# Computing for Geographers

# Computing for Geographers

John A. Dawson   Senior Lecturer, Department of Geography
                 St. David's University College, Lampeter
and
David J. Unwin   Lecturer in Geography, Department of
                 Geography, University of Leicester

**David & Charles**
Newton Abbot   London   Vancouver
**Crane, Russak & Company, Inc.**
New York

ISBN 0 7153 7196 7 (UK)
ISBN 0 8448 0571 7 (USA)
Library of Congress Catalog Card Number 75-37400

© John A. Dawson and David J. Unwin 1976

All rights reserved. No part of this
publication may be reproduced, stored
in a retrieval system, or transmitted,
in any form or by any means, electronic,
mechanical, photocopying, recording or
otherwise, without the prior permission
of David & Charles (Publishers) Limited

Printed in Great Britain
by A. Wheaton & Co., Exeter
for David & Charles (Publishers) Limited
Brunel House   Newton Abbot   Devon

Published in the United States of America
by Crane, Russak & Company, Inc.
347 Madison Avenue, New York   New York 10017

Published in Canada
by Douglas David & Charles Limited
1875 Welch Street   North Vancouver   BC

910.8285
D325c

# Contents

PREFACE 7

CHAPTER 1 THE GEOGRAPHER AND THE COMPUTER 11
Two Revolutions. Quantification in
Theory and Practice. Calculators and
Computers. The Revolutions Converge.
Why Program? Worksheet. Further Reading

CHAPTER 2 THE BASIC ELEMENTS OF A COMPUTER SYSTEM 25
What the Computer Does. The Central
Processing Unit. Input/Output Devices.
Backing Store. Software. Worksheet.
Further Reading

CHAPTER 3 WRITING AND RUNNING A COMPUTER PROGRAM 46
Problem Identification. Algorithmic
Description and Flowcharts. Coding and
Punching. Compiling and Debugging.
Program Testing and Running.
Documentation. Worksheet. Further
Reading

CHAPTER 4 THE INFRASTRUCTURE OF FORTRAN 56
Giving Orders to the Computer. FORTRAN
Statements. Card Layout. Turning the
Symbols into a Language. Worked
Examples. Worksheet

CHAPTER 5 INPUT AND OUTPUT INSTRUCTIONS 71
The Purpose of Input/Output. The READ
Statement. The WRITE Statement. FORMAT
Specifications. STOP and END. Summary.
Worked Examples. Worksheet. Further
Reading

CHAPTER 6 TRANSFER OF CONTROL 94
Program Control. GO TO Statements.
IF Statements. CONTINUE Statements.
Summary. Worked Examples. Worksheet.
Further Reading

CHAPTER 7 THE DO STATEMENT 121
The Need for Loops. The DO Loop.
Standard Functions. Summary. Worked
Example. Worksheet. Further Reading

| | | |
|---|---|---|
| CHAPTER 8 | THE USE OF ARRAYS | 147 |
| | The Concept of Arrays. The DIMENSION Statement. Use of Arrays. Two-dimensional Arrays. Summary. Worked Examples. Worksheet. Further Reading | |
| CHAPTER 9 | RECAPITULATION | 183 |
| | Introduction. Questionnaire Analysis. Determination of a Resultant Vector for Orientation Data. Simple Factory-Warehouse Allocation. Worksheet. Further Reading | |
| CHAPTER 10 | FURTHER INPUT/OUTPUT COMMANDS | 215 |
| | Additional FORMAT Specifications. Repetition of FORMAT Specifications. Line Control Specifications. Alternative Output Channels. Input and Output Arrays. The DATA Statement. Alternative Reference to FORMAT Specifications. Summary. Worked Examples. Worksheet. Further Reading | |
| CHAPTER 11 | PROGRAM SEGMENTATION | 249 |
| | Subprograms. Statement Functions. Function Subprograms. Subroutine Subprograms. The COMMON Statement. The EQUIVALENCE Statement. Worked Examples. Worksheet. Further Reading | |
| CHAPTER 12 | COMMON ERRORS AND GOOD PROGRAMMING STYLE | 289 |
| | Programming Style. Program Integrity. DOUBLE PRECISION, COMPLEX and LOGICAL Data. Program Design and Portability. Presentation and Documentation. Efficiency and Use of Store. Summary. Further Reading | |
| CHAPTER 13 | MODERN COMPUTING | 313 |
| | Introduction. The Use of Files. Program Libraries and Data Banks. Computer-graphics. Other Languages. The Future of Computing in Geography. Further Reading | |
| APPENDIX | STANDARD FUNCTIONS OF ANSI FORTRAN | 354 |
| INDEX | | 357 |

# Preface

A glance at the shelves of most university bookshops will show no shortage of books on how to write computer programs; why have we produced yet another?  Almost all the existing books were written with science-based students in mind and use examples and techniques that relate to the problems met in disciplines such as engineering, mathematics, physics and psychology.  For the geographer who is learning to program, the well-stocked shelves take on the appearance of a desert.  Either he can opt for a highly technical language manual or for books that encourage him to develop programs to calculate stresses in bridges.  In either case, the temptation is to conclude that computing has little to offer except in very specialised branches of the discipline.

   This book is written to encourage the geographer to learn computer programming using the FORTRAN language. Its character is determined by four major considerations. First, we have both taught computer use to geography students at all levels and have found that existing texts not only use unfamiliar examples but also demand a level of mathematics higher than that possessed by the majority of our students.  In this book the examples concentrate upon simple geographical problems of a type that occur at all levels within the subject and whose substantive background ought to be well known.  We have tried particularly to avoid complex statistical manipulations of data matrices.  Secondly, it is our considered opinion that computing is best taught to geographers by allowing the student to program the machine for himself in a reasonably high-level programming language.  The bulk of the book is thus made up of a simplified account of one such language, that can be used as a class text, for self instruction, or for reference.  Practical considerations dictated that the language used be FORTRAN.  Thirdly, we feel that all geographers should be aware of computer potential and have therefore added chapters that deal with the development of computers and modern methods including graphics. Finally, the aims of the book are practical rather than theoretical. The computer scientist may find strange emphases placed upon certain programming methods and an absence of terminology and symbolism of the type that is generally used in programming texts. By his training, the geographer approaches problem definition and solution in particular ways and the examples used in the following chapters show how the computer can be used to assist in these processes, not how they can be twisted to fit into a computing environment. Nevertheless, we hope that the techniques adopted will prove to be of more general applicability. As with any jointly written book, the reader who enjoys guessing games will be able to speculate as to who wrote what; as a clue we offer the observation that each of us wrote exactly half but hope that the joins do not show too much.

All the listed programs have been produced as lineprinter copies of existing card decks that have been run successfully on at least two computer systems but it should be remembered that they are not intended to be used as a series of 'geographically useful' library items. Listings for such programs are widely available elsewhere (*see Chapter 13*). Rather, they are intended to illustrate particular programming methods and applications. Any errors that are present in both the programs and the text are entirely our responsibility.

By their tolerance of exercises that have subsequently been refined and included in this volume, students of a number of universities have played an invaluable part in its production. The text and approach owes much to courses we have taught since the mid 1960s at Aberystwyth, Lampeter, Nottingham and Western Australia. We acknowledge with thanks the feedback provided by students on these courses, particularly Robert Berry, Michael Love, Darrel Nutter and Michael Wright who have worked through most of the exercises. Computing acknowledgements are due to Alan Gilmour (University College Swansea Computing Centre), Alan Rogers (St David's University College Computer Unit) and David Fisher (Leicester University Computer Laboratory) for providing suitable computing environments and to Professors David Thomas (Lampeter),

Clarence Kidson (Aberystwyth), Norman Pye (Leicester), and David Murray (Western Australia) for providing suitable geographic ones. Our profound thanks are also due to Ceinwen Jones who has punched large numbers of cards, Rhiannon Mercer who typed an often very difficult manuscript, Margaret Walker and Christopher Lewis who produced the camera-ready copy and David Orme and Terry Garfield who produced excellent diagrams from exceedingly scruffy bits of paper. Finally, a debt is due to Jo Dawson and Kathy Unwin who at times have been computer widows but who still gave us considerable encouragement in the project.

D.J.U.  J.A.D.
*Christmas 1974*

# Chapter 1
# The Geographer and the Computer

1.1 Two Revolutions
During the 1950s a 'quantitative revolution' spread through geography fundamentally changing the philosophy and outlook of much geographic thought. The course of this revolution and its consequences have been studied by a number of writers, all of whom agree that it changed the subject from an academically isolated and essentially qualitative discipline into one which increasingly concerned itself with the quantitative testing of general statements about spatial distributions. Historians of science and society will probably note that this was the period in which the electronic computer became widely available, sparking off a computer revolution which has already led to far-ranging changes in virtually all the sciences, in industry and in society as a whole. The computer has proved to be a powerful tool to alleviate repetitive work, increase efficiency and enormously speed calculations. In common with many recent technological developments it enables the individual to be released from many dull or error-prone activities. On the other hand, computers can be used unthinkingly to the detriment of mankind by causing increased redundancy in industry and commerce, by facilitating the invasion of our privacy and by allowing mindless 'number-crunching' to be substituted for genuinely creative thought.

In this book we attempt to set out guidelines by which the geographer both as an academic and as an individual in society can become acquainted with computer use, can develop a facility to write simple but labour-saving programs in a commonly-used computer programming language called FORTRAN and, most important of all, can gain a sensible attitude towards computer use not only in geography but also in everyday life.

Explicitly our concern is with geography and computer

use; the wider context of computer appreciation is implicit in many of the following pages. We begin by reviewing some of the background to computer use in geography by looking first at the demand generated in the subject by the quantitative revolution and secondly at the supply afforded by the development of today's highly sophisticated machines. A final section will examine some of the uses to which the computer has been put in geography during the past twenty years.

1.2 Quantification in Theory and Practice

In a short paper entitled *The Quantitative Revolution and Theoretical Geography*, Burton (1963) outlined the radical transformation in spirit and purpose which had taken place in geography during the 1950s and which has been called 'the quantitative revolution'. Although this revolution had its ultimate origin in the fields of mathematics and physics, important and direct antecedents are to be found in more familiar disciplines such as economics, geology and psychology. Some geographers began to look for quantitative techniques to apply to geographic problems while non-geographers began to bring new methods to bear on old geographic questions. Sometimes the same problem was approached by both groups, as is illustrated by the early application of multivariate statistical methods to regionalisation. Prominent among the early quantitative geographers was a school of geomorphologists whose inspiration was drawn from the work of the American geologist A. N. Strahler and the engineer R. E. Horton and who worked on the quantification of landform geometry (Horton, 1945; Strahler, 1950); a group of economic geographers led by W. L. Garrison in the University of Washington (Berry and Marble, 1968), and in Britain a group of geographers spurred on by R. J. Chorley and P. Haggett (Chorley and Haggett, 1967). The application of new techniques to geographic questions by non-geographers can be illustrated by the branch of statistical analysis called centrography which is concerned with the distribution, dispersal and association of spatial distributions and social physics with its work on population interaction using analogies borrowed from theoretical physics (Stewart, 1950; Neft, 1966; Wilson, 1974).

The introduction of quantitative methods into geography was not without its opponents; during the early 1950s there were many debates conducted formally in the literature and at some informal conferences about the desirability or otherwise of the new methods but by 1956 quantification appears to have been accepted, and the quantifiers turned to debate some of the serious technical and methodological problems which geographic use of these methods entailed. The initially 'new' ideas were accepted as a part of the conventional wisdom and by 1970 were in turn being challenged by a generation of post-quantitative geographers. If this revolution had simply been a fad or fashion then it would have run its course and died. That this did not happen is evidence of a deeper influence. The revolution was an attempt to bring the subject into the mainstream of science by consciously seeking general theories to explain spatially distributed phenomena. Prior to 1950 geography had retreated into an idiographic viewpoint from which geographers were content to make only limited generalisations about areas of the earth's surface which in themselves were regarded as being unique. This attitude reached its most extreme form in the doctrine of exceptionalism which argued that because no two places can have the same absolute location on the earth, any discipline concerned with location must deal with the intrinsically unique. It follows that description and interpretation cannot be accomplished by reference to any general laws and that quantification is only useful as a refined form of description.

Quantification plays an important role in modern theoretical and applied geography and it is this role which has generated a considerable demand within the subject for access to rapid information-processing facilities. In theoretical geography, as in any science, quantification implies that three basic requirements are met. First, any measurements must take place on an *operationally definable process* otherwise the resulting data will have no meaning. In practice this requirement usually implies that the quantifier has generated a hypothesis that is to be tested. Secondly, and in order to test any hypothesis, quantification must produce reliable results in that there is a *reproducibility of outcome* of the measurement operation. Repetition of the analysis produces the same results and directly comparable results are produced when different data are used. Thirdly, in order to evaluate a tested hypothesis, quantification must be *valid* - there must be true or accurate measurement. This differs from reproducibility in that it is possible persistently to reproduce an error. Thus quantification allows the clear, unambiguous statement of theory and the standardised precise data collection essential to theory testing. As Burton (1963) observed, 'The need to develop theory precedes the quantitative revolution, but quantification adds point to the need, and offers a technique whereby theory may be developed and improved'.

Quantification has not only had an impact on the theoretical constructs of geography. The range of application of geographic ideas and methods to problems of environmental planning was greatly extended by the quantitative movement. Many of the applications of geography in the 1950s rested on the provision of information for policy makers and for monitoring the performance of existing plans. Techniques for simulating the results of different plans and plan evaluation methods have followed this information stage from the early 1960s onward. (*See Chapter 13 for a further discussion on simulation.*)

As information-providers, geographers were influenced in at least three ways by the change in geographical philosophy. As a result of the rapidly expanding amount of data available from official and unofficial census-taking agencies and the application of remote sensing techniques to data gathering, the need arose for greater quantification simply to make sense of this wealth of information. The data explosion meant that faster, more rigorous data analyses were required. It has been noted, for example, that the manual analysis and description of land-use patterns takes on average ten times as long as the initial field survey. Secondly, within the data gathering exercise it was realised that considerable care was needed in experimental design, sampling method and data analysis. The more rigorous approach necessary for these exercises both stimulated quantification and benefitted from it. Thirdly, the quantitative revolution aided the presentation of material for assimilation by policy makers. Map design, a major building block in all branches of geography, adopted sophisticated numerical techniques to extend the methods for the visual presentation of material. Even by 1960 applied geography had thus been influenced by quantification but during the 1960s an even stronger wind of change was felt as the implications of the new philosophy became fully apparent.

It can be seen that by the late 1950s internal philosophical and methodological developments had led to the existence within the subject of a substantial body of

theoretical and applied geographers who had a major
interest in quantitative methods, and it is fortunate that
at about the same time a second revolution - the widespread availability of the electronic computer - was
beginning to make itself felt.

1.3 Calculators and Computers

A notable feature of many calculations in geography is that
they involve large numbers of often simple operations
which if attempted by hand are inherently very tedious and
error-prone. The idea of making machines to do such
calculations is probably as old as calculation itself, and
the antecedents of the modern electronic computer can be
found in devices that are centuries old. The familiar
abacus is probably the oldest and most important of man's
early attempts to mechanise calculation and in experienced
hands is very fast indeed. Improvements on it are recent,
dating in the main from the seventeenth century when
devices such as the logarithm (1614), Napier's Bones,
Oughtred's slide rule and Pascal's adding machine (1642)
were developed. These devices are all non-automatic in
the sense that they need the frequent attention of
a human operator, a method of working which sets severe
limits on their speed, and it is to Charles P. Babbage
that we owe much of the theory of truly automatic machines.
Babbage had the unusual blend of a practical mind and an
interest in theoretical mathematics and in 1812 conceived of a machine to produce mathematical tables
reliably and rapidly. His so-called 'difference engine'
was exhibited in 1822 and was followed by ambitious but
unfulfilled plans for a complex general purpose 'analytical
engine' which, if it had worked, would have incorporated
many of the principles of today's electronic computer.
At about the same time as Babbage worked, silk weavers in
Lyons were using a primitive form of punched card to control the patterns produced on their looms, a system
invented by J. M. Jacquard. Jacquard's loom introduced
the crucial ideas of using a pre-set program to control
a complex machine and of storing the necessary control
information on a punched card.

The story of the electronic computer now moves to
America where the next advance was made. This was
Herman Hollerith's invention of electromechanical punched-card sorting and tabulating equipment for the United
States' Census Bureau. In 1886 the returns of the 1880

census were still being sorted and counted and it was clear that this job would not be finished before 1890 when the next census was due. Hollerith saw that the answer to this problem lay in some form of mechanisation, and realised that a punched card of the Jacquard type could be used to store numerical information. From this it was a short step to devise machines to sort and count these cards and the way was open to the automatic processing of large amounts of information.

Although at the turn of the century the twin ideas of data storage using punched cards and automatic calculation using mechanical devices were well established, further development of these ideas to produce a full-scale automatic computer has gone hand-in-hand with developments in electronic engineering. In 1937 the Bell Telephone Laboratories produced an electromechanical calculator and in 1944 a crudely programmable calculator was produced at Harvard University, but the first true universal computer is usually held to be the Electronic Numerical Integrator and Calculator (ENIAC) built by Eckert and Mauchly in 1946 at the University of Pennsylvania. Very shortly afterwards prototype computers were built in a number of University centres in both the USA and Britain with commercial development of these machines following in the early 1950s.

Computers have developed in response to parallel improvements in electronics, systems engineering and operating methods, but conventionally are said to have passed through four distinct generations of machines and at the time of writing may be on the threshold of a fifth. The characteristics of each generation are shown in Fig 1.1. The first generation of machines was based largely upon valves and other standard electric components. They were generally slow, of low memory capacity and difficult to program, so that computer programming began as an obscure and esoteric craft tailored more to the needs of the electronics engineer than to those of the computer user. In the late 1950s a second generation of machines based upon transistor technology was introduced and gave a great increase in speed and memory capacity. At the same time the machines were made much easier to program by the provision of a number of programming languages tailored to the existing conventions of specific types of computer user, such as normal mathematical notation for scientific and English

*Fig 1.1* Characteristics of the first four generations of computers

| | Generation | | | |
|---|---|---|---|---|
| | 1 | 2 | 3 | 4 |
| Date of introduction | 1951-2 | 1958-60 | 1963-5 | 1969-on |
| Dominant use | business and science | data processing | information processing | on-line information processing |
| Technology | vacuum tubes and standard components | transistors | medium to large scale integrated circuits | large-scale integrated circuits |
| Method of communication | machine code and assemblers | high-level languages, executives | complex operating systems, multi-programming | conversational systems with hardware subprograms |
| Core store (bits) | $10^7$ | $10^9$ | $10^{11}$ | $10^{13}$ |

language type statements for accounting applications. The earliest geographic applications used these second generation machines but most use post-dates the arrival of third generation machines in the early 1960s. These machines were characterised by items such as integrated circuits, large and fast internal memories, a multitude of so-called peripheral devices and complex systems programs provided by the manufacturer to control automatically the flow of information through the machine. The college student in the 1970s is most likely to have access to a large third generation machine such as the System 360 machines introduced by IBM and the Series 1900 introduced by ICL, but some second generation products may still be found. Such is the speed of progress that most first- and second-generation computers are either on scrap heaps or in museums. The fourth generation is still being introduced and is characterised by a trend towards medium- and large-scale integrated circuitry with semiconductor memories; many often use subprograms provided by the electronics rather than the programmer (Vacroux, 1975), automatic error-checking and have a multitude of remote terminals.

Alongside this development of the large general-purpose machine, the past ten years have also seen the development of small low-cost computers commonly referred to as mini-computers and more lately micro-computers. A basic micro-computer now (1974) costs less than £200. Although definitions vary widely as to what constitutes a mini-computer, these machines are often no bigger than a type-writer, have a fairly small but expandable memory capacity, require no special environment and nothing more than a plug and socket to connect them to the mains power supply. Mini-computers owe their success to an ever-widening range of applications in which they are dedicated to a single or limited range of tasks such as occur in process control, as data shufflers in data communication where they are often linked to a more powerful machine, in instrument control and in research for mathematical calculations.

By 1970 the average computer was very different from the pioneer machines. The internal speed of calculations has increased from times measured in tenths of a second to those measured in thousand-millionths of a second and the storage capacity has increased in a similar proportion. At the same time miniaturisation has allowed the physical size of the machine to be greatly reduced without any loss of processing power so that in 1975 it is possible to buy, off the peg, a machine of desk-top proportions that is capable of doing all that the thirty-ton ENIAC could do only twenty-five years before. Beside this computer revolution the quantitative revolution in geography is mere catapult warfare!

## 1.4 The Revolutions Converge

Although the quantitative revolution in geography had its origin significantly before computers became widely available there can be little doubt that computer use has greatly accelerated its course and acceptance. As geography became more numerical in its approach it was inevitable that some geographers would grasp the consequences of the computer revolution for their work. Not surprisingly American geographers were the first to use computers, in the late 1950s; in Britain early users were Moser and Scott in their 1961 study of British towns and Coppock in his *Agricultural Atlas of England and Wales* (1964).

Figure 1.2 is an attempt to chart the course of the introduction of computer methods into the mainstream of geography. On it is plotted the proportion of computer-based papers of all major papers published in the *Annals* of the Association of American Geographers and the *Transactions* of the Institute of British Geographers. Remembering that a time lag of perhaps three to four years often passes between the initiation and the completion of a research project, there has been a very rapid increase in computer use. The first applications were reported in the period 1957-1962 and in 1963 there was a pronounced peak as the 'primary adopters' reported their work. From 1965 onwards the proportion rose to over 40 per cent in 1973.

A second way of looking at the rate of acceptance of computer methods in geography is to examine the increasing familiarity of geographers with these methods. During the early 1960s use was confined to a few academics and graduate students, but in the 1970s it is by no means unusual for all graduate students and most of the academic staff in a typical university geography department to have carried out computer work and there is an ever-increasing body of numerate undergraduates competent in computer use (Utting and Hall, 1973). More progressive secondary schools are beginning to introduce computer methods into the sixth-form geographical curriculum.

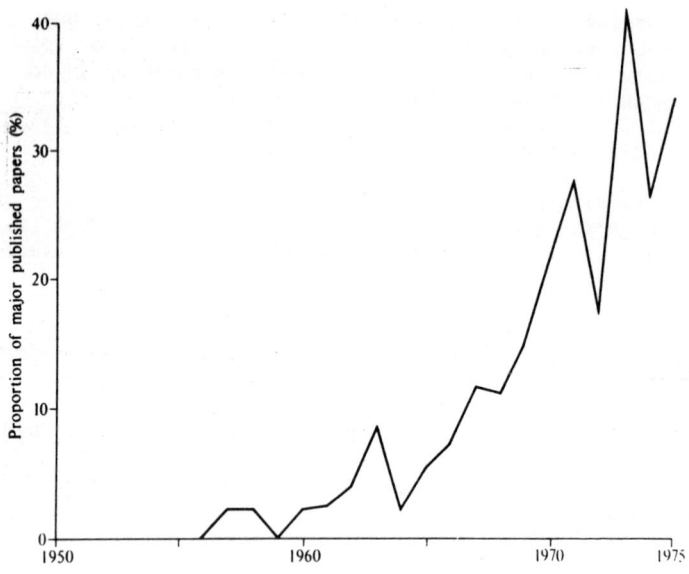

*Fig 1.2* Proportion of major papers in two geographical
journals definitely making use of computers

Not only has the amount of use on geography grown rapidly, it also seems that there have been shifts in the nature of the problems tackled at the research frontier. Initially, computer use involved only a small band of workers in a limited number of centres and use was largely exploratory. Inevitably, the new machines were used to solve existing problems using well-tried methods and the major attraction was the computer's facility for rapid error-free work. Essentially such use is of a *book-keeping* nature in which the computer is used as an efficient adding machine. Data are grouped and re-grouped according to different criteria, and simple descriptive statistics are calculated and occasionally mapped (Kao, 1963). From the early 1960s onwards a second phase of use involved many more workers and the larger, faster, third-generation machines. In this phase the use of the machine was characteristically *computational* as new analytic methods borrowed from other disciplines were applied to familiar geographic problems. Techniques such as factor-, discriminant- and trend-analysis were used to solve problems of regionalisation, classification and spatial variation (King, 1969). Dissatisfaction with these methods coupled with a much greater awareness of computer potential has led to a third phase of use in which the machine is used to solve new kinds of problems which in the past would never have been attempted because of the labour involved. Frequently this work is of a *predictive* nature. The computer has been used to simulate or feign real world processes (*see Chapter 13*), and it has thus proved possible to explore geography not only as it is and was but also the many alternative geographies of the future (Haggett and Chisholm, 1970).

1.5 Why Program?
Despite the many problems that their use might create it is obvious that computers are here to stay and will increasingly affect our life-styles and academic outlook. An understanding of the computer therefore can be thought of as an essential part of any education. Computing is not an esoteric or specialised activity; it is a versatile tool in any work that has a factual or intellectual content. As we have seen, the early machines were difficult to use and were mainly concerned with processing numerical information, but modern machines are much easier to use and as memory devices have improved, so a much more general information-storage role has become important. To this extent computers will usurp many of the traditional functions of the library, book and census schedule and their use will be regarded as essential to virtually all scholarly activity.

It is often suggested that a knowledge of computing can be obtained by being a passive computer-user and depending upon a specialist programming consultant and/or pre-prepared general-purpose package programs. In our experience there are a great many pitfalls in this approach. We have observed, for example, that a researcher who cannot program may spend more time explaining his problem to a consultant than he would have taken to learn to program the problem himself. . All too frequently a standard 'package' will not do precisely what is required to solve the problem and the problem is then tailored by the

21

**non-programmer** to suit the program rather than vice versa.
**Alternatively** the package requires its data to be presented in a manner which does not suit the needs of the **researcher**. Such use is intrinsically conservative - **analyses** are restricted only to those methods for which programs already exist and the full potential of the computer is never realised. It is only by learning to program and run one's own problems, no matter how simple, that a full appreciation of the uses and limitations of the computer can be obtained. A major difficulty in this is that students have widely different backgrounds and mathematical abilities. In the remainder of this book we will introduce computer programming from the point of view of a background in geography, selecting illustrative problems familiar to geography students with only a limited mathematical competence and in general avoiding 'difficult' areas such as multivariate statistics that require relatively advanced mathematics for their full understanding.

1.6 Worksheet
At the end of each of the succeeding chapters is a worksheet similar to that below. The worksheets may involve library work, short comprehension tests, writing a program, or even correcting errors in parts of a program.

1 Try to find at least five examples in the recent geographic literature of studies that make extensive use of the computer and for each comment briefly upon the type of operations undertaken, whether or not the study would have been contemplated without computer assistance, and the extent to which this assistance sheds light on the basic research problem.

2 Find out what makes and types of machine are available for use in your institution and what arrangements are made for student use.

1.7 Further Reading
Adams, J.M. and Haden, D.H. *Computers: appreciation, applications, implications* (1973)
Bernstein, J. *The analytical engine: computers - past, present and future* (1963)
Berry, B.J.L. and Marble, D.F. (eds) *Spatial analysis: a reader in statistical geography* (1968)
Borko, H. *Computer applications in the behavioural sciences* (1962)

Burton, I. 'The quantitative revolution and theoretical geography', *Canadian Geographer*, 7, 151-62 (1963)
Chorley, R.J. and Haggett, P. (eds) *Models in geography* (1967)
Coppock, J.T. *An agricultural atlas of England and Wales* (1964)
Desmonde, W.H. *Computers and their uses* (1971)
Gotlieb, C.C. and Borodin, A. *Social issues in computing* (1973)
Hagerstrand, T. 'The geographer and the computer', *Transactions, Institute of British Geographers*, 42, 1-19 (1967)
Haggett, P. 'On geographical research in a computer environment', *Geographical Journal*, 135, 497-505 (1969)
— and Chisholm, M.D.I. (eds) *Spatial forecasting: some problems in small area projections* (1970)
Harvey, D. *Explanation in geography* (1969)
Horton, R.E. 'Erosional development of streams and their drainage basins', *Bulletin, Geological Society of America*, 59, 275-370 (1945)
Kao, R.C. 'The use of computers in the processing and analysis of geographical information', *Geographical Review*, 53, 530-47 (1963)
King, L.T. *Statistical analysis in geography* (1969)
Moser, C.A. and Scott, W. *British towns: a statistical study of their social and economic differences* (1961)
Neft, D.A. *Statistical analysis for areal distributions* (1966)
Penney, G. *Computers in the social sciences* (1973)
Pylyshyn, Z.W. *Perspectives on the computer revolution* (1970)
Stewart, J.Q. 'The development of social physics', *American Journal of Physics*, 18, 239-53 (1950)
Strahler, A.N. 'Davis's concept of slope development viewed in the light of recent quantitative investigations', *Annals, Association of American Geographers*, 40, 209-13 (1950)
Tobler, W.R. 'Automation in the preparation of thematic maps', *Cartographic Journal*, 2, 32-8 (1965)
University Grants Committee *Teaching computing in universities* (1970)
Utting, J. and Hall, J. *The use of computers in university social science departments* (1973)
Vacroux, A.G. 'Microcomputers', *Scientific American*, 232 no 5, 32-40 (1975)

Weizenbaum, J. 'On the impact of computers on society', *Science*, 612-4 (12 May 1972)

Wilson, A.G. *Urban and regional models in geography and planning* (1974)

# Chapter 2
# The Basic Elements of a Computer System

2.1 What the Computer Does
Before considering in detail how the various parts of
a computer system work, it is best to give an outline of
what a computer does.  This can be most easily visualised
if a simple task is broken down into its components and
each component is given a name.
   A student has to draw a proportional circle map to show
town population in his study area by locating at each
town-site a shaded circle whose area will be proportional
to the town population.  In order to do this he needs to
calculate the radius of each circle using the formula
$$R = c\sqrt{P}$$
in which P is the town population whose square root is
taken, c is a constant used to ensure that the circles
have the correct range of sizes and R is the required
radius.  The sequence of operations the student goes
through could be as follows:

  (i) Read off the town population from a census schedule
      and locate the town on the map.  We can call this
      the INPUT to the procedure and the input material
      its DATA.
 (ii) Look up the value of the constant c which may have
      been noted on a sheet of paper alongside.  We can
      call this a reference to background material or
      STORAGE.  Such information could be stored in the
      student's head or on a piece of paper.
(iii) Use the equation to calculate the value of the
      radius R; this is the ARITHMETIC.
 (iv) Draw in the circle at the appropriate location;
      this is the OUTPUT of the procedure.
  (v) Return to step (i) and process a further town.

There are five distinct elements in the exercise, input,

data, storage, arithmetic and output, but for the results
to make sense these elements must be correctly ordered so
that we can then add a sixth element:

(vi) The sequence of instructions which we call the
PROGRAM.

Just as there may be alternative ways of ordering the
sequence of operations - for example, all the radii could
be calculated and stored before any drawing is performed,
or each calculation could be followed by the corresponding
drawing - so there can be alternative programs to produce
the same result. Finally, it would be unusual for any
student to draw a map without some guidance from his tutor
to ensure that the calculations were correctly made and
the map properly drawn. There must be also:

(vii) An element of CONTROL over the whole exercise.

A computer system works in much the same way as our
hypothetical student. The original *data* prepared in some
acceptable form are fed into the machine as *input* and the
completed work returned as *output*. Between input and
output is the *Central Processing Unit* (CPU) which has
a number of functions. Within it reference can be made
to stored data and program; this is the *storage* function.
A further part of the CPU, the *arithmetic* unit, can perform the necessary calculations and a further part still
exerts *control* over the job as a whole. The reference
of input and output to the CPU is logically ordered by the
*program*. For work that involves storing large quantities
of data the internal CPU store may not be large enough,
in which case reference can be made to external or *backing
store* (the piece of paper containing the value of c in our
student example).

Figure 2.1 is a schematic drawing of such a simple system.
It consists of a CPU surrounded by a number of specialised
devices known as *peripherals*, including input devices to
transfer data to it, output devices to accept information
from it and backing store to hold additional data. So
far our computer will merely be a collection of circuits
wired together and packed into metal cabinets. This is
the computer *hardware*, but for it to function usefully it
also needs co-ordinating programs to control the flow of
work through the CPU and its peripherals and to determine
what operations are needed in each particular job. These

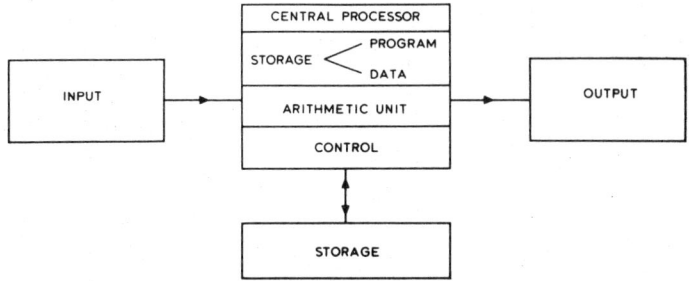

*Fig 2.1* Basic elements of a computer system

programs comprise the *operating system* and are collectively known as the computer *software*.

2.2 The Central Processing Unit
The CPU is the heart of the computer where the functions of storage, arithmetic and control are carried out. Because of the variety of types and sizes available it is difficult to talk of a typical CPU, but the major distinguishing feature of all is the capacity to hold and manipulate large volumes of data at great speed. In practice it is not economical to design CPU's that are capable of storing *all* the information they need, all that need be stored are the data and instructions for immediate use.

How is information stored within the CPU? Physically, many modern units make use of small ferrite rings or *cores* some of 12 to 50 thousandths of an inch outside diameter which can be magnetised by passing an electric current along wires threaded through them. Figure 2.2 shows how each individual core is wired and how these are built into larger blocks. Each core element is capable of being magnetised in two directions and can thus take on one of two 'states' which in turn can be represented by the binary digits 0 or 1. A number of these core elements are arranged in a square and are connected by wires running along the rows and columns. A third wire, called the sense wire, is threaded diagonally across the lattice and

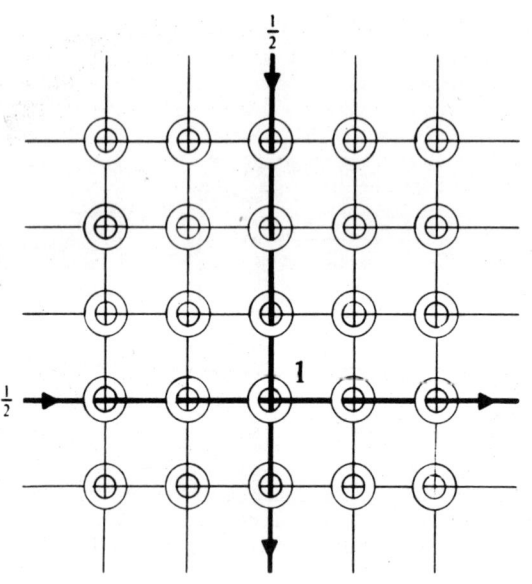

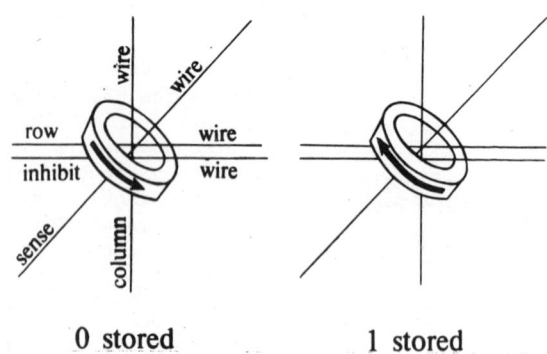

*Fig 2.2* Sections of magnetic core store

enables reading of the binary representation to take place
and a fourth wire, the inhibit wire, is used to restore
information after it has been read.  Only one element can
lie at the intersection of any particular row and column
and this allows information to be selectively written to
particular elements.  The current necessary to alter the
direction of magnetism in a core element must be of
a certain minimum value, so that if a current of just half
this is sent down a row wire and half down a column wire
it is only at their intersection that the minimum will be
reached and only one core element will have its magnetism
changed.  To enable simultaneous access to a number of
core elements, several squares of core are assembled
vertically giving three-dimensional blocks of storage.
More recently, computer manufacturers have begun to use
faster and cheaper stores based upon the use of thin films
and semi-conductors that will eventually replace ferrite
core as the principal storage medium.

Storage within the CPU is thus of individual, indivisible
units of information called binary digits or *bits*, but
a store containing many thousands of bits would be completely meaningless unless there were some way of assembling these bits into the numbers, symbols and letters we
use in everyday life.  It is usual for the computer store
in the CPU to be divided into groups of bits whose
patterning enables particular characters and numbers to be
represented.  A 6-bit grouping would enable the 64 combinations possible from the six digits 000000 to 111111 to
be created and this is more than sufficient to store in
code-form all the 26 letters of the alphabet but it is
inadequate if we try to store numbers in the same way.
The highest number that can be represented using a 6-bit
group is only 63, for an 8-bit group it is 255 and so on.
Obviously even an 8-bit group is too small to accommodate
the range of numbers we would like to use.  One possible
solution to this problem is to retain a 6-bit grouping to
accommodate a single character and to build up large
numbers expressed in their binary form as *words* each of
which is a number of 6-bit groups.  A 24-bit word would
then be capable of storing 4 characters each of 6 bits
or one 24-bit number.  The subdivisions of the core into
character locations necessary to do this are sometimes
called *bytes*.  On modern machines the total *word-length*
can vary from as little as 12 to as much as 60 and is an
important factor when comparing machine sizes which are

quoted in both words and bytes. The symbol 'K' is used to denote 1024 items. A 16K CPU has 16,384 locations. Hence a 16K machine using a 24-bit word length has as many core elements as a 32K machine which uses a 12-bit word. Most medium-sized processors will have at least 32K, 24-bit word storage, while a few machines allow the user to select the required word length.

We have not yet considered the various ways in which numbers can be coded as sequences of binary digits. When counting in the familiar decimal system we carry over a digit every time a ten is reached by moving a one to the left and inserting a zero:

In *decimal*          6 + 4 = 10

Values increase by powers of ten as we move from right to left as for example in the number 128:

*Place Value*          $10^2$   $10^1$   $10^0$
*Counter*            1    2    8
128 =    ( 1 x $10^2$ ) + ( 2 x $10^1$ ) + ( 8 x $10^0$ )

In the binary system used by the CPU the carry-over occurs at two and the place values increase by additional powers of two:

*Place Value*     $2^7$ $2^6$ $2^5$ $2^4$ $2^3$ $2^2$ $2^1$ $2^0$
*Counter*         1   0   0   0   0   0   0   0
128 =    (1 x $2^7$) + (0 x $2^6$) + (0 x $2^5$) + (0 x $2^4$) +
         (0 x $2^3$) + (0 x $2^2$) + (0 x $2^1$) + (0 x $2^0$)

This means that the pure binary representation of the number 128 is 10000000. The decimal equivalent of the binary number 11101 is:

$$\begin{aligned} 1 \times 2^0 &= 1 \\ 0 \times 2^1 &= 0 \\ 1 \times 2^2 &= 4 \\ 1 \times 2^3 &= 8 \\ 1 \times 2^4 &= \underline{16} \\ \text{decimal} & \quad 29 \end{aligned}$$

Although the computer performs most of its arithmetic using the binary representation, conversion between the two systems is easy and the machine will always accept decimal numbers as input and output. *Other than as an aid in understanding how a computer works there is no reason why the geographer-user should ever bother himself with the details of binary notation and arithmetic.*

Alternative systems are sometimes used to code each decimal number. For example, a string of 4-bit groups each of which is the binary code for a decimal digit may be used. This is a form known as *Binary Coded Decimal* (BCD):

| Decimal | Binary |
|---------|--------|
| 1 | 0001 |
| 2 | 0010 |
| 3 | 0011 |
| 4 | 0100 |
| 5 | 0101 |
| 6 | 0110 |
| 7 | 0111 |
| 8 | 1000 |
| 9 | 1001 |
| 0 | 0000 |

In BCD the decimal number 128 would be stored as 0001 0010 1000. This coding takes up more store than pure binary and makes arithmetic more difficult but it can be a very efficient way of storing characters and is frequently used in backing-store devices. When needed the BCD coded information is converted into pure binary before it is manipulated by the machine.

The computer must also be able to distinguish between positive and negative numbers; this is usually done using a single bit in a fixed position within each word to indicate one or other of the two possible states. Finally, computations involving very large or very small numbers are awkward and accuracy is lost if long strings of zeros conclude or lead the significant part of each number. In science this difficulty is often avoided by writing such numbers in two parts, a *mantissa* which contains only the significant digits and an *exponent* which places the decimal point. For example the number 186,000 can be written $0.186 \times 10^6$ in which the mantissa is 0.186 and the exponent is 6. In computer storage a similar convention is adopted. One word is used to hold the binary representation of the mantissa, another its exponent. Although this form of storage enables the computer to deal with very large and very small numbers to greater accuracy than most of us would use in a pencil and paper calculation, it does restrict the accuracy of the representation to a fixed number of significant digits. With a 24-bit word it is doubtful whether more than 8 significant digits would be retained and in certain types of calculation this can lead to very considerable rounding errors in the computer.

The CPU we have discussed is fundamentally a *digital* device which deals with discrete (0 or 1) entities and which takes a finite time to perform calculations and output results. It can be contrasted with an *analogue* machine in which numbers are represented by physical forces such as voltages and resistances. Generally, analogue computers are inaccurate if compared with digital machines but they have the advantage that output is simultaneous with input and are normally used as simulators to serve special purposes. Other than to note that analogue computation seems to have a considerable future in geographical research, it will not be discussed further in this book.

2.3 Input/Output Devices

Data in their original form cannot be fed directly into and out from the CPU. At input they must be machine-readable, recorded on a medium that is physically acceptable and in a manner that is recognisable by the machine. At output information can only be presented in a limited number of ways depending upon the particular devices that are available at a particular computer centre. The most important input/output media are punched cards, paper-tape and print-out. More specialised media are the faces of cathode ray tubes, typewriter-like consoles, magnetic tapes and moving pens.

*Punched Cards*. A punched card is a piece of thin, high-quality cardboard of exact size and thickness. In the past many forms of card have been used but today almost all card processing uses the 80-column form illustrated in Fig 2.3. The card is divided into 80 vertical columns each of which has 12 sites into which holes may be punched so that a single column may record a symbol, digit or letter by means of a pattern of holes unique to that character. The exact codes used vary slightly but almost all manufacturers use the pattern illustrated in Fig 2.3 for the decimal digits and letters of the alphabet. Digits present no difficulties and can be represented by punching just one hole-site per column whereas letters need two such holes and symbols such as those used for currency (£ or $) have a still more complex code. Cards are said to be *interpreted* if they carry along their upper margin a printed version of the characters corresponding to the punched hole sites. Punched cards are prepared for input by a *keypunch* machine consisting of a keyboard, a hopper containing a supply of unpunched cards, a punching head and a stacker to accept the cards after they have passed the punch. Each card is fed from the hopper to the punching head at which its column-by-column progress is controlled from a keyboard. As the required information is typed on the keyboard so a series of very small knives

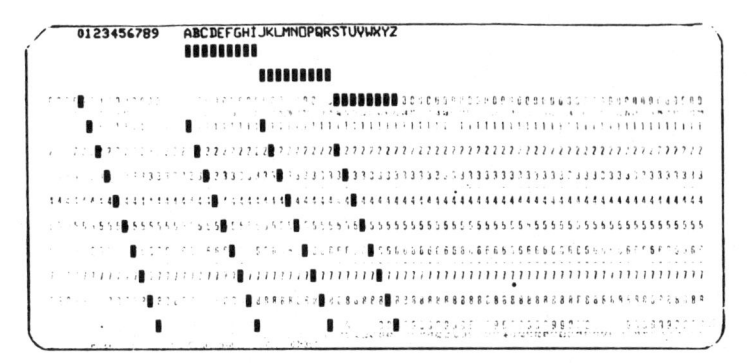

AN 80-COLUMN PUNCHED CARD

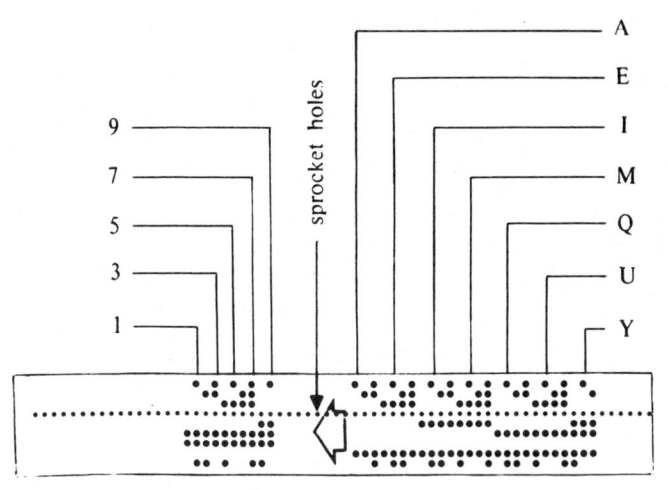

8-TRACK PUNCHED PAPER TAPE

*Fig 2.3* An 80-column punched card and 8-track punched paper tape

cut holes into the card corresponding to that information and the completed cards are then stacked ready for input to the machine. This card input is accomplished using

33

a *card reader* which is an electromechanical or optical
device that can recognise the hole patterns on the cards
by means of either small brushes or photo-electric cells
and convert this pattern into appropriate pulses for
transmission to the CPU.  Card readers vary in speed from
less than 100 to over 1500 cards per minute, which is very
much faster than the 100 or so cards per hour that a skilled
keypunch operator can prepare.  Some computers may also
have a facility to output decks of punched cards using
a card punch.  By modern standards this is a very slow
process.

*Punched Paper Tape.*  The punched card has the inherent
feature of a natural break at the end of each card whereas
a punched paper tape, a continuous ribbon of paper tape
onto which data are recorded by punching holes across its
width, does not have this limitation.  Each row of holes
represents the code for a single character but the number
of sites available is less than on a card and varies from
manufacturer to manufacturer.  Originally 11/16 inch
5-track tape was used, which allowed a maximum of five
hole-sites per character, but modern machines now use 7-
or 8-track tape (Fig 2.3).  Paper tape can be prepared
at a keyboard device similar in its operation to the keypunch and is read into the machine using a *paper-tape
reader*.  It can also be obtained as output using a high-
speed *paper-tape punch*, or as output from a great many
automatic recording devices.  At input, tapes can be read
at speeds up to about 1000 characters a second which is
roughly equivalent to a card reader speed of 750 cards
a minute, so that in terms of speed of operation there is
little to choose between the two;  but (volume for volume)
paper tape is cheaper and easier to store than cards.
It is far less flexible when in use.  Once prepared, the
order of the data items on the tape cannot be changed,
alterations are difficult to make and physically they can
be easily torn.  With cards individual data cards may be
extracted or new cards added into the pack.  Damaged
cards are easily replaced but disaster can befall if the
cards are dropped.

*Specialised Input Media.*  The preparation of both cards
and tape involves gathering data onto a source document
and then subsequently transferring these data into
machine-readable form but in some applications, notably
bank cheques and cash-register tapes, data can be collected
in a standardised form suitable for direct computer input

and a variety of more or less specialised input-media are available. In *Optical Character Recognition* the aim is to produce a document which is optically scanned by the machine. The input medium must be of a standard shape and size and like the punched card it must have its information recorded in pre-determined positions. One method uses a machine capable of recognising the shapes of characters but a more common system simply gives the various positions on the document a value and records data at these positions by a black pencil mark. Students in North America are familiar with the type of document used to record responses to multiple choice examinations shown in Fig 2.4. Although the possibility of errors due to badly marked forms is quite high it seems likely that these specialised media will become more widely used. Social survey by questionnaire often demands the standardised recording of multiple-choice information and there is no reason why document reading techniques should not be used for the direct input of the basic field-data.

A *digitiser* automatically records the rectangular co-ordinates of points on a map or graph and punches them onto paper tape or cards. A pencil-like instrument is manually moved across the surface of a map or graph which is laid on a special table and the movement of this pencil is tracked electromechanically by arms underneath the table enabling the current position of the pencil to be translated directly into x,y co-ordinates which are then output in machine-readable form. Because they can rapidly process mapped information, digitisers are of considerable application in geography, and are beginning to be regarded as part of the standard equipment of geography departments.

*Magnetic Tape*. The quantity of raw data stored on cards, paper tape or document reader forms is limited by its bulk and comparatively low speed of operation. Bulk storage of large quantities of data is better accomplished using a medium such as magnetic tape. A magnetic tape is a $\frac{1}{2}$in- to 1in-wide strip of tough, flexible plastic coated on one side with iron oxide and magnetisable at spots designated as storage locations in much the same way as the holes are used on paper tape. The length of a tape may be as much as 3600ft and information may be recorded at densities of up to 6250 characters to the inch. *Magnetic tape decks* linked permanently to the computer are used to read the information from these tapes at speeds of up to

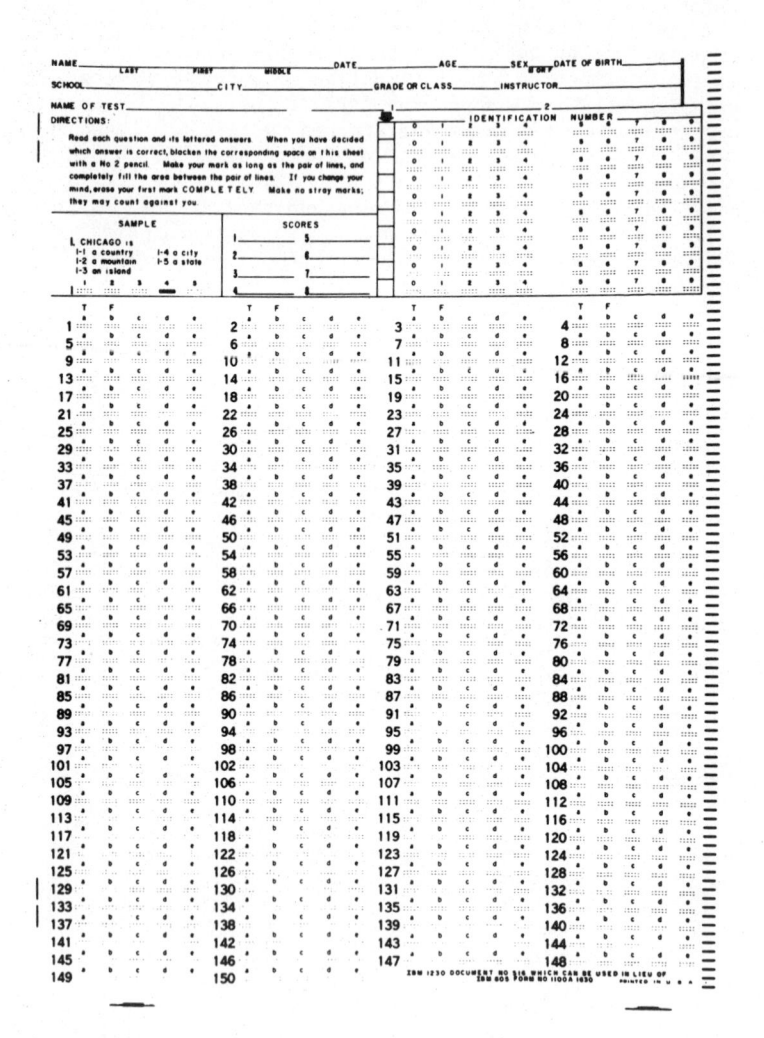

*Fig 2.4* A document reader card

150,000 characters per second. Although it is possible to prepare magnetic tapes using a special keypunch device, most tapes are produced in the first instance as computer output. Typically a large number of data cards will be read into the machine, edited in some way, and then their contents copied onto a magnetic tape which can subsequently be used in lieu of the original cards. Reading and writing magnetic tapes are relatively rapid and generally error-free operations and most computer installations will use such tapes as a backing store. The major disadvantage is that access to individual items on a tape is *serial*, one item succeeding another in a series along the tape's length, so that to read any single item it might prove necessary to scan the entire tape length in a very inefficient manner.

*Teletypes.* The input/output media considered so far involve the preparation of some machine-acceptable objects onto which data are recorded. It is also possible to communicate directly with the computer (called on-line) by way of a teletype keyboard. This is a typewriter-like device directly linked to the CPU. The teletype can be used for both input and output and is normally used in one of three ways:

(i) As the main input/output medium in a time-sharing system that allows several users to communicate with the machine more or less simultaneously.
(ii) As a control console concerned with the machine operating procedures to allow the computer operator to communicate with the various control programs.
(iii) As the input/output medium for small low-cost mini-computers.

*Print Outs.* The printer is the most important computer output device. It has the ability rapidly to produce large volumes of output in a form that is immediately intelligible to most people at speeds of up to 1200 lines per minute with each line containing 120 characters. This output, called *print out*, is now a familiar part of our daily lives as more and more commercial concerns use it for invoicing, sales records and so on. Several types of printer are used. Single character printers are really typewriters under computer control and can operate at speeds up to 10 characters a second. Chain printers use a continuously moving chain of type slugs, which has on it several repetitions of the basic character set, together with hammers fired against the paper pressing it onto an inked ribbon and the required type slug. This type of *lineprinter* normally operates at speeds of up to 600 lines a minute. Barrel printers also use a continuously moving type-face but this is embossed onto the surface of a rapidly revolving metal cylinder. They are faster than the older chain printers but are now being superseded by printers which rely upon non-mechanical methods such as optical and thermal processes similar in many ways to xerography.

*Graphical displays.* Virtually all the examples and exercises given in this book involve printer output but in Chapter 13 attention is drawn to programming methods that use a number of specialised peripherals to generate a graphical output from the machine. A Chinese proverb reputedly says that *a picture is worth a thousand words*, and it is certainly true that a great deal of computer output would be best presented as diagrams, graphs and maps. Although the standard symbols used on the lineprinter can be shuffled to produce very crude pictures (*see Chapter 10*) the *cathode ray tube* and *graph plotter* are far more flexible and easy to use. A cathode ray tube is similar to a television screen on which can be displayed numbers, letters, characters and lines. Because this process is entirely electronic it is also very rapid but unless the screen is photographed there is no permanent record of the output produced and the small screen can limit some applications. Certain types of system also allow the user to input information back through the tube using a *light pen* whose movements are tracked by the machine as it crosses the tube's face. Geographers and land-use planners are beginning to use this type of display to show the results of simulation studies in map and graph form and to allow the user to interact with the computer to modify his program or data as the job proceeds. The graph plotter is a pen driven under computer control across the face of a recording medium such as paper. Although an inked pen and paper are normally used other possible media are a light source on photosensitive paper and a cutting tool across metal. Two examples of plotter output are shown in Fig 2.5, one of which shows some computer art, the other the results of a more serious study. Plotters can be of two types. In a *drum plotter* the pen moves in one plane while the paper is moved in the other whereas a *flat-bed* plotter produces generally higher-quality output by keeping the paper in a fixed position and moving the pen in both the x and y planes across its surface. Plotter output has obvious uses in automatic

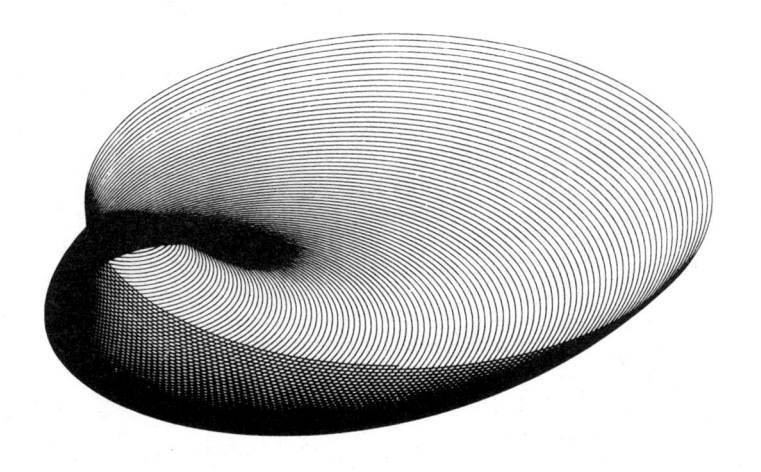

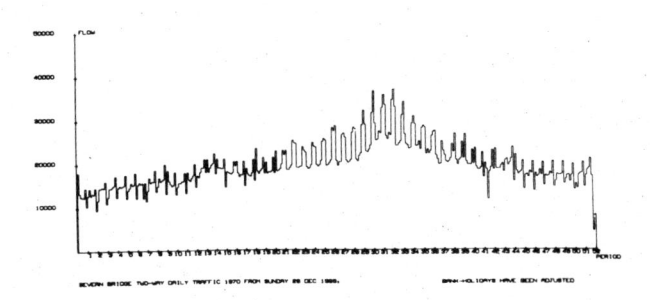

*Fig 2.5* Examples of graphical output

cartography but it is very slow when compared to the CPU speed and it is normal practice to operate it independently of the CPU using a previously prepared magnetic tape containing plotter commands.

From this brief survey it will be apparent that the range of possible input/output peripherals is very great.

A typical computer installation will have a selection of devices and probably a preferred mode of working. Fortunately it is possible to learn a great deal about computing and get extremely useful results using only the simplest and most common method in which cards are input and the output is by lineprinter and for the greater part of this book we have used this method.

## 2.4 Backing Store

Our computer system now has the hardware to enable it to input, process and output information.  Backing store in one form or another enables the computer system to store very large quantities of information in a manner that is readily accessible to the CPU.  Magnetic tapes are probably the most widely used of such storage media but their serial access means that the time taken to retrieve a specific datum depends upon where the read/write heads are positioned relative to where on the tape the required information is located.  If we remember that a tape can be over half-a-mile long, then it is obvious that this could be a lengthy process perhaps of the order of minutes rather than milliseconds.  In contrast, core storage is *random* in that one item can be accessed as quickly as any other, but is very much more expensive than magnetic tape and could not economically be used to hold information that is used only infrequently.  As a compromise between these two types of store, magnetic drums, discs and cards have been developed which have some of the characteristics of both.

*Magnetic drum* store was used in the early computers and consists of a copper-sleeved cylinder coated with a magnetic substance.  This working surface is divided into a number of tracks and each track into sectors which can hold a single word or character.  The track density along a drum might be of the order of 30 per inch with perhaps 100 bits of information per track.  Every storage location can thus be addressed by its track and sector numbers and as the drum is rapidly rotated past a series of fixed read/write heads items can be accessed very quickly.

*Magnetic disc* store has a number of discs each with two recording surfaces mounted on a central spindle rather like a stack of gramophone records.  Each surface has a number of concentric recording rings or tracks again divided into data sectors and can be read or written to by heads located on moveable arms above and below each disc.  As the discs rotate so each datum passes a head and can be accessed if

required.  Although disc sizes vary greatly a typical pack might hold 100 million or more bytes.
*Magnetic card* devices have an even greater capacity up to several thousand million characters.  As the name suggests, these are small magnetised cards and are stored in a magazine.  The required card is first selected from the stack and then individual data accessed from it.

## 2.5 Software

A computer system is much more than the physical components which make up the hardware, it includes the sets of instructions known as the *software* that allow effective use of this equipment and nowadays a manufacturer will spend as much time and effort in developing software as he does in designing hardware.  To understand the various software elements we must revert to our student cartographer introduced at the beginning of this chapter.  It will be remembered that we introduced the idea of control over the exercise and that this control had two distinct components:

(i) A series of operations devised by the particular student to enable his map to be drawn which would almost certainly differ from student to student. This is the *program* for the specific task.
(ii) A control function exerted by his tutor which would be much the same for all students doing the exercise and which has its computing equivalent in the *operating system* or *system software* of the machine. This system software is built up of instructions to control the work flow, organise the input/output peripherals and ensure that each program is efficiently handled.

Operating systems can exert control in many ways.  Perhaps the simplest system is that used in early machines in which the computer processed one job at a time.  Jobs followed sequentially into the machine and while in it had available to them all the CPU's store.  This type of operating is intrinsically inefficient.  Many jobs will require only a fraction of the store available and will waste expensive CPU time while the fast processor waits for input and output to be transferred to and from the extremely slow peripherals.  To see how serious this delay can be, imagine a slowed-down CPU which performs one instruction per second.  This slowed system would receive one character *per day* from an input teletype, one character every five minutes from a fast card reader and one character every ten seconds from a very fast magnetic tape drive!

It is apparent that, unless special arrangements are made, the fast CPU will spend most of its time waiting on its slow peripherals so that almost all modern operating systems try to cut down this slack time. Systems vary greatly, but at their heart is usually a special executive program for managing the physical resources of the machine such as CPU time, main memory, disc space or on-line peripheral time in order to achieve some desired objective such as the maximum throughput of work. Modern machines have a direct memory access facility in which, in addition to the CPU, there are a series of controllers for the fast peripherals that can communicate directly with the memory and transfer blocks of data without wasting CPU time. When the CPU encounters an instruction to transfer data it gives the peripheral controller the first location, or address, of the memory to or from which data are to be transferred. This transfer is initiated but the CPU leaves the controller to get on with it and moves to the next instruction. At the end of the transfer the controller issues a signal, or 'interrupt', which forces the CPU to check that the transfer has been successful and to allow it to process the transferred data.

A second way of keeping the CPU occupied is in multiprogramming where direct memory access is used but in addition the memory contains data and instructions for more than one program at any one time and the CPU switches between programs doing whatever useful work it can.

The nature of individual operating systems varies greatly from small machines which do not have such a system and rely upon the supervision of an operator to large and powerful machines that have extremely complex systems capable of performing a great many functions more or less simultaneously A system that is widely used in British universities with ICL 1900 series computers is called GEORGE and some four or five basic versions of this system are now operational. The very popular IBM System 360 machines also have a range of available operating systems. The novice programmer need not know very much about the workings of his particular operating system, simply enough to identify his work and request that the facilities his programs require be available, but the more advanced computer-user may spend a lot of time and effort manipulating the operating system to gain most advantage from the computer.

Of more importance to the novice are the sequences of instructions needed to instruct the machine to perform specific tasks.  Just as data must be presented in machine-readable form, so must these instructions be presented in a form that the computer can recognise and translate into appropriate actions.  The language used by the computer is called *machine code*, and because the machine communicates in numbers this code consists of a series of digits to specify:

   (i) The operations to be performed,
  (ii) The quantities on which they are to be performed, and
 (iii) What to do with the result.

Programming in machine code is extremely difficult even for a skilled programmer and very early in the development of computers it occurred to computer scientists that the machine itself might be used to translate from a simple programming language into machine code and then execute that code as before.  In order to do this the computer is supplied with a special program called an assembler or compiler that can be brought into operation from the backing store.  The simplest is the *assembler* which permits the use of assembly languages using mnemonics such as ADD and MULT for the operations and labels such as X and Y for the quantities to be used.  The computer now takes over the responsibility of ensuring that all the required quantities are correctly stored.  Assemblers work in two ways.  Some operate as *interpreters* whereby each individual assembly-language instruction is translated into machine code and then executed immediately whereas others act as *translators* which convert the entire program into machine code and then execute that code.

In using assembly languages we have progressed from basic machine codes, but programming is still a difficult and tedious business.  Fortunately we can go at least one stage further and allow one program statement to generate a whole series of machine instructions and rely upon a special program called a *compiler* provided by the machine to perform the necessary conversions.  In doing this some efficiency will be lost but computer languages using macro-instructions can be written in forms that resemble a natural language and use standard mathematical notation. A number of such 'high-level' languages have been developed.

*Fig 2.6* Five forms of the same calculation

Mathematical expression to be evaluated:  x = b + c

| | |
|---|---|
| In binary | 101011-001101001 |
| | 110111-001011101 |
| | 101101-010110110 |
| In assembler | F-C41 |
| | A-C54 |
| | S-D06 |
| In ALGOL | X := B + C ; |
| In BASIC | LET X = B + C |
| In FORTRAN | X = B + C |

Figure 2.6 illustrates how the same calculation might be written in machine code, assembly and high-level languages.
   The most common high-level languages are COBOL (Common Business Orientated Language) for commercial use, and ALGOL (Algorithmic Language), BASIC (Beginner's All Purpose Symbolic Instruction Code) and FORTRAN (Formula Translation) for scientific use. Recently, in an attempt to produce a multi-purpose language, PL/1 has been developed to combine features from all four. Of these languages FORTRAN is the oldest and most widely used. Since its introduction in a fairly primitive form in 1954 it has been developed into a highly practical programming language for scientific use and the vast majority of scientific programs are written using it. Although it cannot equal the elegance of ALGOL or the simplicity of BASIC it has proved easy to learn and use and virtually all college and university programming environments allow its use. It is for this reason that FORTRAN is used in all the examples and exercises but there is no reason why the principles used should not be utilised in any other high-level language.

2.6 Worksheet
1 computers are often seen to be large, impersonal grey cabinets or giant tape recorders full of mystery, an attitude that often seems to be encouraged by computer personnel. To dispel this attitude we suggest that you

visit your computer centre in order to examine the hardware and watch the operators at work.

2 Acquire and examine examples of punched cards, paper tape, print-out and digital plotter drawings.

3 Write down the following decimal numbers in their binary representation: 2 4 8 32 3 27 1024 1111.
Write down the decimal equivalents of the following binary numbers: 10111 11 100101 1111 111 10001 11001.
Write down the following decimal numbers in binary coded decimal form: 42 987 66 1230.

4 Find out which high-level language compilers are available at your computer installation.

2.7 Further Reading

Barry, R.G. 'The punched card and its application in geographic research', *Erdkunde*, 15, 140-2 (1960)
Chandler, K. *A dictionary of computers* (1970)
Cutbill, J.L. (ed) *Data processing in biology and geology* (1970)
ECU *Automated cartography and planning* (1971)
Fenichel, R.R. and Weizenbaum, J. (eds) *Computing and computation* (1971)
Fry, T.F. *Computer appreciation* (1970)
Gregory, K.J. and Brown, E.H. 'Data processing and the study of landforms', *Zeitschrift dur Geomorphologie*, 10, 237-63 (1966)
Hathaway, P. 'Some applications of computer graphics to transportation planning', *Greater London Intelligence Quarterly*, 26, 31-8 (1974)
Hollingdale, S.H. and Toothill, G.C. *Electronic computers* (1965)
Laver, M. *Introducing computers* (1973)
Rosing, K.E. and Wood, P.A. *The character of a conurbation* (1971)
Sawyer, J.S. 'Geographical output from computers, and the production of numerically forecast or analysed synoptic charts', *Meteorological Magazine*, 89, 187-90 (1960)
Sprunt, B. 'Geographics: a computer's eye view of terrain', *Area*, 4, 54-9 (1970)

# Chapter 3
# Writing and Running a Computer Program

3.1 Problem Identification

The view most people have of computers is derived from the mass media. Often the image is of an all-powerful, unfeeling machine which poses some nebulous threat to personal privacy or individual jobs. Alternatively the computer is the latest of man's great achievements and is capable of solving any problem presented to it. It cannot be stressed too strongly that whether or not a computer appears to be unfeeling, producing 'miracle' results, or is just another piece of office furniture depends not only upon the computer but also on the person using it. The machine is told what to do and what not to do by its programmer and the results will only be as good or as bad as the programmer and program allow them to be. Unless extremely advanced methods are used the computer has no common sense and can do no extra thinking for its programmer; every step in the computer solution must be spelt out in advance by the computer program. There is thus much more to the art of problem-solving by computer than the work carried out by the machine, and in this chapter the complete process of setting up a problem for computer solution is examined. This can be subdivided into six sequential and related steps (Fig 3.1).

Initially the programmer must *identify* a worthwhile problem and then define it in a way that is capable of solution. A great many worthwhile problems may prove to be either insoluble with the available resources or capable of solution without computer assistance, but in the past twenty years the range of problems tackled by computer methods has broadened enormously. In applications such as the calculation of summary statistics and simple indices, problem identification and definition may be very simple, but in others, such as complex resource allocation

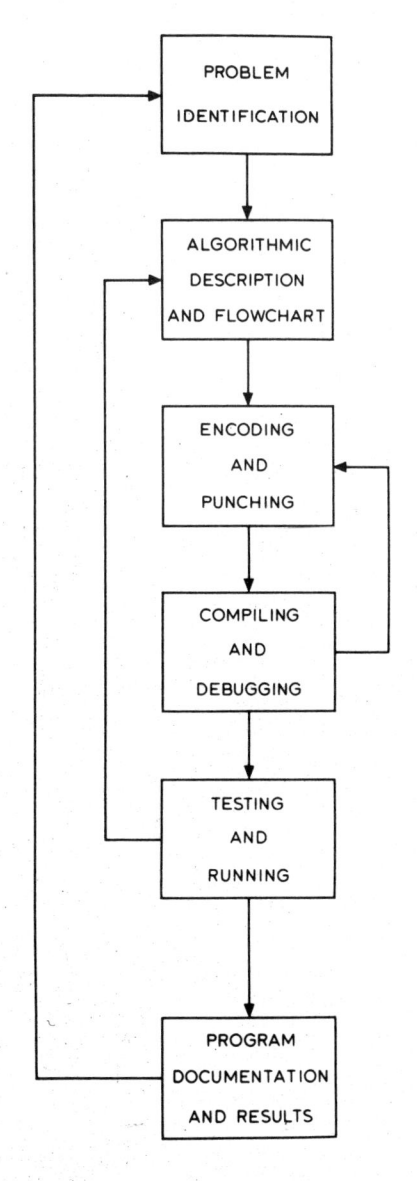

*Fig 3.1* Stages in problem solving by computer

models (Cicchetti, 1973; Hamilton *et al.*, 1969), it may
be very difficult and occupy a large proportion of the
programmer's time.  The identification and definition of
problems is beyond the scope of this book, and must
obviously depend upon the intelligence and background of
the student himself.  A number of books concentrate on the
identification (Abler, Adams and Gould, 1971; Morrill, 1970;
Haggett, 1972) and techniques of analysis of geographical
problems (Isard, 1960; Cole and King, 1968; Dornkamp and
King, 1971).

3.2 Algorithmic Description and Flowcharts
From the identification of a problem for which computer
assistance is required in its solution, the programmer now
moves to a description of that problem as an *algorithm*.
An algorithm is a precise and complete statement of a procedure which if followed will lead to the desired solution;
the conversion of good ideas into sound algorithms is at
the heart of effective computer use.  There are two points
to bear in mind when devising an algorithm:

  (i) Everything must be specified in advance, including
      procedures to be followed in every circumstance
      that could possibly arise.
 (ii) The machine cannot think logically unless it has
      been given absolutely explicit rules on which to
      base decisions.  Unless programmed appropriately,
      the machine will not have the wit to spot invalid
      data and nonsensical results.

Let us consider a very simple problem which might occur as
part of a geographic investigation, and attempt to devise
an algorithm for it.  In a study of map projections we
wish to know whether or not three places, A,B and C lie
upon a straight line in the particular projection under
review.  If they are not on a straight line we also want
to know the extent of their deviation from it.  Mathematically, there are a number of possible ways to solve
this problem, but one of the most efficient is to specify
the spatial co-ordinates of the places and to use these as
data to calculate the area of the triangle ABC that they
enclose (Fig 3.2).  A convenient formula to use is:
AREA = $\frac{1}{2}$ $(x_1.y_2 + x_2.y_3 + x_3.y_1 - x_2.y_1 - x_3.y_2 - x_1.y_3)$
where $(x_1,y_1)$, $(x_2,y_2)$ and $(x_3,y_3)$ define the rectangular
co-ordinates of the three places.  Inspection of Fig 3.2
makes it obvious that the points lie on a straight line if,

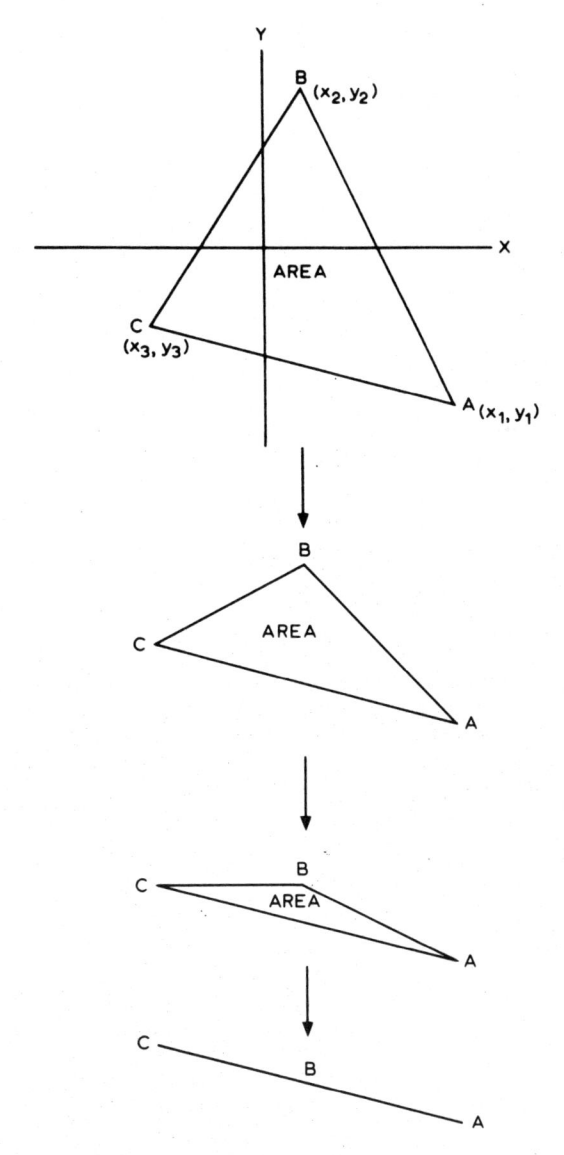

*Fig 3.2* Test of collinearity for three points

and only if, the area of the triangle is zero. If the
points are not 'collinear' the area of the enclosed
triangle provides a measure of the amount of deviation.
The equation forms the basis of an algorithm in that it
suggests a procedure for solving the problem. If we sub-
stitute numerical values for the unknown co-ordinates
$(x_1,y_1)$, $(x_2,y_2)$, $(x_3,y_3)$ we can calculate AREA and use this
knowledge to test whether or not the points are collinear.
The equation is not in itself a complete algorithm because
we also need a precise statement of the sequence of
operations and a precise statement of what to do with the
input data and results. A full algorithm might consist
of the following steps:

  (i) Start the computer.
 (ii) Input the values $x_1$, $y_1$, $x_2$, $y_2$, $x_3$, $y_3$ which will
      probably have been read off a base map.
(iii) Use the equation to calculate the area ABC.
 (iv) If the area ABC is zero then output a message to
      say that the points are collinear. If this area
      is non-zero output its numerical value.
  (v) Stop the computer.

An important tool in algorithmic description for FORTRAN
programming is the *flowchart* which provides a visual representation of the sequence of operations (ICL, 1971;
Schrieber, 1969). Flowcharts enable complex problems to
be readily appreciated and greatly facilitate communication
between different programmers. Each is made up of
a series of boxes, the shapes of which have been standardised
to indicate the type of operation, together with connecting
lines and arrows to show the 'flow of control' between the
operations. The range of possible operations and hence
shapes is now very large but for simple programs six basic
symbols are used:

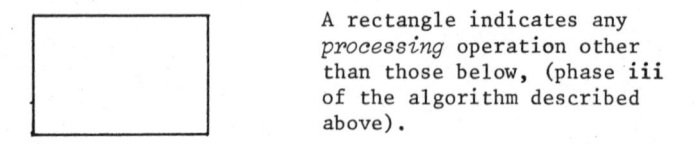

A rectangle indicates any *processing* operation other than those below, (phase **iii** of the algorithm described above).

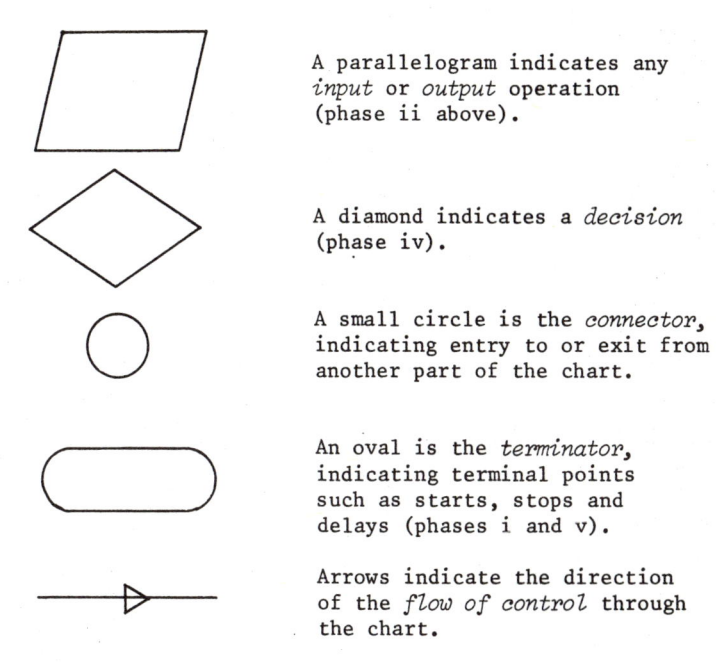

A parallelogram indicates any *input* or *output* operation (phase ii above).

A diamond indicates a *decision* (phase iv).

A small circle is the *connector*, indicating entry to or exit from another part of the chart.

An oval is the *terminator*, indicating terminal points such as starts, stops and delays (phases i and v).

Arrows indicate the direction of the *flow of control* through the chart.

A flowchart for the collinearity algorithm is given as Fig 3.3. A seventh symbol, the *annotation* box, has been used to provide additional descriptive information at the point where AREA is calculated.

Once an algorithm has been devised it is possible to move on to produce a computer program from it, but even at this stage it is useful and sound practice to run mentally through the flowchart using some artificial, but plausible, data in order to ensure that it will cater for all eventualities.

3.3 Coding and Punching

If the algorithm appears to be sound we can proceed to restate it as a precisely defined series of instructions to the computer coded in a particular programming language. Figure 3.4 shows a FORTRAN program for the collinearity problem written out onto special sheets, or *coding forms*, to indicate what the various instructions are and where they should be punched on the standard 80-column punched card. At this stage the precise details of each

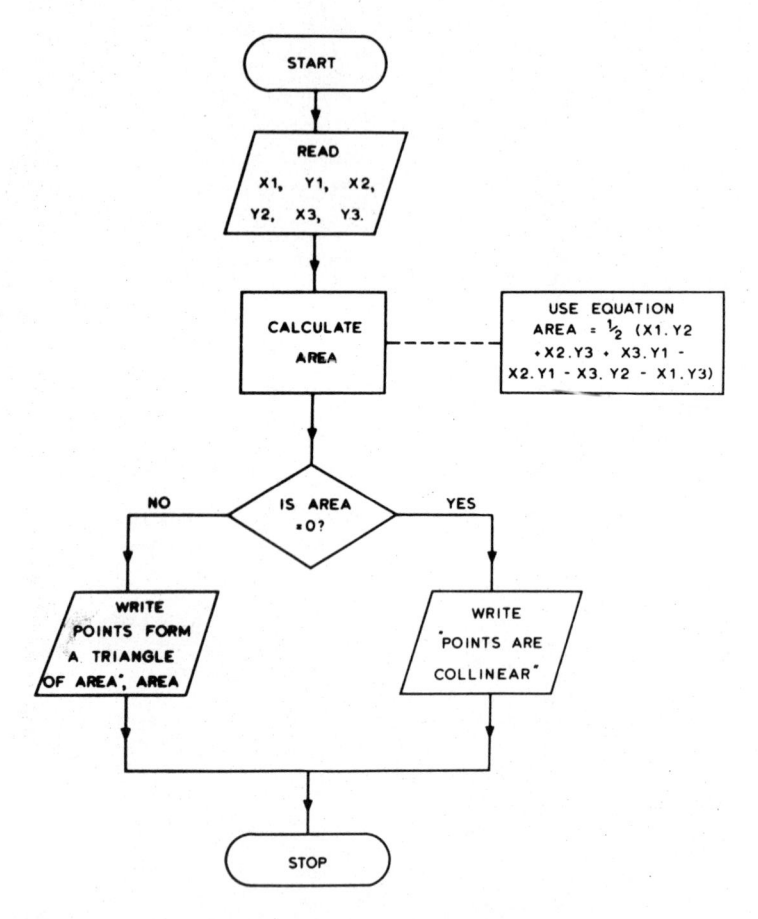

*Fig 3.3* Flowchart for collinearity algorithm

instruction and the various rules for positioning these on punched cards need not worry us and will be dealt with in Chapter 4. With the coding forms as a guide, the complete program can now be punched onto cards. In some institutions the computer laboratory will operate a punching service for its users and all that need be done is to hand

```
C TEST FOR COLLINEARITY OF THREE POINTS                    10
C                                                          20
      NIN = 5                                              30
      NOUT = 6                                             40
C READ IN THE X,Y COORDINATES OF THE POINTS ON INPUT CHANNEL 5
      READ(NIN,100), X1,Y1,X2,Y2,X3,Y3                     60
C CALCULATE AREA USING GIVEN FORMULA
      AREA = 0.5*( X1*Y2 + X2*Y3 + X3*Y1 - X2*Y1 - X3*Y2 - X1*Y3)  80
      IF (AREA) 20,10,20                                   90
   10 WRITE(NOUT,110)                                     100
      GO TO 30                                            110
   20 WRITE(NOUT,120), AREA                               120
   30 STOP                                                130
C FORMAT STATEMENTS SPECIFY HOW TO INPUT THE DATA AND OUTPUT THE ANSWER 140
  100 FORMAT( 6F6.0)                                      150
  110 FORMAT( 26H THE POINTS ARE COLLINEAR )
  120 FORMAT( 36H THE POINTS FORM A TRIANGLE OF AREA ,F10.3)
      END
```

*Fig 3.4* A FORTRAN coded program for the collinearity problem

in the completed forms, but at others it will be necessary to learn how to use a keypunch machine.

3.4 Compiling and Debugging

When the cards have been punched the programmer is ready to submit his deck of cards to the machine for the first time and at this stage it will be necessary to add special *job control* cards to the deck in order to provide the operating system with information about the job to be processed, identify the particular user and so on. These cards are punched in much the same way as FORTRAN program cards but they are usually unique to the particular operating system and information about them should be obtained from the particular computer laboratory being used.

With the job control cards in place the program can now be run for the first time, but there are so many chances to make mistakes in the programming or punching that most programs do not work when first tried. These errors must be identified, located and removed by changing the statements that are in error and replacing the faulty cards. Fortunately the compiler will have built-in error detection procedures and will politely inform the user on the print out that errors are present. It will usually also give *diagnostics* which locate and describe each individual error. The corrected program is now re-submitted to the machine until it is completely error-free. This process is called *debugging* and can be very simple, but for a long and complex program it might take many weeks of effort.

## 3.5 Program Testing and Running

Although the debugged program now makes sense to the compiler and is written in perfectly acceptable FORTRAN we still do not know for certain that it will give the desired results when executed. This is because errors may have been made in the logic of the program or conditions not anticipated in the original problem definition may appear and cause erroneous calculations. At this stage it is vital that the program is tried using data for which the correct answers are known, a procedure known as *program testing*, before it is used with real data to solve real problems. The user must also take great pains to ensure that his program is numerically sound and will produce as accurate a set of answers as is possible from a variety of data inputs.

## 3.6 Documentation

A final step in the preparation of a computer program, that of *documentation*, is one which is often forgotten but which is very important. The programmer should produce an explanatory version of the program describing its purpose, the algorithm used (with a flowchart) and the results which might be obtained together with a copy of the program statements and an example of the output. Program specifications are of use in at least three ways. They enable the programmer to keep an accurate record of the work he has done, allow communication with other workers in the field and enable other users to make use of the program. Frequently program documentation is very badly done with the excuse that 'it is only intended for my own research', but this is a very short-sighted attitude - if only because in one year's time you may have forgotten how the program was constructed. Some good examples of program specifications are to be found in the serial publication of geological programs from the University of Kansas (Merriam, 1966-71).

All this sounds a long process. As Fig 3.1 shows it may be necessary to backtrack repeatedly to correct errors made earlier, to improve the algorithm and perhaps even to modify the problem. Once a reliable program has been developed, however, the benefits can be enormous. It can be used rapidly to process very large quantities of data without error or to solve a large number of similar problems.

## 3.7 Worksheet

1 Find out how you can get cards punched at your local computer centre.

2 Familiarise yourself with a card keypunch machine, punching a card to show the representation of the characters ABCDEFGHIJKLMNOPQRSTUVWXYZ 0123456789
= + - * / ( ) , . $.

3 Obtain permission to use your local machine, acquire a 'job number' for use in identifying your work to the machine and find out the correct procedures for submitting and collecting work.

4 Ask at your computer centre for details of the correct job-control cards necessary to run a straightforward FORTRAN program that has card input and lineprinter output.

5 Identify the major activities you undertake in a typical day and devise a flowchart representation of them.

6 The arithmetic mean of a series of numbers is given by dividing their sum by the number of items. Draw a flowchart to illustrate these operations.

7 Draw a flowchart for a program to control, with total safety, the operation of the barriers at a railway level-crossing.

## 3.8 Further Reading

Abler, R., Adams, J.S. and Gould, P. *Spatial organisation. The geographer's view of the world* (1971)
Cicchetti, C.J. *Forecasting recreation in the United States* (1973)
Cole, J.P. and King, C.A.M. *Quantitative geography* (1968)
Doornkamp, J.C. and King, C.A.M. *Numerical analysis in geomorphology* (1971)
Haggett, P. *Geography: a modern synthesis* (1972)
Hamilton, H.R., et al. *System simulation for regional analysis: river basin planning* (1969)
I.C.L. *Thinking as a programmer. Basic principles of flowcharting* (1971)
Isard, W. *Methods of regional analysis* (1960)
Merriam, D.F. (ed) *Computer contributions* (1966-71)
Morrill, R.L. *The spatial organisation of society* (1970)
Schrieber, T.J. *Fundamentals of flowcharting* (1969)

# Chapter 4
# The Infrastructure of Fortran

4.1 Giving Orders to the Computer
Earlier chapters have shown that there are many ways of informing the computer what to do and in what sequence to perform the various commands. The main method of communication is by a computer program written in a *language* capable of being absorbed and understood by *system software* (*see Chapter 2*) within the computer. There are many languages available for this purpose and the most commonly used is FORTRAN, with which almost all of this book is concerned.

FORTRAN was created in 1954 by a group of men working for International Business Machines. The name FORTRAN is derived from FORmula TRANslation because the original reason for developing the language was to make the use of computers easier for mathematicians and engineers. In the last 25 years FORTRAN has developed rapidly with improved versions incorporating a wider range of facilities and extending the working vocabulary of the language. As with most languages, 'dialects' have emerged and many university computer facilities offer several versions of FORTRAN each with their own individual linguistic nuances. A standard basic version of FORTRAN does exist - rather like *Queen's English* - and this is termed ANSI FORTRAN. This version was recommended by the American National Standards Institute in 1966, and should be available on virtually all the computers capable of understanding some form of FORTRAN. A revision of this standard is to be published but at the time of writing it has not appeared. Mastery of ANSI FORTRAN can be used as a foundation for learning other versions of the language specific to particular computer installations.

4.2 FORTRAN Statements
A program is often split into several distinct parts and

the more complex the problem being solved by the program the greater number of parts there are to program. One part, however, is the *main program* and this acts as controller for the other *subprograms*, calling them up and using them as necessary. Execution of the whole program starts at the beginning of the main program. A program or a subprogram is composed of *statements* each of which defines a particular instruction or specifies the nature of the data being dealt with. Generally there is only one statement on each punched card. Not all statements are directly executable by the computer, but provide instructions to do something with other statements or to set up the program to enable it to deal with executable statements.

Executable statements are of one of four types:

(i) *Arithmetic statements* which allow calculation to take place and may be used to evaluate arithmetic expressions (*see later sections of this chapter*).
(ii) *Input-output statements* which are used to transfer data into the program and to extract results from the program (*see Chapter 5*).
(iii) *Transfer of control statements* which are used to control the sequence in which the statements are executed (*see Chapter 6*).
(iv) *Testing statements* which are used to apply tests of various types as for example for logical or arithmetic equality between values within the program (*see Chapter 6*).

Executable statements are dealt with in the sequence in which they are written and this sequence is only changed by a transfer of control statement or the main program calling up a subprogram. To allow changes to take place in the sequence of execution it is sometimes necessary to be able to identify individual statements so that control may be transferred to a particular point in the program. Statements may carry *labels* and the label is used as an index system to identify statements.

FORTRAN thus consists of a sequence of statements, each roughly equivalent to a natural language sentence, using symbols somewhat similar to those used in algebra. For example, part of a FORTRAN statement may contain an expression similar to:

$$(X+Y)/Z$$
or $$A+B-C$$

The symbols available are those which have been discussed already in respect of card punches. The symbols are:

| | |
|---|---|
| alphabetic letters | A to Z (capitals only) |
| numerical digits | 0 to 9 |
| = | equals |
| + | plus |
| - | minus |
| * | asterisk |
| / | slash |
| ( | left parenthesis |
| ) | right parenthesis |
| , | comma |
| . | decimal point |
| $ | currency symbol |
| | blank (which in this book is conventionally written as b̲) |

There are thus 36 alphanumeric characters and 11 special characters. Care should be taken when writing some of these characters and it is useful to adopt a clear cut convention. A frequently used one is:

| alphabetic | numeric |
|---|---|
| O | ∅ |
| I | 1 |
| Z̷ | 2 |

4.3 Card Layout
Usually each statement may be transferred to one punched card (*see Chapter 3 and Fig 3.4*) and the card is sufficiently large to hold the whole statement in a single line. Sometimes however a particularly long statement needs more than one card and this is possible by indicating on the second card that the statement is a continuation from the previous card. The exact positioning of the statement on a card is governed by some simple rules.
A card is made up of 80 columns each of which is a potential character position. Column 1 is at the left and column 80 at the right.
   The *rules* of card layout divide the card into five groups of columns:

          Column   1
          Columns  1 -  5
          Columns  7 - 72
          Column   6
          Columns 73 - 80

*Column 1.* Any card having a C in column 1 is treated as a *comment* card and is not processed in any way. By using this facility it is possible to annotate a program and place comments within the program to remind the reader of the purpose of particular statements at various places in the total program. Plentiful use of comment cards makes the program more easily understandable to other people and also may help the original programmer to understand the program if he uses it again in the future (*see Fig 4.1*).

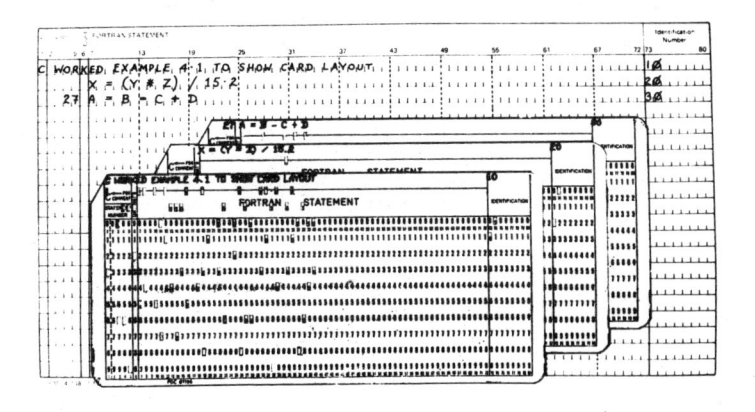

*Fig 4.1* The layout of characters on a FORTRAN statement card

*Columns 1-5.* If a statement is to be labelled (*see above*) then the label number is placed in columns 1 to 5. The label consists of from 1 to 5 digits (eg 1074 or 3 or 22222) and the digits may be placed anywhere within columns 1 to 5. Blanks, even within a number, are ignored. The value of a label number is not significant for it is used only to index the particular statement for cross-reference by another statement. For this reason the same label number may not be used more than once within each section of the program. If a statement is not to carry a label then columns 1-5 are left blank. The third statement in Fig 4.1 carries the label 27.
*Columns 7-72.* These columns contain the main body of the FORTRAN statement which can be punched anywhere in this column block. With one exception noted in Chapter 5

blanks will be ignored and can be used to space out
symbols and make them more easily readable. For example,
spaces are sometimes placed either side of arithmetic
symbols such as:
$$A + B - C$$
Different conventions are adopted by different programmers
so that as a programmer becomes more adept at writing
programs he will develop an individual style.

*Column 6.* If a statement is particularly long and needs
more than one card to contain the symbols then
a *continuation* card is needed. To turn an ordinary card
into a continuation card a symbol other than 0 (zero) or
blank is placed in column 6. If this is done then
column 7 is treated as if it followed directly on from the
previous column 72. This procedure can be repeated for up
to 19 continuation cards but only rarely is it ever
necessary to use more than one continuation card. A continuation line may, of course, only follow an initial line
or another continuation line.

*Columns 73-80.* Any symbols punched in the last eight
columns of a card are not processed. Columns 73-80
therefore may be used to identify the cards making up the
program. Often some identification of the program and
a serial number sequence representing the order of the
cards in the program is punched in these columns. If
a pack of cards is dropped it is much easier to place them
in correct sequence again if the cards carry some numbering
system - and accidents do happen! Figure 4.1 shows three
statements written onto a coding form with the correct card
layout. These statements are transferred to cards as
shown in this figure.

4.4 Turning the Symbols into a Language

Up to this point the symbols in the language of FORTRAN
and the physical layout of these symbols has been discussed. Now it is possible to consider what the symbols
actually represent in a mathematical and logical sense.

Within FORTRAN statements which carry out mathematical
calculations there are two basic types of symbol sequences.
One represents *constants* and the second represents
*variables*. As the name suggests, constants are specific
numerical quantities which remain fixed during the
execution of the program. Constants thus have explicit
values, for example 23 or 714.417. Variables contain
numerical values which may be altered within the program
and during its execution. A variable is referred to by

a name rather than an explicit value. Thus for example, the symbol A or Z or JI (and many others) may be used to refer to a numerical quantity which may, if desired, be allowed to vary through the program.

For example the expression (15.4 + 12.7)/19.2 uses constants but (A + B)/C uses variables and at different points in the program the values in A, B and C may be altered.

*Constants*. In FORTRAN programming two types of numbers are distinguished and these are called *integer* and *real*. Integer and real numbers may effectively be defined by the absence or presence of a decimal point. Integer numbers are whole numbers without decimal points. Thus 7 is an integer constant. Real numbers contain a decimal point, for example 7.1 or even 7. This distinction between real and integer is basic and important to the FORTRAN language because the tyo types of numbers are handled differently by the computer. The different ways of dealing with real and integer numbers will be referred to constantly both in the remainder of this chapter and in subsequent chapters. For most statistical calculations which the geographer needs, it is easiest and often necessary to use real numbers, but integer numbers are useful if a program is required to count values such as in questionnaire analysis.

Constants in a FORTRAN program may be, therefore, integer or real. - Integer constants are written simply as a whole-number. The only non-numeric characters allowed are the minus and plus signs. If no sign is present then the constant is assumed positive. Commas must not be used to separate blocks of numbers. Examples of integer constants are:

1
99
+99
-99
3774624

The maximum positive (minimum if negative) size of an integer constant allowable depends on the particular computer being used. On the large computers generally used by geographers, usually we can safely assume there are no limits. For example, on the IBM 360 series the largest value permissible is $2^{32} - 1$ and this is a computer with a relatively small allowable maximum value. For all geographical calculations the size limit on integer numbers need not worry the programmer.

Real constants are written with a decimal point and may have an optional sign indicated by the + or - character. When no sign is written the number is assumed to be positive. Examples of real constants are:

 1.0
 +2.1345
 -0.001
 635629.9
 2.

Unlike normal practice the comma is *not* used to indicate thousands and millions. Like integer constants the absolute maximum or minimum real numbers allowed can be very large or small but this does not mean that the number of digits calculated and retained is equally large. In fact, most computers will retain between 7 and 16 significant digits and it is worthwhile checking with the individual computer centre how many digits the machine calculates. In most applications in geography this accuracy-limit need not worry us too much but there are situations in which it becomes very important and special techniques to increase the precision of calculations should be adopted.

Often it may be necessary to specify very large or very small real constants. It is obviously very cumbersome to write such numbers with a great many noughts on either side of the decimal point so that FORTRAN allows us to use an exponent mantissa notation. A large number, such as 1,632,500,000 can be written as $1.6325 \times 10^9$, and in FORTRAN this is expressed using the letter E as 1.6325E9. The E and the integer number that follows (ie 9) indicate the exponent, or power of ten, to which the real number, or mantissa, must be multiplied. Both exponent and mantissa can have a positive or negative sign. Examples of this E-form are:

 1.0E6      equivalent to 1,000,000
 1.0E-6                   0.000001
 -2.3245E2                -232.45
 -0.345678E-4             -0.0000345678

*Variables*. Both real and integer constants have one value only during the life of the FORTRAN program. Variables, on the other hand, are identified by a symbolic name and, if desired, may be given a new value several times in the program. In the use of variables the actual number which is being manipulated is stored in a memory cell which is referenced by a symbolic name, for example A, B, C or Z.

Calculations can then be performed referring only to the symbolic names, for example:
$$(A+B)/C$$
This expression may be encountered several times in the program but each time different values lie in the memory cells referenced as A, B and C and so effectively different calculations are carried out although the same expression is used.

In similar manner to the use of constants, real or integer numbers may be dealt with by variables. But real and integer variables have slightly different rules governing the symbolic names which refer to the variables. The general rule for naming variables is:
*The name must consist of from 1 to 6 letters or numbers with the first character being a letter. No other symbols are allowed in the name.*
Additionally for integer variable names:
*For an integer variable, the initial letter must be one of the following: $I, J, K, L, M, N$.*
Additionally for real variable names:
*For a real variable, the initial letter must be one from the sequences A to H or O to Z.*

The first letter of the variable name implicitly determines whether the number referenced by that name is real or integer. Other than the three rules outlined above, the names are left to the invention of the programmer but it can be useful to use names which to some extent reflect the meanings of the values they represent. Although any sequence of letters could be used it is sensible, for example, to use variable names such as AREA to represent a calculated area, DIST a distance and so on.

Examples of acceptable variable names which are implicitly of type INTEGER are I, INDIA, MEAN, NAME, N1974, MUM, J, L4MY, LOUISE, KAKE, M, JUMBO, KLM, IBM and INTEGE. Examples of acceptable real variable names are A, ZOOM, PDP11, POP2, BASIC, BEER, STD, STDEV, QWERTY, VARIAB, ENG6NM, V and even FORTRA. There are ways of overriding the rules governing the naming of integer and real variables but these can be left until later as a good deal of care is required if it is decided to tamper with the normal conventions.

*Arithmetic expressions and operations.* Having defined the types of numbers which can be handled it is necessary to see how the numbers themselves can be manipulated. Five simple operations are possible on expressions involving constants or variables. The operations, with the FORTRAN symbol and examples are:

|   |   |   |
|---|---|---|
| add | + | A+B |
| subtract | - | A-B |
| multiply | * | A*B |
| divide | / | A/B |
| raise to a power | ** | A**3 |

These five symbols allow the evaluation of quite complex expressions. Brackets may be used in expressions and the same rules of evaluation apply as in simple arithmetic with inner brackets evaluated first. If no brackets are included then a convention of evaluation is applied in which the powers are calculated first, followed by multiplication and division, followed by addition and subtraction; evaluation of operations of equal status is from left to right.

In using arithmetic expressions no two operators may appear next to one another in the same expression. Thus, for example A*-B is not permissible. This problem can be overcome by the use of brackets, so A*(-B) is acceptable; alternatively this particular expression could be written as -B*A. While over-use of brackets may slow down execution of a program by a fraction of a second, under-use of them may well result in errors either in the syntax of the program or in the result of the calculation. The latter may be difficult to pick out; if the expression is complicated then an error in the answer may not be readily seen. *In general, if in doubt, use brackets.* Frequent use of brackets has the added advantage that the sequence of evaluation is more clearly understood by someone (other than the programmer) studying the program.

Arithmetic operations may be carried out using either real or integer quantities but, with the exception of raising a real number to an integer power (2.0**5), not upon a mixture of both in the same expression. In integers, multiplication, subtraction and addition are straightforward but divisions, as in the expressions NUM/NS and 3/2, present problems. In performing the second of these divisions the computer will truncate the answer to give an integer result of 1 and not the 1.5 we might have expected. This is a common but easily avoided source of programming error. By similar logic we can raise a real number to a real power, as in 7.0**3.1428, but are not allowed to raise integers to real powers as this would usually give a real number.

A final point to note in dealing with arithmetic expressions is that problems again can arise with division. If, for example, the expression
$$3.0 * (1.0/3.0)$$
is calculated and the program is then required to compare the result with 1.0 then the two numbers would not be equal. Because calculations involve decimal numbers not fractions the result of the expression would be .99999999 (assuming the calculations are to 8 decimal places) which is not equal to 1.0. Fortunately, there are ways of getting around this but these can wait until later chapters. The possibility of such problems has to be anticipated in computer programming.

*Arithmetic assignment statements.* From the previous subsection it can be seen that the computer can be told to deal with complex arithmetic expressions and to evaluate them. But the computer has no common sense. What then has to be done with the evaluated expression? Clearly, the machine must either remember the answer for further use or write it out as an end product. To remember the answer it must be stored under another variable name so it can be easily referenced. This transfer of the result of an evaluation to a variable name is carried out using the symbol =.

The FORTRAN statement
$$A = 17.4 * 12.3$$
will evaluate the right hand side and will reference the answer (214.02) to real variable name A. Such a statement is called an *arithmetic assignment statement* and its general form is
$$variable = expression$$
The statement is an order to calculate the value on the right and to give that value to the variable named on the left.

The use of = in this statement is not equivalent to its general use in most of mathematics. In FORTRAN, = means 'make the variable equal to the value of the expression.' If variables B and C contain 2.1 and 2.2 respectively then the statement A = B+C will assign the value 4.3 to variable A. The right hand side of the statement may be an arithmetic expression (B+C) but equally well it may be a constant or a variable name. For example:
$$X = 74.2$$
$$I = J$$
are perfectly acceptable.

65

The use of the = sign in its FORTRAN meaning can lead to some strange looking statements which would be mathematically peculiar.  Statements such as
$$X = X + X$$
or $\quad X = X + 1.\emptyset$
which in the algebraic sense of = are meaningless, are quite common in FORTRAN.  The first statement simply means double the value of variable X and the second adds 1 to its value.

4.5 Worked Examples
(i) We can illustrate the use of these assignment statements by looking at two examples.  The first is drawn from demography where tedious numerical calculations ideally suited to computer arithmetic are often performed.
Suppose that we are given the total male population, the female population in age groups 0-14, 15-49 and over 50, the total live births, the total deaths in the age group under 1 and the age group 1 and over, and the number of marriages.  In all there are 8 variables.  We are required to calculate the following summary demographic measures:

| | | |
|---|---|---|
| crude birth rate | = | births x 1000/total population |
| crude death rate | = | deaths x 1000/total population |
| **cru**de marriage rate | = | marriages x 1000/total population |
| general fertility rate | = | live births x 1000/female population in 15-49 age group |
| infant mortality rate | = | deaths under 1 x 1000/total live births |
| male/female sex ratio | = | male population/female population |

The sequence of FORTRAN arithmetic assignments to perform the required calculations is shown in Fig 4.2.

Each line of program represents one card of a possible computer program.  The cards are all identified by the key word DEMO carried at the end of each line and the cards are numbered in sequence in columns 78-80.  A series of introductory comment cards containing C in column 1 to signify they contain non-executable statements are followed by eight cards which carry the all-important arithmetic

```
C LET THE FOLLOWING NAMES STAND FOR THE GIVEN AND REQUIRED VARIABLES -   DEMO  10
C POPM - MALE POPULATION, POPF0, POPF15, POPF50, - FEMALE POPULATIONS BYDEMO 20
C AGE GROUPS, POPF - TOTAL FEMALE POPULATION, BL - LIVE BIRTHS, D1, D2 -DEMO 30
C DEATHS IN TWO AGE GROUPS, MED - NUMBER OF MARRIAGES, BR - BIRTH RATE, DEMO  40
C DR - DEATH RATE, CMR - CRUDE MARRIAGE RATE, RIM - RATE OF INFANT       DEMO  50
C MORTALITY, SR - MALE TO FEMALE RATIO, POPTOT - TOTAL POPULATION, FR -  DEMO  60
C FERTILITY RATE                                                         DEMO  70
C                                                                        DEMO  80
C THE FIRST STEP IS TO CALCULATE TOTAL FEMALES AND TOTAL POPULATION      DEMO  90
C THE OTHER REQUIRED VALUES MAY THEN BE CALCULATED IN TURN               DEMO 100
C                                                                        DEMO 110
      POPF = POPF0 + POPF15 + POPF50                                     DEMO 120
      POPTOT = POPF + POPM                                               DEMO 130
      BR = BL * 1000.0 / POPTOT                                          DEMO 140
      DR = (D1 + D2) * 1000.0 / POPTOT                                   DEMO 150
      CMR = MED * 1000.0 / POPTOT                                        DEMO 160
      FR = LB * 1000.0 / POPF15                                          DEMO 170
      RIM = D1 * 1000.0 / LB                                             DEMO 180
      SR = POPM / POPF                                                   DEMO 190
```

*Fig 4.2* A FORTRAN statement sequence to calculate summary demographic statistics

expressions. The first two calculate total populations and the remaining six deal with each index in turn. Thus the death rate for which real variable location DR has been used is equal to the sum total of deaths (D1 + D2) multiplied by a thousand (to turn it into a rate) and divided by the total population (POPTOT). The brackets around (D1 + D2) are necessary to ensure that this total will be calculated first. In a later chapter we will return to these statements to use them as part of a full FORTRAN program.

(ii) The second example is taken from pedology and shows a case where a complex calculation has to be performed on a small amount of data. Casagrande estimated that the depth of freezing in a soil is proportional to the square root of the number of day-degrees below freezing point and according to the equation:

$$z^2 = (1.73 \times 10^5 \frac{k}{mpL}) H$$

where z is the depth of freezing into the soil (cms)
      k is the thermal conductivity of the wet soil (cal/deg C/cm/s)
      p is the density of water (g/cm$^3$)
      m is the volumetric moisture and ice content of the soil (cm$^3$/cm$^3$)
      L is the latent heat of fusion of ice (80 cal/gm)
      H is the number of accumulated days-degree C below the freezing point of the water.

A FORTRAN statement to calculate z from this equation could be:

  Z = ((1.73 * (10.0**5*(AK/(AM*AP*AL))))*H)**0.5

The square root is found by raising the whole expression to the power 0.5. In the expression inner brackets are evaluated first so the term mpL is calculated in the first instance (AM*AP*AL) and it is then divided into K (AK). Unless indicated otherwise by the use of brackets exponentiation is carried out before multiplication so that the next step of evaluation of the expression is 10.0**5 followed by multiplication of this result by the result of earlier steps in the evaluation. With evaluation in brackets from left to right, the next step is a multiplication by 1.73 followed by a multiplication by H. Finally the square root of the whole evaluation is taken and the result entered into real variable location Z. The evaluation is shown in the following diagram:

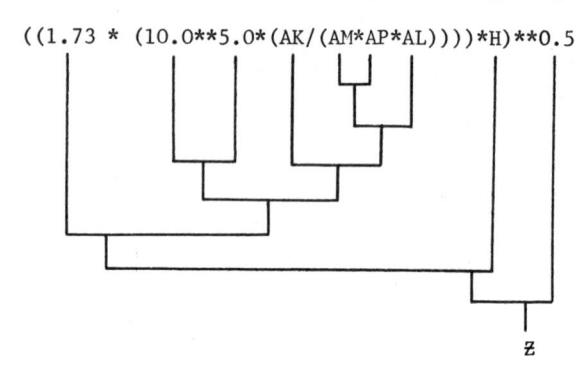

((1.73 * (10.0**5.0*(AK/(AM*AP*AL))))*H)**0.5

Z

## 4.6 Worksheet

1 Transfer the program statements shown in Fig 4.2 to cards.

2 Which of the following numbers are not acceptable as FORTRAN integer constants?
+1760  +12.0  -273  6,080  1E20

3 Write down the following as FORTRAN real constants:
1760  63,360  $10^6$  -760  1/3

4 Using E notation write down the following very large or very small numbers as FORTRAN real constants:
Earth's mass         5,980,000,000,000,000,000,000,000 kgm
Sun's mass           1.99 x $10^{37}$ kgm

Mean distance of earth from sun   149,598,500,000 m
Land area of earth   $1.45 \times 10^{12}$ m$^2$
Mass of an electron   $9.1078 \times 10^{-28}$ gm
Mass of a proton   $1.6725 \times 10^{-24}$ gm

5 Which of the following are not acceptable as FORTRAN real constants and why?
33,000   0.394   3E10   +57.296   -96.2   3.1416   E+7

6 For each of the following variable names state which are acceptable in FORTRAN and to which implicit type they refer:
I, EARTH, GEOG, MOON, COMPUTER, I*, FORTRAN, 2TON, MANAGER, ME-OJ, DEL(2), RAD, NOR21, LOCQUO, ?D

7 What is wrong with each of the following FORTRAN statements:
  (i) ZERO = 19,561.323/M
 (ii) X = ((A+B)/C)-D)
(iii) POPCH = POP1971 - POP1961
 (iv) Y-Z = (Y*-Z)-(Y/-Z)
  (v) MEAN = SUM/N
 (vi) MEAN = MEAN**∅.5

8 Write down the following formulae as FORTRAN arithmetic expressions:

  (i) Area of triangle   $A = \tfrac{1}{2}b.p$   b = base
  p = perpendicular height

 (ii) Slope   $s = v/h$   v = vertical interval
  h = horizontal equivalent

(iii) Wien's Law   $L_{max} = 0.2893/T$   $L_{max}$ = wavelength of maximum emission
  T = temperature (°K)

 (iv) Outmigration rate $= \dfrac{O}{P} \times 100$   O = number of out-migrants
  Immigration rate $= \dfrac{I}{P} \times 100$   I = number of immigrants
  Net migration rate $= \dfrac{I-O}{P} \times 100$   P = total population

69

Gross migration rate $= \dfrac{I + O}{P} \times 100$

(v) Stefan's Law $\quad E = e\sigma T^4$

$E$ = rate of emission of radiant energy
$e$ = emissivity
$\sigma$ = a constant
$T$ = temperature ($^{\circ}K$)

(vi) Newton's Law

$$F = G \dfrac{m_1 \cdot m_2}{d^2}$$

$G$ = acceleration due to gravity
$m_1, m_2$ = mass of objects
$d$ = distance

(vii) Net radiation

$$R = (Q+q)(1-a) + i_d - i_u$$

$Q$ = incident direct radiation
$q$ = diffuse radiation
$a$ = surface albedo
$i_d$ = counter radiation
$i_u$ = outgoing radiation

(viii) Temperature conversion
$$C = 5/9 \times (F-32)$$

(ix) Kinetic energy $\quad E = \tfrac{1}{2} m.v^2$

$m$ = mass
$v$ = velocity

(x) Rossby's relationship for the Zonal Westerlies

$$c = u - B\,(L/2\pi)^2$$

$u$ = velocity of zonal current
$B$ = local rate of change of Coriolis parameter
$L$ = wavelength

(xi) Drag on a sand grain (0.22mm diameter) in air (ie critical velocity when just beginning to move)
$$V = A \sqrt{\left(\dfrac{\delta-P}{P}\, g.d.\right)}$$

$A$ = constant (0.1)
$\delta$ = density of sand
$P$ = density of air
$g$ = gravitational acceleration
$d$ = diameter of the grain

# Chapter 5
# Input and Output Instructions

5.1 The Purpose of Input-Output
So far, only the basic building blocks of the FORTRAN language have been considered. While we now know enough to write statements that solve simple problems, we still cannot write a complete computer program. The purpose of this chapter is to provide the extra bits and pieces to enable simple programs to be written. Two groups of program statements are missing from our working knowledge of FORTRAN. These are commands to write out the answers and commands to transfer data into the computer and are commonly termed output and input instructions.

The data used in the examples in the last chapter have all been in the form of constants, either real or integer. Statements such as
$$X = 7.42$$
$$J = 14$$
were used to allocate constant values to named locations and effectively to introduce data into the program. In these statements X and J remain fixed throughout the program unless reset in similar fashion. It is more common for most of the storage locations used in a program to have data introduced into them from data cards placed at the end of the program. From earlier chapters it is seen that a complete submission to the computer usually consists of three main types of cards:

  (i) Statement cards to control the program on its passage through the computer (job control cards - specific to computer installations).
  (ii) Program statement cards in sequence.
  (iii) Data cards (which usually follow immediately after the program statements) and marked as such by a further job control card.

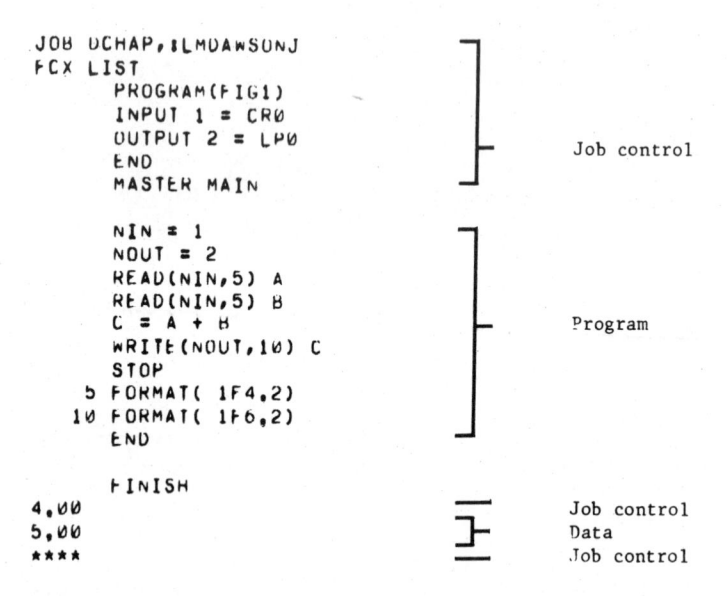

```
JOB UCHAP,ILMDAWSONJ     ⎤
FCX LIST                 |
     PROGRAM(FIG1)       |
     INPUT 1 = CR0       |
     OUTPUT 2 = LP0      ⎬   Job control
     END                 |
     MASTER MAIN         ⎦

     NIN = 1             ⎤
     NOUT = 2            |
     READ(NIN,5) A       |
     READ(NIN,5) B       |
     C = A + B           ⎬   Program
     WRITE(NOUT,10) C    |
     STOP                |
   5 FORMAT( 1F4.2)      |
  10 FORMAT( 1F6.2)      |
     END                 ⎦

     FINISH                  Job control
4.00                         Data
5.00
****                         Job control
```

*Fig 5.1* A listing of a simple program to add two numbers

Figure 5.1 shows a typical series of statements, to perform a trivial addition of 5 and 4 using a particular computer system. It can be seen that the data are placed at the end of the whole program and are *automatically* cross-referenced with the variables named in a READ command. The first READ command, READ (NIN,5), refers to the first datum, ie 4.00. It causes the value 4.00 to be automatically placed in real variable location named A. Similarly the value 5.00 is placed into B. It is now quite simple to replace the two data cards with ones containing different numbers and so to use the program with other data. There are several statements in the program in Fig 5.1 which have not been dealt with in the previous chapter.

5.2 The READ Statement
In the READ instruction, data are transferred to storage locations from a variety of input sources (*see Chapter 3*) of which the most frequently used is punched cards fed

through a card reader. The computer must be informed of the source of data to be dealt with by the READ instruction and its layout on the particular input medium. Each type of input peripheral, eg card reader, paper tape reader, magnetic tape deck etc, is assigned a number. Sometimes this is done by the local computer centre, sometimes in the job control cards at the beginning of a computer job. In the program in Fig 5.1 it is done in the job control cards with the command

INPUT 1 = CRØ

meaning assign the number 1 to the card reader (CRØ). This is not the case with all computers, however, and it is advisable to check what numbers are ascribed to the different types of input and output machines. Because of this variation in numbers used, a convention has been adopted in this book whereby the number of an input device is held as an integer variable and is set by a constant as the first operation in a program. The location is called NIN. In running any program in this book it will be necessary to check the number of the input device being used - the card reader - and then if necessary to alter the integer assignment statement involving NIN. In Fig 5.1 the statement NIN = 1 corresponds with the order in the job control setting the card reader to respond to the number 1. In some computer centres NIN = 5 may be the rule, in some it may be NIN = 15, or in others a different number and yet again in others it may not be necessary specifically to assign these numbers at all.

The READ statement consists of the command READ followed by three specified pieces of information:

(i) The number of the input device to be used (in this book we use the number currently held in location NIN)

(ii) A description of the type and arrangement of the data as it exists at the end of the program. The description is termed the FORMAT of the data and is specified on a separate card which is cross-referenced by a label from the READ statement. In Fig 5.1 the cross-reference is made by the label reference number 5.

(iii) The name of the storage location or locations into which the data are to be transferred, for example

READ (NIN, 15) X, Y, Z, K, J
15 FORMAT (.................)

The read statement begins in column 7 of the standard
computer card. The input device number and FORMAT
reference number are separated by a comma and together
are enclosed in brackets. If data are to be transferred
to several storage locations then the names of the locations
are themselves separated by commas and if several READ
statements are included *each refers to a new card containing data*. A second READ statement cannot be used to
transfer data from a data card previously referenced by an
earlier READ statement. In the example:
```
          READ (NIN, 20) A,B,C
          READ (NIN, 20) X,Y,Z
       20 FORMAT (..........)
```
two data cards are necessary, the first containing three
numbers and the second containing three more numbers.
One data card with six numbers on would be incorrect. If
the data are held on one card containing six numbers then
the first READ statement would have to be changed to:
```
          READ (NIN, 25) A, B, C, X, Y, Z
       25 FORMAT (..........)
```
and the second READ statement left out.

   The second element of the READ command is the FORMAT
specification which describes the type and layout of the
data on the data card – whether for example the data are
numbers or letters, integer or real, as well as the position
of the numbers or letters on the card. The FORMAT statement is not operational in the sense of other statements
encountered so far. It is referred to by the READ statement using a label (*see Chapter 4*). There are a wide
variety of ways the data may be positioned on the data
card and consequently great variety in FORMAT statements
exists. Here we propose to deal with only two simple
input or output formats and one specific output format.
More complicated FORMAT types are discussed in Chapter 10,
where we show how advanced FORMAT types can be used to
produce maps and graphs of geographical variables.

   Return now to the example in Fig 5.1. The FORMAT
referenced by both READ statements is labelled 5.
```
                 5 FORMAT ( 1F4.2)
```
It consists of:

1. A label, in this case 5 punched in columns 1 to 5.
2. The term FORMAT punched from column 7 onwards.
3. A specification conventionally beginning with a space
   and enclosed in brackets. The specification consists
   of a mixture of numbers and letters and indicates at

least three characteristics of the material being considered:
(i) The number of times a *field* is to be repeated sequentially. A *field* is a block of characters which make up a datum.
(ii) The *mode* of the data to be transferred. The *mode* is the type of data - integer, real, alphanumeric - and is indicated by a single letter.
(iii) The *number* of characters in the field.
In this case the specification is ( 1F4.2) giving one field of data in the F-mode with the number of characters signified by 4.2.

A *field* may be repeated any number of times. If so many fields are specified that, in total, they add up to a greater number than the available character spaces on a data card then the next data card is considered as referring to the one FORMAT specification. Because the last column of a card is automatically the end of a field, a particular datum may not be split between two cards. In the example above the 1 refers to one datum (field) requiring transfer from the data card to the variable store. In the first READ statement this number is 4.00 and is transferred to A; in the second READ statement, which uses the same FORMAT statement, 5.00 is transferred to B. In each case only 1 field is transferred.

The *mode* of the data is real. F stands for floating point (real) in which the number of characters (including one for the decimal point) is specified together with the number of characters to the right of the decimal place. Thus F4.2 means the field contains a real number containing 4 characters (including the decimal point) two of which lie to the right of the decimal point. Other examples would be:

F7.1 (7 characters in total - one after the decimal point - eg 12345.6)
F15.9 (15 characters in total - 9 after the decimal point - eg 12345.123456789)

In general the F format is of the form
$$x \, F \, y.z$$
where $x$ is the number of real numbers for transfer
   $y$ is the number of characters in each number
   $z$ is the number of places after the decimal point.

Therefore, $y$ is always greater than $z$ by at least 1. The value of $z$ must also be conditioned by the working accuracy of the computer. There is little point in transferring data with 10 decimal places from data cards to variable locations if the arithmetic manipulations on the variables are accurate to only 8 decimal places.

## 5.3 The WRITE Statement

As well as transferring data from cards to variable storage locations it is also vital to transfer answers from storage locations to a form in which they can be assimilated by the programmer usually as print-out on the lineprinter (*see Chapter 3*). The WRITE statement is the opposite of the READ statement and is governed by broadly similar rules. In Fig 5.1 statements directly relating to the WRITE command are:

1. OUTPUT 2 = LPØ (in job control sequence)
   This assigns the number 2 to the lineprinter and is directly analagous with INPUT 1 = CRØ.
2. NOUT = 2. This sets the integer location NOUT to the necessary value to enable the WRITE statement to access the lineprinter (cf NIN = 1).
3. WRITE (NOUT, 1Ø) C. This statement includes the command WRITE followed by a similar sequence of specifications to those explained for READ. In effect the value in real variable location C is to be output on the lineprinter (device 2) using the format as specified in the FORMAT statement labelled 10.
4. 1Ø FORMAT ( 1F6.2). One number is to be output and it is to contain two places of decimal within a maximum possible total of six characters.

## 5.4 FORMAT specification

The above account of the READ/WRITE statements in Fig 5.1 serves to illustrate the general principles of input-output. Remembering that the more advanced features are being kept in reserve for Chapter 10, there are other points of possible difficulty which need to be mentioned at this stage.

Apart from the labels, statements dealing with input and output all start in column 7, but the data must be punched on the data card in accordance with the format specification associated with it and thus the rules of card layout for FORTRAN programs (*see Chapter 4*) do not apply to data.

The FORMAT specifications in Fig 5.1 all deal with
positive real numbers. If negative numbers are input
as data then the sign becomes one of the characters of the
field length. Thus F4.2 would not be sufficient for
-4.0̸0̸ as the minus sign has increased the field length to
F5.2. Similarly in the WRITE statement sufficient field
length must be given for the inclusion of the sign
irrespective of whether the output numbers are positive or
negative. After a complex calculation the exact field
length of the number to be output or its sign is not
always known and so the exact specification of the field
length in the WRITE format is not possible. In the
F format at least one digit will be printed to the left of
the decimal point even if this digit is 0̸. If the field
is too large, blanks will be inserted to fill out the
field. The number 4.00 output in F10.2 would appear with
b indicating the blank spaces :
bbbbbb4.00 and in the same format -4.00 would be
bbbbb-4.00 and again in the same format .40 would be
bbbbbb0.40

In F mode, output fields of the general form $xFy.z$ must
allow for the sign, a possible 0̸, and the decimal point
characters so that the total field length $y$ must exceed
the number of decimal places specified ($z$) by at least
three. It is usually well worthwhile to *overestimate* the
total field length.

The input and output forms of F mode considered so far
have only dealt with a single datum and the fields have
not been repeated. To repeat a field, the value $x$ in the
general form $xFy.z$ is used to indicate the number of fields
to be input or output. With input there are no problems.
Data in 3F5.2 might appear on the data card thus:

12.1213.1308.10

What appears to be a continuous sequence of digits and
decimal points is broken up automatically into the three
pieces of data 12.12, 13.13 and 8.10. On output it is
relatively easy to write a FORMAT that produces a sequence
of numbers looking like gibberish and it is difficult for
the reader to break the sequence of digits into meaningful
numbers. Again there is a good case in output formats for
stating a longer overall field length than the minimum
required. In this way the blanks used to fill the field
become spaces separating meaningful numbers.

The F mode is only one of many modes each of which may
be used to deal with different types of data. For
example, if integer values are being transferred either

from data cards to variable locations or from variable
locations to lineprinter output then I mode is used. All
the problems over decimal points in F mode are absent in
I mode. The general form of I mode is
$$xIy$$
where $x$ is the number of times the field of length $y$ is to
be repeated. Acceptable I mode formats are:
 ( 5I4) ie five integer numbers of four characters taking
   up columns 1-20 of the data card.
 ( 2I3) ie two integer numbers each of 3 characters.
 ( 20I8)
In the last case two data cards would be required to carry
the data. On the data cards the number of columns needed
to carry the values is the multiple of the field length
and number of times it occurs. Again, as in F mode,
negative signs occupy separate positions.

Both I and F mode formats may occur in multiple and
mixed forms. In multiple forms each format block is
separated by a comma so that
              10 FORMAT ( 2F4.1, 3F9.3, 1F5.2)
would be quite permissible and would relate to data of the
following form:
         11.1-3.344444.44455555.55566666.66677.77
Similarly ( 3I2, 4I6) or other multiple forms are equally
allowable. From these multiple formats it is a short
step to mixed formats, for example
                  ( 3I2, 4F9.2, 2I2)
In this manner it is possible to input or output both real
and integer values in the same READ or WRITE sequence. If
a multiple format mode specification is used, care must be
taken to ensure that the real numbers are read into, or
written from, real variable locations and that integer
mode specifications relate to integer mode variable names.

In this first look at input-output methods and simple
formatting it is necessary to consider only one final
format mode. This is the H mode and for the moment it
relates only to output. In F and I modes the output of
a program is limited to numbers, and there can be no
annotation of any of the answers produced. All that is
written onto the lineprinter paper are numbers. Often,
however, it is useful to have pieces of text which
annotate particular numbers. If, for example, we are
calculating birth rates and death rates from a set of
figures it would be more useful if the answers came out as:
         BIRTH RATE IS 20.4 PER THOUSAND
         DEATH RATE IS 18.7 PER THOUSAND

than if we merely got:
               20.4
               18.7
Format specifications using H mode allow us to add text to answers in this fashion.  The piece of text is inserted as part of the program, not as data, and so it is not placed in a storage location but is kept constant throughout the program.  In similar way to other FORMAT statements the H mode consists of

   (i) A label (occupying columns between 1 and 5 on the punched card).
  (ii) The word FORMAT (usually starting in column 7).
 (iii) The specification enclosed by brackets (usually starting in column 13 or 14).

The specification first states the number of characters to be printed in the piece of text (characters can be special symbols, numbers or letters), followed by the letter H, followed by a blank (which is counted as a character), followed by the string of characters which are required to be printed.  For example:
         15 FORMAT ( 14H BIRTH RATE IS).
The spaces between words are counted as characters in similar fashion to the space after the H.  To use this FORMAT statement a WRITE statement of the form
               WRITE (NOUT, 15)
would be used where NOUT has previously been set to the output device number.  No variable location needs to be specified as no value is in effect output;  only the specified characters in the format are written.
  Mixing of modes within a single format specification is permissible so that an annotated answer would be produced from the statements:
 15 FORMAT ( 14H BIRTH RATE IS, 1F5.1, 13H PER THOUSAND)
       WRITE (NOUT, 15) BR
when real storage location BR held the calculated value to be output.  In this case a value is to be output, between two H mode specifications, and so in the WRITE statement a variable name must be stated.  Within the H mode the full character set is available for use.  No matter what it is, each character including blanks is counted separately in the specification of the field length, as for example in:
            (23H POP. CHANGE. 1961-71 =, 1F6.2)

79

The length of an H mode field is only limited by the width of the lineprinter paper which is normally 120 characters but which does vary from machine to machine.

The three format modes, F, I and H allow for most of the input and output needs of simple programs and as such will be sufficient for the moment. Some other modes will be considered in Chapter 10 when we have acquired the knowledge to use them. The position of FORMAT statements in the program does not materially affect the program in any way but clearly the position of READ and WRITE statements will. A program cannot contain orders to WRITE an answer before it has calculated the value to be written out, nor can it read cards it does not have. The labelled FORMAT statements do not need any information, other than that in the statement itself, and the method of cross-referencing READ and WRITE to FORMAT by the use of labels enables them to be placed anywhere in the statement sequence. Their position differs with different programmers' styles. Some like to keep the FORMAT statement close to the input-output statement to which it refers. Others prefer to collect them together at the end or beginning of the program.

5.5 STOP and END
Before we can write a program and effectively produce answers a statement sequence is required to stop the program when it ends. Two statements are available. These are:
STOP
END
and both begin in column 7 of the punched card. The last program statement must always be END, for this signifies that the program is now complete. Its use is illustrated in Fig 5.1. The statement STOP merely stops execution of the program and is used in programs in which some critical condition in the program has been reached indicating that no further processing is necessary. The STOP command then can come anywhere within the program but is usually associated with some additional statement which checks the value of a storage location. Having seen how to end a program we are now in a position to write a program, run it and produce answers intelligible to the programmer and other people.

5.6 Summary
In this chapter an introduction to the use of statements to transfer data cards into the computer store and from the

store to the lineprinter has been presented.  The form of
the READ and WRITE statement has been considered and the
method outlined for (i) indexing the device number and
(ii) labelling the FORMAT statement within the read and
write instructions.  Three types of FORMAT mode have been
considered:

   (i) F mode, for real numbers, with the general form
       ($x$F$y$.$z$)
  (ii) I mode, for integer numbers, with the general form
       ($x$I$y$)
(iii) H mode, for any symbols, with the general form
       ($y$H$\underline{b}aaaa....a$)
where $y$ is the field length of the format
    $x$ is the number of times the field length is to be
       repeated
    $z$ is the number of digits after the decimal point
       in the real value
    $\underline{b}$ is a blank
    $\overline{a}$ is an acceptable symbol, numeric, alphabetic or
       special.

Finally the commands STOP and END have been mentioned to
indicate the means of halting execution of the program.

5.7 Worked Examples
(i) *Problem.* In the previous chapter an example of a small
scale demographic analysis was used to illustrate a sequence
of programming instructions.  This example will now be
extended into a full program to calculate the various demographic ratios for Italy.
*Algorithm and flowchart.* The values are calculated as laid
out in worked example 4.5(i) and a very simple flowchart
for the program is shown in Fig 5.2.
*Program.* A listing of the program is shown below (Fig 5.3)
and is basically the same as that in the previous chapter,
but with the additions of input-output statements.
The input statement is:
READ (NIN,1∅) POPM, POPF∅, POPF15, POPF5∅, BL, D1, D2, WED
and is cross-referenced to both
                NIN = 5
          1∅ FORMAT ( 8F1∅.∅)
The statement instructs the computer to transfer the eight
data elements from a single data card to the eight named
locations.  The layout of the data on the card is

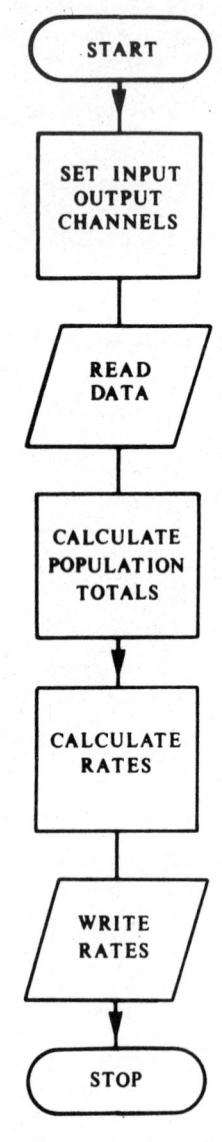

*Fig 5.2* Flowchart for worked example of demographic analysis

```
C  LET THE FOLLOWING NAMES STAND FOR THE GIVEN AND REQUIRED VARIABLES -  DEMO   10
C  POPM - MALE POPULATION, POPF0, POPF15, POPF50, - FEMALE POPULATIONS BYDEMO   20
C  AGE GROUPS, POPF - TOTAL FEMALE POPULATION, BL - LIVE BIRTHS, D1, D2 -DEMO   30
C  DEATHS IN TWO AGE GROUPS, WED - NUMBER OF MARRIAGES, BR - BIRTH RATE, DEMO   40
C  DR - DEATH RATE, CMR - CRUDE MARRIAGE RATE, RIM - RATE OF INFANT       DEMO   50
C  MORTALITY, SR - MALE TO FEMALE RATIO, POPTOT - TOTAL POPULATION, FR -  DEMO   60
C  FERTILITY RATE                                                         DEMO   70
C                                                                         DEMO   80
C  READ IN DATA                                                           DEMO   81
       NIN = 5                                                            DEMO   82
       NOUT = 6                                                           DEMO   83
       READ (NIN,10 ) POPM,POPF0,POPF15,POPF50,BL,D1,D2,WED                DEMO   84
C  CALCULATE VALUES                                                       DEMO   89
C  THE FIRST STEP IS TO CALCULATE TOTAL FEMALES AND TOTAL POPULATION      DEMO   90
C  THE OTHER REQUIRED VALUES MAY THEN BE CALCULATED IN TURN               DEMO  100
C                                                                         DEMO  110
       POPF = POPF0 + POPF15 + POPF50                                     DEMO  120
       POPTOT = POPF + POPM                                               DEMO  130
       BR = BL * 1000.0 / POPTOT                                          DEMO  140
       DR = (D1 + D2) * 1000.0 / POPTOT                                   DEMO  150
       CMR = WED * 1000.0 / POPTOT                                        DEMO  160
       FR = BL * 1000.0 / POPF15                                          DEMO  170
       RIM = D1 * 1000.0 / BL                                             DEMO  180
       SR = POPM / POPF                                                   DEMO  190
C  WRITE OUT ANSWERS                                                      DEMO  200
       WRITE ( NOUT,20)                                                   DEMO  210
       WRITE (NOUT,25) BR,DR,FR                                           DEMO  220
       WRITE (NOUT,15) RIM,CMR,SR                                         DEMO  230
    25 FORMAT ( 14H BIRTH RATE IS,1F8.3,14H DEATH RATE IS,1F8.3,18H FERTI DEMO  240
      *LITY RATE IS,1F8.3)                                                DEMO  250
    15 FORMAT ( 28H RATE OF INFANT MORTALITY IS,1F8.3,1/H MARRIAGE RATE IDEMO  260
      *S,1F8.3,13H SEX RATIO IS,1F8.4)                                    DEMO  270
    20 FORMAT ( 25H RESULTS FOR ITALY ARE -- )                            DEMO  280
    10 FORMAT ( 8F10.0)                                                   DEMO  290
       STOP                                                               DEMO  295
       END                                                                DEMO  300
```

*Fig 5.3* Program listing for worked example of demographic analysis

specified by the FORMAT statement and is such that each datum takes 10 card columns. The various totals and rates are then calculated in statements DEMO 120 to DEMO 190 inclusive. Output is by the lineprinter (device number 6 - NOUT = 6) and the answers are prefaced by an H mode FORMAT statement giving the name of the country to which the data refers. The FORMAT statements are collected together at the end of the program sequence.

*Data and run.* The raw data for Italy in 1968 are:

|  |  |  |
|---|---|---|
| Male population |  | 26,188,500 |
| Female population aged | 0-5 | 6,394,000 |
|  | 15-49 | 13,202,000 |
|  | over 50 | 7,707,000 |
| Live births |  | 930,641 |
| Deaths aged less than 1 |  | 31,511 |
| aged over 1 |  | 478,611 |
| Marriages |  | 375,074 |

The results, in demographic rates per thousand population,

83

```
RESULTS FOR ITALY ARE --
BIRTH RATE IS  19.760 DEATH RATE IS  10.831 FERTILITY RATE IS  70.492
RATE OF INFANT MORTALITY IS  33.859 MARRIAGE RATE IS   7.964 SEX RATIO IS  1.2525
```

*Fig 5.4* Results for Italy of demographic analysis program

are shown in Fig 5.4. The birth rate is greatly in excess of the death rate but natural increase is tempered by a relatively high rate of infant mortality. To obtain comparative data for other countries the program needs to be run several times each time with data for a different country.

(ii) *Problem.* When a geomorphologist analyses a sample from a sedimentary deposit one of the first operations is to sort the contents of the sample into size groups using various laboratory techniques. The grain sizes are most conveniently referred to the $\phi$ (pronounced 'phi') scale on which $\phi$ is $-\log_2$ of the particle diameter expressed in mm, and the weights in each size group are shown by a graph on which the vertical axis represents the cumulative percentage by weight of the sample and the horizontal axis the $\phi$ sizes from very coarse boulder-sized particles ($\phi$ less than $-8$) through to very fine clay-sized material ($\phi$ greater than $+9$). Such a graph is shown in Fig 5.5.
*Algorithm and flowchart.* Inman (1952) and Folk and Ward (1957) suggest a series of statistics to characterise these curves and so differentiate types of deposit which depend upon the interpolated $\phi$ values at cumulative percent values of 95, 84, 75, 50, 25, 16, and 5 (*see Fig 5.5*). There are three measures of the average particle size, two of its sorting or dispersion about this mean, three of its 'skewness' or departure from the normal curve and two of its kurtosis or peakedness. The measures are:

| | |
|---|---|
| Median diameter | $Md\phi = \phi 50$ |
| Inman mean diameter | $M\phi = \frac{1}{2}(\phi 16 + \phi 84)$ |
| Folk and Ward mean diameter | $= \frac{1}{3}(\phi 16 + \phi 50 + \phi 84)$ |
| Inman sorting measure | $\sigma\phi = \frac{1}{2}(\phi 84 - \phi 16)$ |
| Folk and Ward sorting | $= \frac{1}{4}(\phi 84 - \phi 16) + (\phi 95 - \phi 5)/6.6$ |
| Inman first skew measure | $= (M\phi - Md\phi)/\sigma\phi$ |

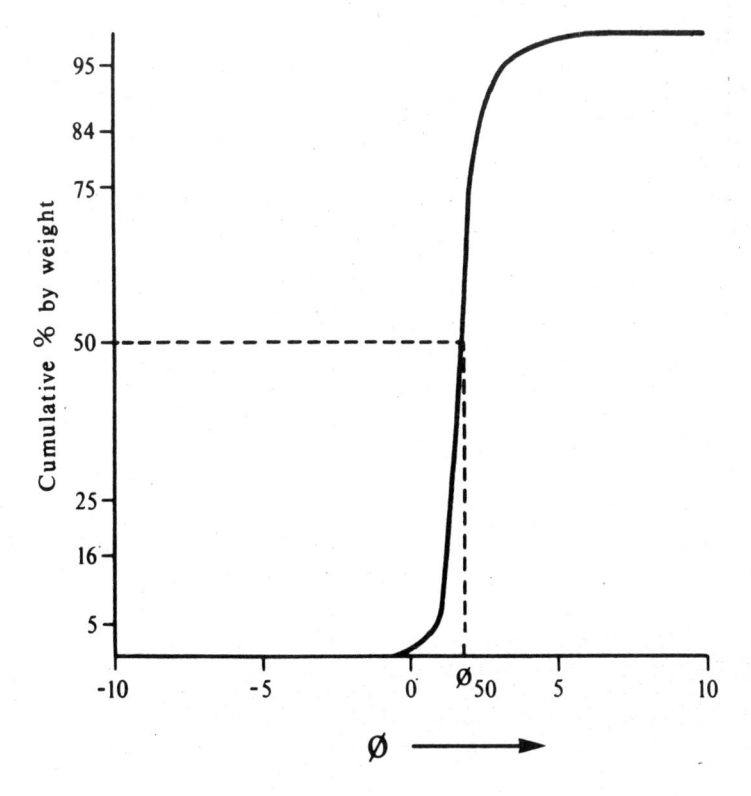

*Fig 5.5* An example of a sediment size distribution curve

Inman second skew measure = $\frac{1}{2}((\phi 5 - \phi 95) - Md\phi)/\sigma\phi$

Folk and Ward skew measure = $\frac{\phi 16 + \phi 84 - 2.\phi 50}{2(\phi 84 - \phi 16)} + \frac{\phi 5 + \phi 95 - 2.\phi 50}{2(\phi 95 - \phi 5)}$

Inman kurtosis measure = $(\frac{1}{2}(\phi 16 - \phi 5) + \frac{1}{2}(\phi 95 - \phi 84))/\sigma\phi$

Folk and Ward graphic kurtosis = $(\phi 95 - \phi 5)/ 2.44 (\phi 75 - \phi 25)$

These calculations are simple enough but when done by hand they are both time-consuming and error-prone; instead we prefer to write a very simple computer program that makes

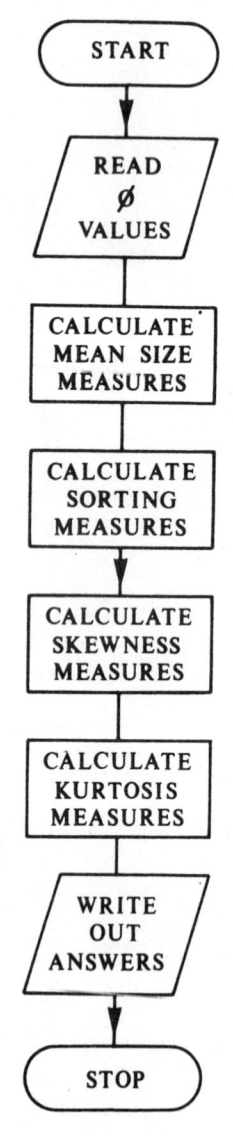

*Fig 5.6* Flowchart for worked example for grain size analysis

```
C PROGRAM TO CALCULATE INMAN AND FOLK AND WARD MEASURES FOR A SIZE           10
C OF SEDIMENT DISTRIBUTION CURVE                                              20
C REFERENCE = KING,C,A,M,(1966) TECHNIQUES IN GEOMORPHULOGY,274               30
C                                                                             40
C SET INPUT AND OUTPUT FOR THIS MACHINE USING VARIABLES NIN AND NOUT          50
C                                                                             60
      NIN = 5                                                                 70
      NOUT = 6                                                                80
C READ IN THE PERCENTAGES GREATER THAN VALUES WHICH HAVE BEEN                 90
C READ OFF THE DRAWN CURVE AT PHI VALUES OF 95,84,75,50,25,16 AND 5          100
      READ(NIN,90) PHI95,PHI84,PHI75,PHI50,PHI25,PHI16,PHI5                  110
C CALCULATE THE MEANS                                                        120
      AMD = PHI50                                                            130
      AINM = 0.5 * (PHI16 + PHI84)                                           140
      FWM = (PHI16+PHI50+PHI84)/3.0                                          150
C CALCULATE THE SORTING MEASURES                                             160
      SINMAN = (PHI84-PHI16)/2.0                                             170
      SFW = ((PHI84-PHI16)/4.0) + ((PHI95-PHI5)/6.6)                         180
C CALCULATE THE SKEWNESS MEASURES                                            190
      SKEWI1 = (AINM-AMD)/SINMAN                                             200
      SKEWI2 = (0.5*(PHI5+PHI95) - AMD )/ SINMAN                             210
      WIGS = ((PHI16+PHI84=2.0*PHI50)/(2.0*(PHI84-PHI16))) +                 220
     1       ((PHI5+PHI95 - 2.0*PHI50)/(2.0*(PHI95- PHI5)))                  230
C CALCULATE THE KURTOSIS MEASURES                                            240
      BINMAN = (0.5*(PHI16-PHI5) + 0.5*(PHI95 -PHI84 ))/ SINMAN              250
      FWGK = (PHI95-PHI5)/(2.44*(PHI75-PHI25))                               260
C WRITE THESE ALL OUT AT ONE GO                                              270
      WRITE(NOUT,92) AMD, AINM, FWM,SINMAN,SFW, SKEWI1,SKEWI2, WIGS,         280
     1 BINMAN, FWGK                                                          290
      STOP                                                                   300
   90 FORMAT(7F5.2)                                                          310
   92 FORMAT( 10F10.3)                                                       320
      END                                                                    330
```

*Fig 5.7* Program listing for grain size analysis

use only of assignment and input/output statements.

A suitable flowchart and program to perform these operations are presented as Figs 5.6 and 5.7. The algorithm is extremely simple; all the computer need do is to read in seven numbers, perform ten specified arithmetic assignments involving these numbers and output the results. As the flowchart shows this may be represented by a straight-line sequence of operations.

The computer reads its data from a single data card on which are punched the seven $\phi$ values that characterise the grain size curve according to the FORMAT statement labelled 90. Each number occupies five columns on the card and has a decimal point with two decimal places to the right of it. These values are thus assigned to the locations named PHI95, PHI84, PHI75, PHI50, PHI25, PHI16, and PHI5. The next ten statements are the assignments needed to form the various statistics and place their values into the locations AMD, AINM, FWM, SINMAN, SFW, SKEWI1, SKEWI2, WIGS, BINMAN and FWGK. The meaning of each assignment in the

FORTRAN program should be perfectly clear if the reader compares them with the arithmetic expressions set out at the start of this section. Finally the ten statistics for the input curve are printed out using the pair of statements:

```
      WRITE (NOUT,92) AMD,AINM,FWM,SINMAN,SFW,SKEWI1,
     1                SKEWI2,WIGS,BINMAN,FWGK
   92 FORMAT ( 10F10.3)
```

*Sample data run.* Two runs have been performed to contrast sediments from two distinct sandy beach environments. Sand samples were collected from a beach cusp trough and from midbeach level and a size of sediment distribution curve derived by laboratory analysis. The interpolated $\phi$ values for each curve were:

| Percentile | Sample A (beach cusp) | Sample B (mid-beach) |
|---|---|---|
| 95 | 2.50 | 3.15 |
| 84 | 2.25 | 3.10 |
| 75 | 2.20 | 3.05 |
| 50 | 1.90 | 2.68 |
| 25 | 1.50 | 2.60 |
| 16 | 1.30 | 2.45 |
| 5  | 1.10 | 1.95 |

The data card for the first run was punched:
    2.50 2.25 2.20 1.90 1.50 1.30 1.10
and for the second
    3.15 3.10 3.05 2.68 2.60 2.45 1.95
The results obtained were:

| Statistic | Run 1 Sample A (beach cusp) | Run 2 Sample B (mid-beach) |
|---|---|---|
| Median diameter | 1.900 | 2.680 |
| Inman mean diameter | 1.775 | 2.775 |
| Folk and Ward mean diameter | 1.817 | 2.743 |
| Inman sorting measure | 0.475 | 0.325 |
| Folk and Ward sorting | 0.450 | 0.344 |
| Inman first skew measure | −0.263 | 0.292 |
| Inman second skew measure | −0.211 | −0.400 |
| Folk and Ward skew measure | −0.203 | 0.038 |
| Inman kurtosis measure | 0.474 | 0.846 |
| Folk and Ward graphic kurtosis | 0.820 | 1.093 |

As these statistics are usually interpreted they would

indicate that the beach cusp sands are of medium size, are
well sorted, but are non-normal with rather more coarse
material than might be expected and having a distribution
that is also slightly more peaked than a normal curve.
Sediment B from the mid-beach is a very well sorted fine
sand with skewness values that indicate that it is very
nearly a normal curve.  The kurtosis values indicate
a curve which is in fact slightly less peaked than a normal
curve.

5.8 Worksheet
1 Correct the following input-output statements.  Assume
the input channel is 1 and the output channel is 2.
   (i)      READ (1,12) A.B.C.
           12 FORMAT 3F9,2
  (ii)     WRITE (2,15) (X,Y,Z)
           15FORMAT( 2F9.2)
 (iii)     READ (1,1) A,J,K,B
        1 FORMAT ( 1I2,2F4.1,1I3)
  (iv)     WRITE (8,1) ONE,TWO,THREE
        8 FORMAT (3F7.9)

2 Because it is a highly variable result of a large number
of factors, aridity is notoriously difficult to define.
In principle it represents the balance between water supply
and loss at the earth's surface but in practice this cannot
easily be measured.  The balance depends mainly upon precipitation and temperature so that many authors have produced indices of aridity that involve some relationship
between these two readily measured variables.  Three such
indices are:

$$\text{Lang's Rain Factor} = p/t$$
$$\text{De Martonne's index of aridity} = p/(t + 10)$$
$$\text{Angstrom's coefficient of humidity} = p/(1.07^t)$$

$t$ = temperature in $°C$
$p$ = rainfall in mm

Given that one inch of rainfall is the same as 25.4mm and
that the conversion between $°F$ and $°C$ is $°C = 5/9(°F - 32)$,
write a computer program that will input the mean annual
rainfall and temperature in inches and $°F$, convert these
to mm and $°C$ and then calculate and print each index.
Run your program three times with the following data for
three European cities and comment upon the results
obtained:

|  | Precipitation (inches) | Mean annual temperature °F |
|---|---|---|
| Bergen, Norway | 78.3 | 46.0 |
| London, England | 23.4 | 50.9 |
| Athens, Greece | 15.7 | 64.0 |

3 The X and Y co-ordinates of London, Cambridge and Bristol on a kilometre grid are:

| London | 430 | 179 |
|---|---|---|
| Bristol | 258 | 172 |
| Cambridge | 446 | 258 |

Write and run a program to determine by Pythagoras (distance = $((x_1 - x_2)^2 + (y_1 - y_2)^2)^{0.5}$) the straight line distances between each pair of places. The Automobile Association provides road distances (in miles) as follows:

| London-Bristol | 116 |
|---|---|
| London-Cambridge | 54 |
| Bristol-Cambridge | 148 |

Write the program so that it will calculate the straight line distance as a percentage of actual road distance for each of the three routes. Annotate your program with suitable comment cards and your output with appropriate H FORMAT statements. A possible flowchart is provided in Fig 5.8.

4 Modern geomorphology has concentrated a great deal of attention upon the relationships between the functioning of drainage basins and measures of their geometric properties, or form. Some properties, such as the length of watershed (p), basin area (a), maximum (H) and minimum (z) elevations, total length of stream channel (L) and number of streams (n), can be measured directly off maps whereas others must be derived by simple calculation. Among the most often used derived measures are:

| Basin relief | r = H - z (ft) |
|---|---|
| Basin relative relief | Rr = 100 r/5280 p (%) |
| Basin circularity | C = a/(area of circle with same perimeter as basin) |
|  | = $4\pi a/p^2$ |
| Drainage density | D = L/a (miles/sq.mile) |
| Stream frequency | f = n/a (no./sq.mile) |
| Ruggedness number | Rn = D.r/5280 |

Write a program to input values for p, a, H, z, L, and n and calculate r, Rr, C, D, f and Rn.
The fourth-order basin of the Soar river south of Sharnford,

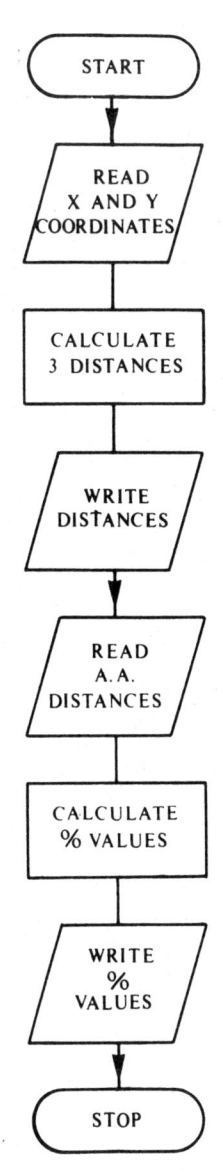

```
          ┌─────────┐
          │  START  │
          └────┬────┘
               ▼
         ╱──────────╲
        ╱   READ     ╲
        ╲  X AND Y   ╱
         ╲COORDINATES╱
               │
               ▼
         ┌──────────┐
         │ CALCULATE│
         │3 DISTANCES│
         └─────┬────┘
               ▼
         ╱──────────╲
         ╲  WRITE   ╱
         ╱DISTANCES ╲
         ╲──────────╱
               │
               ▼
         ╱──────────╲
        ╱   READ     ╲
        ╲   A.A.     ╱
         ╲DISTANCES ╱
               │
               ▼
         ┌──────────┐
         │ CALCULATE│
         │ % VALUES │
         └─────┬────┘
               ▼
         ╱──────────╲
         ╲  WRITE   ╱
         ╱    %     ╲
         ╲  VALUES  ╱
               │
               ▼
          ┌─────────┐
          │  STOP   │
          └─────────┘
```

*Fig 5.8* Flowchart for program to compare straight line and and road distances between three pairs of X Y co-ordinates

Leicester has values p = 12.71 miles, a = 25.66 sq.miles, H = 395ft, z = 217ft, L = 211.6 miles and n = 19. Test your program using these values as data and collect similar data for a fourth-order basin elsewhere in Britain and re-run your program. How effectively do the indices characterise these basins?

5 In the drawing of pie-diagrams it is necessary to calculate the proportion of the full circle accounted for by each of the items that together comprise the total. The areas (in hectares) of crop types in the major regions of Denmark are:

|  | Zealand | Bornholm | S.Islands | Funen | Jutland |
|---|---|---|---|---|---|
| Cereal | 232835 | 24973 | 162262 | 153364 | 1183319 |
| Pulse | 5463 | 273 | 4966 | 3840 | 10745 |
| Roots | 20283 | 1821 | 34879 | 25067 | 191767 |
| Green fodder | 46856 | 7383 | 22426 | 46398 | 627537 |
| Other crops | 30293 | 2638 | 22436 | 14545 | 27918 |
| Fallow | 700 | 82 | 192 | 386 | 855 |
| Horticulture | 2494 | 11 | 1186 | 3903 | 2373 |

*Source:* Landsbrugsstatistick 1971, Danmarks Statistik (1972).

Write and run a program to calculate the percentage of Zealand devoted to each crop and from this calculate the degree divisions for a pie-diagram. Re-run the program four times, once for each of the remaining regions.

6 Given the data below on new dwellings completed in Wales and the United Kingdom as a whole, compare the constructional activity in the two areas by calculating the following measures:

   (i) Ratio of private to public dwellings completed in 1962 and 1972.
  (ii) Population per new dwelling completed in 1962 and 1972.
 (iii) Percentage of national completions (private) accounted for by Wales in 1962 and 1972.
  (iv) Percentage of national completions (public) accounted for by Wales in 1962 and 1972.

|  | Population (1962) | Dwellings completed private | public | Population (1972) | Dwellings completed private | public |
|---|---|---|---|---|---|---|
| Wales | 2653000 | 7501 | 7609 | 2735000 | 10635 | 4135 |
| UK | 53314000 | 178211 | 135432 | 55798000 | 200568 | 130179 |

## 5.9 Further Reading

Folk, R.I. and Ward, W.C. 'Brazos River bar: a study in the significance of grain size parameters', *Journal of Sedimentary Petrology*, 27, 3-26 (1957)

Inman, D.L. 'Measures for describing the size distribution of sediments', *Journal of Sedimentary Petrology*, 22, 125-45 (1952)

# Chapter 6
# Transfer of Control

6.1 Program Control
The program instructions introduced in the last two
chapters enable a very simple sequence of computer operations to be performed. Instructions will always be
executed in the same order as they appear on the punched
cards. Once started, a program can only be stopped using
STOP and END. Often it is necessary to alter this rigid
sequence by programming the computer to perform instructions
in a sequence other than that dictated by the card order,
for example to repeat a particular part of a program many
times, or to repeat a complete program with many different
data sets. To do these more complex operations, we need
to have methods for the *transfer of control* from one part
of a program to another. FORTRAN specifies two families
of statement which effect this transfer. These are:
> GO TO statements
> IF statements

6.2 GO TO Statements
The simplest transfer of control can be achieved by the
unconditional GO TO statement which consists of the
command GO TO followed by a label which defines the next
statement to become operational. An example of such
statement use is:

$$A = B+C$$
$$GO\ TO\ 99$$
$$X = Y/100$$
$$Z = W*(X*X)$$
$$99\ \ D = A-B*C$$

The statement GO TO 99 serves to transfer control directly
to the statement labelled 99 and the two assignment statements involving storage locations X and Z will be ignored.
In this example, control is transferred to statements
lower in the sequence, but this is not a rigid rule and

the labelled statement can occur before the GO TO command.
In such cases control is looped back to an earlier part of
the program thus allowing part of the program to be repeated.
One common example of this type of loop is used when control is returned to the start of the program to enable the
next data block to be considered. The GO TO command may
be labelled to any statement without an existing label.
Transfer of control with the simple GO TO is *unconditional*;
transfer will occur under all circumstances.

   A second type of GO TO statement is one which allows
conditions to be attached to a GO TO statement so that it
transfers control only if particular provisions are met,
otherwise the operational sequence of statements is undisturbed. In this form the statement consists of GO TO
followed by a string of labels enclosed in brackets and
an integer storage location name preceeded by a comma.
The storage location must be of integer form. If the
content of the storage location is 1, control is transferred to the first label in the string, if the storage
location is 2, control passes to the second label, and so
on. For example:

```
         GO TO (15, 24, 3, 420), KEY
      15 A = B/C
         GO TO 1
      24 A = B*C
         GO TO 1
       3 A = B+C
         GO TO 1
     420 A = B-C
       1 A = A*A
```

Depending on the integer value in storage location KEY,
which can take on values from 1 through to 4, location
A will contain a different arithmetic relationship between B and C. An unconditional GO TO is required after
each labelled statement to stop sequential operation of all
the labelled statements below the chosen one. This form
of the GO TO statement is called a *computed* GO TO. If
in this example for some reason, KEY had a value greater
than 4 or less than 1 the program would execute the next
statement in sequence, in this case that labelled 15;
usually this would not be what the programmer had intended.

6.3 IF Statements

Transfer to control is also effected by the *conditional*
IF *statement*. The IF statement can allow a wide and
logically complex set of conditions to be tested before

control is transferred. There are two basic forms of the
IF statement. In the simpler form, the contents of
a storage location are checked to determine whether they
are negative, zero or positive. If negative, control is
passed to one label; if zero, to a second label; and if
positive, to a third. For example the statement
IF (ANS) 9,99,999
checks the contents of ANS. If it is negative then control is passed to labelled statement 9, if zero to 99 and
if positive to 999. The storage location in the IF
statement may be of either real or integer form. An extension of this IF statement allows the storage location
to be replaced by an arithmetic expression, for example
(A+B) or (A+100.0) or of even more complex form and containing brackets. In this case it is the result of the
expression which is tested as negative, zero, or positive.
This type of IF statement (termed *arithmetic* IF) allows
a three way condition to be tested.

An alternative form of the IF command allows more types
of condition to be tested. Generalised, this second IF
form is:
IF(e)S
where e is a relational expression and S is any executable
statement, except another IF of the same form or a DO
(*see Chapter 7*). This definition of the *logical* IF
requires considerable expansion.

The *executable statement* may be any of those dealt with
in this, earlier and the following chapters. Thus it may
be a simple assignment statement such as:
A = B+C
or an input/output statement
WRITE (NOUT, 15)
or a control statement
GO TO 44
The *relational expression* introduces a new concept.
A question may be asked about a relationship between two
expressions and the answer is either yes (true) or no
(false). In the IF statement if the answer is yes then
the executable statement is performed, if the answer if no
then the next statement in sequence is performed.
A possible question may be: 'Is the content of location
A equal to 10.00?' This question becomes the expression
IF (A.EQ.10.0)
An executable statement, say B = D+E, then follows and will
be executed only if A equals exactly 10.0. In full the
statement becomes

                    IF (A.EQ.1∅.∅) B = D+E
Such a statement has considerable use in many geographical
problems.  In classification procedures it is often
necessary to check whether two items are identical;  if
they are then they belong to the same class, otherwise
they are placed in different classes.  In other problems
it may be necessary to test whether a value equals some
critical score, and depending on whether it does or does
not then different actions are taken.
    The .EQ. relationship is one of several possible logical
questions which may be asked.  The other ones are:
                .LT. - less than
                .LE. - less than or equal to
                .GT. - greater than
                .GE. - greater than or equal to
                .NE. - not equal to
    By analogy with the *arithmetic operators* introduced in
Chapter 4 these statements are called *relational operators*
and provide quite a powerful armoury for testing logical
relationships.  Two simple examples serve to illustrate
their use.  A body of data on the housing characteristics
of British towns in 1971 has been read into the computer.
Part of the analysis consists of a study of towns with
a population between 10,000 and 25,000 inclusive and we
wish to know how many of these there are.  If we assume
that the populations are placed successively into storage
location POP the following statements could be used:

    KOUNT = ∅
      .   .   .
      .   .   .
 1∅ (Assign population of next town into POP)
      .   .   .
      .   .   .
    IF (POP.LT.1∅∅∅∅.∅) GO TO 1∅
    IF (POP.GT.25∅∅∅.∅) GO TO 1∅
    KOUNT = KOUNT + 1
      .   .   .
      .   .   .
    (remainder of analysis on towns in size class
    10,000-25,000)
      .   .   .
      .   .   .
      .   .   .
    GO TO 1∅

The tests of population size combine the conditional statement with an unconditional GO TO. The unconditional GO TO returns control to the beginning of the program to allow repetition of the analysis for each town in turn. In order to count the number of towns in the sample a simple counting procedure has been used. When the answer to both the IF statements is yes then 1 is added to the existing value in storage location KOUNT. By setting an initial value of zero in this location a cumulative count procedure is set up. KOUNT is not continually reset to zero because the label 10, the statement to which control is passed by the GO TO command, follows KOUNT=∅ in the statement sequence.

A commonly used model of demographic evolution serves as a second example. The model states that there are three main stages in the demographic transition of a community. These may be summarised as:

Stage 1   high birth rate (crude birth rate over 30‰)
          high death rate (crude death rate over 15‰)
Stage 2a  high birth rate (crude birth rate over 30‰)
          rapidly declining death rate (crude death rate 30‰ and under)
Stage 2b  moderate birth rate (crude birth rate 30‰ and under)
          moderate death rate (crude death rate over 15‰)
Stage 3   low birth rate (crude birth rate 30‰ and under)
          low death rate (crude death rate 15‰ and under)

Stages 2a and 2b are alternatives. Placing a whole series of communities into their appropriate stage is a repetitive exercise well suited to automation by computer. In a program to categorise communities a series of IF statements are necessary to decide whether or not the birth rate is above or below 30 per thousand and the death rate above or below 15 per thousand. A flow chart for such a program could be based on Fig 6.1. Each of the three questions would be programmed as logical IF statements.

The power of logical IF statements may be increased by using them in combination and using the operators .OR. and .AND. The .OR. operator is satisfied if either of the expressions it joins is true. The .AND. operator requires both expressions to be true. Thus in the first example above the two IF statements can be combined to read

    IF(POP.LT.10000.∅.AND.POP.GT.25000.∅) GO TO 10

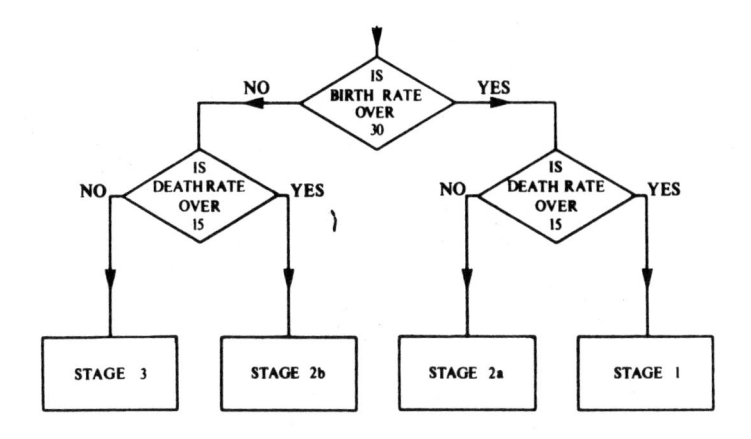

*Fig 6.1* Flowchart for classifying communities on the
basis of their demographic stage

More than one condition can be tested in a single statement. The conditions furthermore need not relate to the same storage location. In the second example above the IF statements also could be combined. For example, assuming that storage locations CB and CD hold the birth and death rates respectively then the statement
$$IF(CB.GT.3\emptyset.\emptyset.AND.CD.GT.15.\emptyset)$$
will isolate stage I communities. Similar combined logical IF statements would isolate the other demographic stages.

A more complex hypothetical, combined logical IF statement could be:
$$IF((A.LE.B+C.AND.X.LT.Y).OR.I.EQ.1\emptyset\emptyset)STOP$$
Real and integer locations are here used in the same IF statement but not in the same relationship within the IF statement. So A.LE.I is inadmissible. The effective use of the .AND. and .OR. is exactly as in the simpler cases; if all the conditions are true the statement at the end of the IF statement is executed, if either of the first two conditions is false and I is not equal to 100 then the next statement in sequence is executed. Evaluation of the statement is .AND. operations followed by

.OR. operations unless altered by the use of brackets.
As in arithmetic expressions, inner brackets are evaluated
first.  The inner brackets in the above expression are
not strictly necessary in this particular case.

In the use of IF statements the execution sequence must
be remembered.  Consider the following table:

| Statement sequence | Execution sequence | |
|---|---|---|
| | if X<10 | if X>10 |
| A = B+C | 1 | 1 |
| IF(X.LT.1Ø.Ø)D=E | 2 | |
| F = G+H | 3 | 2 |

If X is greater than 10 no value is placed in storage
location D, furthermore and more importantly the third
statement is *always* executed, no matter what the value of
X.  In some instances there may be a critical value with
items scoring a lower value requiring treatment in one way
(for example addition) and items scoring above the critical
value requiring a different treatment (for example subtraction), the statements:

        IF(X.LT.1Ø.Ø)D=E+Z
        D = E-Z

are incorrect.  What is required is:

        IF(X.LT.1Ø.Ø)D=E+Z
        IF(X.GT.1Ø.Ø)D=E-Z

Both IF statements are necessary.  If the second IF is
not included D will always equal E-Z even when X is less
than 1Ø.  This is a similar situation to that which arose
in the discussion of the computed GO TO when an unconditional GO TO had to be added in the program sequence.

6.4 CONTINUE Statements

In the transfer of control, labels are frequently attached
to statements.  The unconditional and computed GO TO
always involve a use of labels.  Sometimes it proves
necessary to label an already labelled statement, at
other times it is inconvenient to label a particular
statement although this is the next statement to be executed.  Under these circumstances what is required is
a statement which has no arithmetic or logical effect in
itself but which merely serves as a peg on which to hang
a label and which can be placed at any point to which
control has to be transferred.  The CONTINUE statement
serves these needs.  CONTINUE is a dummy statement which
causes no action but which may be labelled.  Quite simply
it allows execution to continue.

6.5 Summary

This chapter has considered some basic statement types used to transfer control in a computer program. Two main types have been outlined:
> GO TO statements
> IF statements

The treatment of neither is exhaustive and discussion has concentrated on their most frequently used forms. The forms explained, in summary, are:

(i) The *unconditional* GO TO of the form GO TO K *where* K is a statement label.
(ii) The *computed* GO TO of the form GO TO(K1,K2,...Kn),i *where* K1-Kn are labels and i is an integer variable referring to the label sequence.
(iii) The *arithmetic* IF of the form IF(e)K1,K2,K3 *where* e is either an integer or real arithmetic expression and K1,K2,K3, are statement labels.
(iv) The *logical* IF of the form IF(e)S *where* e is a logical expression and S is an executable statement.
(v) The CONTINUE statement.

The chapter has also introduced the concept of repeating instructions by the use of a series of conditional and unconditional statements and the related concept of an incremental counter. The following examples and exercises indicate practical aspects of the use of control transfers.

6.6 Worked Examples

(i) *Problem*. In the study of economic activity one of the first questions a geographer asks is in which areas is economic activity concentrated and which particular industries are local to specific areas? A very simple quantitative way of partially answering these questions lies in the *location quotient*.

The location quotient is a statistical measure which shows the degree to which a specific district has more or less than its share of a particular activity. The measure can be used to show *either* the activity mix of a single region or town by comparing it with the national mix *or* differences in the location of a single activity over a set of town or regions. The amount of an economic activity present at a place can be measured in many ways, by its value added, by the number of employees, by the number of man-hours worked, and several other equally meaningful measures. The location quotient algorithm may be used with any of these measures.

The lowest value of the location quotient is zero which indicates that no activity of the particular type occurs in the area. Complete concentration of all the nation's activity of a particular type into one region will be shown by a quotient of 100/X where X is the percentage share of the activity in the national total of all activities. Between these extremes occurs the point, a location quotient of 1.0, at which a district has a similar share of an industry as is present in the nation. Thus a quotient above 1.0 indicates a district which has a greater proportion of the activity than the national average. One of the earliest users of this measure was Florence (1948). Examples of more recent uses of the location quotient are found in Britton (1967) and Martin (1966) while some problems in its use are considered by Isard (1960).

The present problem is to determine the extent to which male employment in Eire was concentrated in Dublin in 1966. The location quotient may be used for this problem and a location quotient calculated for each employment group.
*Algorithm and flowchart.* The location quotient, using employment as a measure of amount of the activity, is defined as:

$$\frac{a/b}{c/d}$$

*where* a is district employment in selected activity
b is district employment in all activities
c is national employment in selected activity
d is national employment in all activities

In the specific problem under consideration, the district is Dublin, the nation is Eire, and the activities are 13 industry classifications. The data are provided below. The location quotient is calculated for each activity group. In these calculations total employment in Dublin and total employment in Eire are constants. The flowchart is shown in Fig 6.2.

*Program.* The first value read is the number of industry groups for which location quotients are required. The two constant values of total employment in Dublin and in Eire are then read into storage locations DUBALL and EIRALL. A counting procedure is set up to enable a check to be made to determine when all the industry groups have been processed. The counter uses the integer storage

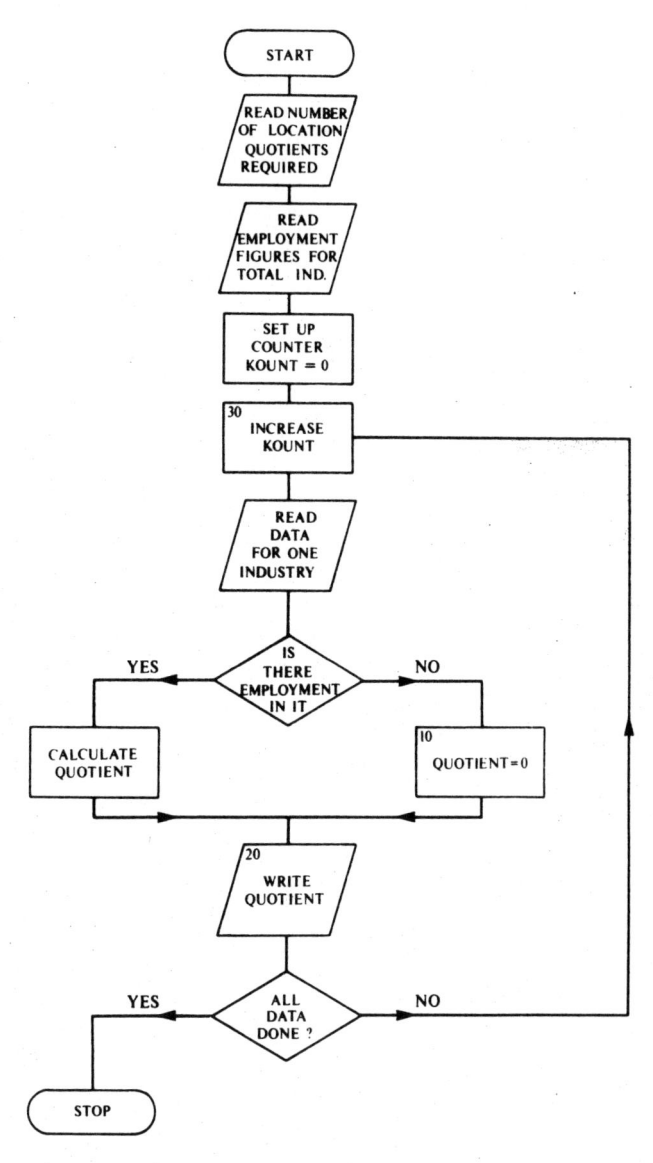

*Fig 6.2* Flowchart for calculation of location quotients

location KOUNT. The counter is incremented and data for an industry group are read. A logical IF statement is used to check whether there is any employment in the particular industry group being processed. If not then the location quotient is immediately made equal to zero (statement label 10) and written out (statement label 20). If a location quotient is to be calculated the calculation is made and the answer is written (statement label 20). A check is then made using a logical IF statement to determine whether all the industry groups have been processed. If some data remain the counter is incremented and the data for the next industry group are read. If all the data have been processed the program stops.
*Data and run.* The data are drawn from the Census of Population of Ireland for 1966 (Central Statistics Office, 1968) and are shown in the table:

Male employment by industry - 1966

| | Code | Dublin | Republic |
|---|---|---|---|
| Agriculture | 1 | 944 | 298014 |
| Fishing | 2 | 156 | 2334 |
| Mining, quarrying and turf production | 3 | 468 | 9079 |
| Manufacturing industries | 4 | 48778 | 135288 |
| Building and construction | 5 | 19027 | 72839 |
| Electricity, gas, water supply | 6 | 4214 | 11182 |
| Commerce | 7 | 31078 | 100419 |
| Insurance, banking, finance | 8 | 4020 | 10420 |
| Transport, communication and storage | 9 | 20458 | 48919 |
| Public administration and defence | 10 | 11979 | 33763 |
| Professions | 11 | 10899 | 37580 |
| Personal service | 12 | 5752 | 15788 |
| Entertainment and sport | 13 | 2472 | 6461 |
| Total | | 160611 | 785196 |

Figure 6.3 shows the program (A) and results (B) obtained from these data. The most heavily concentrated industry, location quotient 2.045, is transport, communication and storage. Such activities would be expected to be concentrated in a capital city such as Dublin. The other tertiary industries show high degrees of concentration. Manufacturing industry with an index of 1.763 is more concentrated than would be expected for the total employment structure of Dublin. The three activities with the

```
C  A PROGRAM TO CALCULATE LOCATION QUOTIENTS                           1M
C                                                                      2M
C  SET INPUT AND OUTPUT CHANNELS USING NIN AND NOUT                    3M
       NIN = 1                                                         4M
       NOUT=2                                                          5M
C  READ NUMBER OF INDUSTRIES FOR WHICH LOCATION QUOTIENT TO BE         6M
C  CALCULATED,  READ EMPLOYMENT TOTAL FOR DUBLIN AND EIRE.             7M
       READ (NIN,100) N                                                8M
       READ (NIN,105) DUBALL,EIRALL                                    9M
C  SET UP  COUNTING PROCEDURE, SO THAT KOUNT STARTS AT ZERO           10M
       KOUNT=0                                                        11M
    30 KOUNT=KOUNT+1                                                  12M
C  READ EMPLOYMENT FOR SPECIFIC INDUSTRY AND CHECK IF ANY EMPLOYEES   13M
C  IN THAT INDUSTRY, IF NOT THE LOCATION QUOTIENT IS ZERO.            14M
       READ (NIN,105) DUBIND,EIRIND                                   15M
       IF (DUBIND.EQ.0.0)GO TO 10                                     16M
C  CALCULATE LOCATION QUOTIENT                                        17M
       QUOT=(DUBIND/DUBALL)/(EIRIND/EIRALL)                           18M
       GO TO 20                                                       19M
    10 QUOT=0.0                                                       20M
C  WRITE ANSWER                                                       21M
    20 WRITE (NOUT,110)KOUNT,QUOT                                     22M
C  CHECK IF ALL DATA HAS BEEN ANALYSED, IF NOT THEN READ DATA FOR     23M
C  NEXT INDUSTRY, IF ALL DATA ANALYSED THEN STOP.                     24M
       IF (N.GT.KOUNT)GO TO 30                                        25M
       STOP                                                           26M
   100 FORMAT ( I3)                                                   27M
   105 FORMAT ( 2F10.2)                                               28M
   110 FORMAT ( 31H LOCATION QUOTIENT FOR INDUSTRY,I3,3H IS,F7.3)     29M
       END                                                            30M
```

LOCATION QUOTIENT FOR INDUSTRY   1 IS   0.015
LOCATION QUOTIENT FOR INDUSTRY   2 IS   0.327
LOCATION QUOTIENT FOR INDUSTRY   3 IS   0.252
LOCATION QUOTIENT FOR INDUSTRY   4 IS   1.763
LOCATION QUOTIENT FOR INDUSTRY   5 IS   1.277
LOCATION QUOTIENT FOR INDUSTRY   6 IS   1.842
LOCATION QUOTIENT FOR INDUSTRY   7 IS   1.513
LOCATION QUOTIENT FOR INDUSTRY   8 IS   1.886
LOCATION QUOTIENT FOR INDUSTRY   9 IS   2.045
LOCATION QUOTIENT FOR INDUSTRY  10 IS   1.735
LOCATION QUOTIENT FOR INDUSTRY  11 IS   1.418
LOCATION QUOTIENT FOR INDUSTRY  12 IS   1.781
LOCATION QUOTIENT FOR INDUSTRY  13 IS   1.872

*Fig 6.3* (A) Program and (B) sample results of the calculation of location quotients

lowest values are, not unexpectedly, those in the primary sector.

(ii) *Problem*. The shapes of the stones in a sedimentary deposit can give valuable insight into its origin. More often than not shapes are described qualitatively using terms such as rounded, angular and irregular but the disadvantage of such terms is that different workers will have different ideas as to exactly what each one means. Some more objective method of analysis is obviously desirable. The method suggested by Zingg (1935) expresses shape in terms of the longest, intermediate and shortest axes of the stone. Further discussion is provided in Miller (1962) and Doornkamp and King (1971). Stones are assigned to one of four categories according to the values of the ratios bb/aa and cc/bb, as illustrated in Fig 6.4, and the numbers in each category are summed to give an idea of the overall character of the deposit.

*Algorithm and flowchart*. Figure 6.5 presents a full flowchart for a computer program to classify stone shapes using the Zingg method. The program, Fig 6.6, makes use of arithmetic and logical IF and unconditional GO TO statements.

*Program*. After writing a heading, the value of NS, the number of stones to be classified, is read and the locations I, NDISC, NSPHER, NBLADE and NRODS set to initial values of either 1 or 0. These locations are used by the program as counters and are thus of type *integer*. Location I simply counts the number of stones analysed while NDISC, NSPHER, NRODS, and NBLADE count the number of stones falling into each shape category. At statement 10 the aa, bb and cc axis lengths for a single stone are read into the machine and assigned to the locations A, B, and C. Arithmetic assignment statements are then used to form the ratios which are required as Y=B/A and X=C/B. The program now continues to assign the stone into a shape category using the arithmetic IF statements numbered 20, 30 and 40. The statement numbered 20 (an arithmetic IF statement) asks if the result of the arithmetic expression Y-0.6666 is greater than, equal to, or less than zero which is equivalent to testing if the current value stored in location Y is greater than, exactly equal to, or less than 0.6666 (ie 2/3). If Y is greater than 0.6666 then the expression has a positive value and control passes to statement 40. The same action occurs if Y is exactly 0.6666 but if it is less than

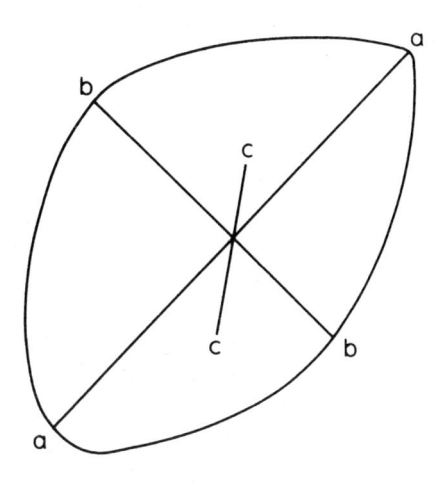

*Fig 6.4* Zingg's shape classification method

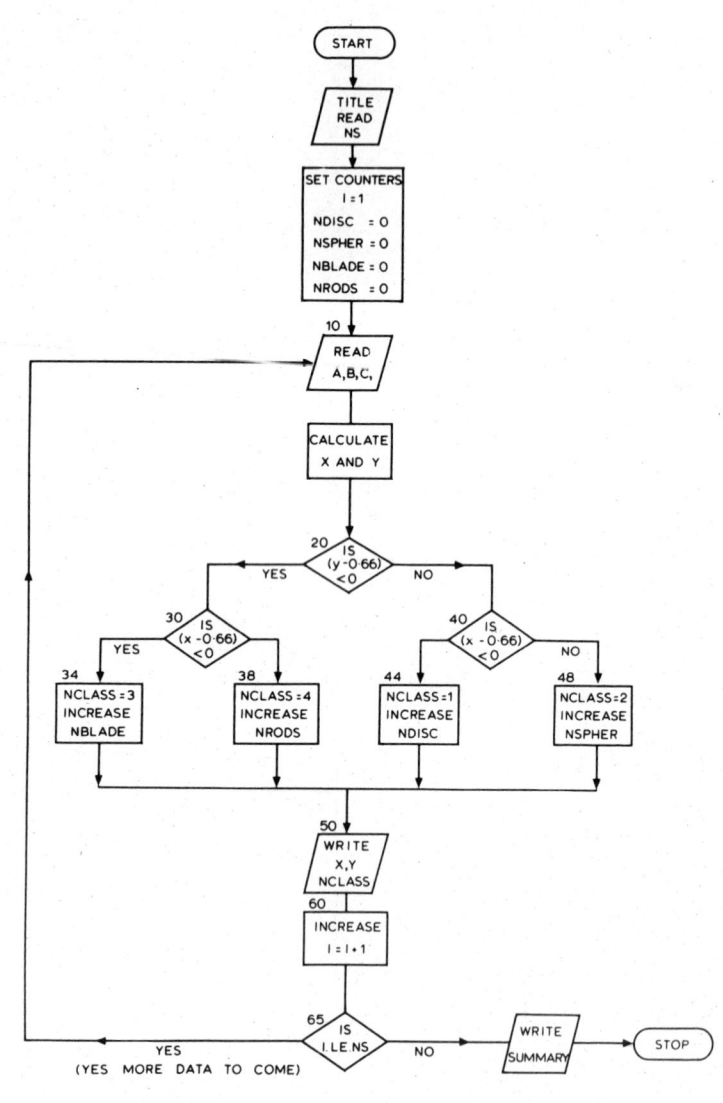

*Fig 6.5* Flowchart for Zingg shape classification program

this the expression gives a negative result and control
passes to statement 30. The program has now classified
the stone into one of two major groups according to the
value of the ratio bb/aa. Statements 30 and 40 make
a similar test to that made at 20 using the ratio cc/bb
which is held in location X and, depending upon the branch
followed, the stone shape class is fully determined. In
each case the variable NCLASS is set at the number of the
category into which the stone falls and the appropriate
counter incremented by one. In all cases control is un-
conditionally passed to statement 50 which prints out the
data and results for the stone and we are now ready to
process a second and successive data cards. At statement
60 the card counter is increased by one and at statement
65 a logical IF statement asks if this new value of I is
still less than or equal to the total number of stones in
the sample. If it now exceeds the total number in the
sample we have reached the end of the data but if it is
less then data remain and control is passed back to statement 10. At the end of the analysis the accumulated totals
in each shape category are printed out and the program ends.
Although this program is only capable of classifying into
four different categories the same principles can be used
to produce much more complex schemes such as those required
by the climatic classifications of Miller, Koppen and
Thornthwaite.

*Data and run.* Figure 6.6 gives an example of the print out
from the program. The data are the aa, bb and cc axis
lengths (inches) measured on 25 beach cobbles taken from
a shingle spit at Benacre Ness, Suffolk, England. The
values of the axis ratios and the category into which each
cobble falls have been printed out as part of the same
table, while the totals in each category are printed
separately. It can be seen that the majority of the
cobbles have the disc-like shape characteristic of beach
sediments.

6.7 Worksheet
1 Correct the following statements:
    (i)    GO TO 1∅
    (ii)       IF (A-B) 1∅, 1∅, 1∅
    (iii)      GO TO, (3,2,1), A
    (iv)       IF (A+C-B) 1,2
    (v)       GO TO A
    (vi)       IF (I=1), GO TO 25
    (vii)    6 IF (I.GT.A) GO TO 6
    (viii)      IF (K.LE.3..OR.K.LE.4..OR.K.LE.5.) K+K=K

```
C CASE STUDY - ZINGGS CLASSIFICATION OF COBBLE SHAPES
C INPUT IS A CARD ,FORMAT I3, GIVING THE NUMBER OF COBBLES
C FOLLOWED BY ONE CARD PER COBBLE , FORMAT 3F6.3 GIVING THE A,B,C AXIS
C LENGTHS
C
C SET THE INPUT - OUTPUT CHANNELS FOR THIS MACHINE
      NIN = 5
      NOUT = 6
C WRITE HEADING
      WRITE(NOUT,100)
C READ NO OF CASES INTO LOCATION NS
      READ(NIN,110) NS
C SET COUNTING LOCATIONS AT INITIAL VALUES
      I = 1
      NDISC = 0
      NRODS = 0
      NBLADE = 0
      NSPHER = 0
C READ IN DATA FOR A SINGLE COBBLE AND FORM AXIS RATIOS X AND Y
   10 READ(NIN,120) A,B,C
      Y = B/A
      X = C/B
C CLASSIFY ACCORDING TO ZINGGS CRITERIA
   20 IF ( Y - 0.6666) 30,40,40
   30 IF ( X - 0.6666) 34,38,38
C STONE IS BLADE LIKE
C INCREMENT APPROPRIATE COUNTING LOCATION
   34 NCLASS = 3
      NBLADE = NBLADE + 1
      GO TO 50
C STONE IS ROD LIKE
   38 NCLASS = 4
      NRODS = NRODS + 1
      GO TO 50
C SECOND BRANCH FOR DISCS AND SPHERES
   40 IF ( X - 0.6666) 44, 48, 48
C STONE IS DISC LIKE
   44 NCLASS = 1
      NDISC = NDISC + 1
      GO TO 50
C STONE IS SPHERE LIKE
   48 NCLASS = 2
      NSPHER = NSPHER + 1
C WRITE ANSWERS FOR THIS STONE
   50 WRITE(NOUT,130) A,B,C,X,Y,NCLASS
C INCREMENT AND TEST FOR DATAS END
   60 I = I + 1
   65 IF ( I .LE. NS ) GO TO 10
C WRITE SUMMARY INFORMATION ABOUT WHOLE SAMPLE
   70 WRITE(NOUT,140) NDISC,NSPHER,NBLADE,NRODS
      STOP
  100 FORMAT(55H     A          B          C          X          Y. CLASS)
  110 FORMAT( I3)
  120 FORMAT( 3F6.3)
  130 FORMAT(5F10.3,I5)
  140 FORMAT( 7H DISCS=,I5, 9H SPHERES=,I5, 8H BLADES=,I5, 6H RODS=,I5)
      END
```

| A | B | C | X | Y | CLASS |
|---|---|---|---|---|---|
| 1.500 | 1.150 | 0.475 | 0.413 | 0.767 | 1 |
| 1.975 | 1.275 | 0.675 | 0.529 | 0.646 | 3 |
| 1.675 | 1.100 | 0.625 | 0.568 | 0.657 | 3 |
| 1.475 | 1.400 | 0.500 | 0.357 | 0.949 | 1 |
| 1.725 | 1.550 | 0.950 | 0.613 | 0.899 | 1 |
| 2.075 | 1.500 | 0.700 | 0.467 | 0.723 | 1 |
| 1.625 | 1.300 | 0.700 | 0.538 | 0.800 | 1 |
| 2.300 | 1.875 | 1.125 | 0.600 | 0.815 | 1 |
| 1.650 | 1.125 | 0.875 | 0.778 | 0.682 | 2 |
| 1.800 | 1.275 | 0.375 | 0.294 | 0.708 | 1 |
| 1.100 | 0.975 | 0.575 | 0.590 | 0.886 | 1 |
| 1.950 | 1.725 | 0.600 | 0.348 | 0.885 | 1 |
| 2.125 | 1.850 | 0.900 | 0.486 | 0.871 | 1 |
| 2.025 | 1.425 | 0.600 | 0.421 | 0.704 | 1 |
| 2.150 | 1.275 | 0.750 | 0.588 | 0.593 | 3 |
| 1.675 | 0.900 | 0.650 | 0.722 | 0.537 | 4 |
| 1.500 | 1.300 | 0.650 | 0.500 | 0.867 | 1 |
| 1.900 | 1.750 | 0.775 | 0.443 | 0.921 | 1 |
| 1.875 | 1.375 | 0.825 | 0.600 | 0.733 | 1 |
| 2.100 | 1.400 | 0.625 | 0.446 | 0.667 | 1 |
| 2.275 | 1.525 | 0.750 | 0.492 | 0.670 | 1 |
| 3.252 | 1.525 | 0.800 | 0.525 | 0.469 | 3 |
| 2.300 | 1.750 | 0.925 | 0.529 | 0.761 | 1 |
| 2.100 | 1.625 | 0.800 | 0.438 | 0.869 | 1 |
| 2.100 | 1.500 | 0.700 | 0.467 | 0.714 | 1 |
| DISCS= 19 | SPHERES= 1 | BLADES= 4 | RODS = 1 | | |

*Fig 6.6* Program and sample results of the calculation of Zingg measures of stone shapes

```
1  (ix)        IF (X+2).LT.Y) STOP
   (x)         2 J=1
               IF (J.EQ.1), GO TO 2
   (xi)        GO TO (3,4,5,) I+K
   (xii)       GO TO 10
               6 WRITE (NOUT, 10)A
               10 FORMAT (IF 4.5)
   (xiii)      IF (I.GE.2.AND.LT.10) GO TO 14
   (xiv)       IF (A), 3,2,1
```

2 Geographical material is sometimes collected as incidence data in which a set of characteristics are either present or absent from a given phenomenon under study. A number of measures of association between phenomena with incidence measurements are available. Many of these measures are suited to computer analysis as they involve a series of repetitive comparisons. If we assume two items are given

111

on which there are measures of presence(1)/absence(0) it is then possible to draw up a table thus:

|  |  | Item 1 (j) 1 | 0 | Total |
|---|---|---|---|---|
| Item 2 (k) | 1 | n(J.K) | n(j.K) | n(K) |
|  | 0 | n(J.k) | n(j.k) | n(k) |
| Total |  | n(J) | n(j) | n |

where n(J.K) is the number of measures present on both items
n(j.k) is the number of measures absent from both items
n(J.k) is the number of measures present on item 1 but absent on item 2
n(j.K) is the number of measures absent on item 1 but present on item 2
the number of matched cells = n(J.K)+n(j.k) = M
the number of unmatched cells = n(J.k)+n(j.K) = U

There are six commonly used measures of association between items 1 and 2:

(i) Jaccard's measure S=n(J.K)/(n(J.K)+U)
the closer the association the nearer S is to 1.0
(ii) Simple ratio R=M/(M+U)
(iii) Rogers and Tanimoto's measure T=M/(M+2U)
(iv) Haman's measure H=(M-U)/n
(v) Yules measure Y=(n(J.K).n(j.k)-n(j.K).n(J.k))/
(n(J.K)n(j.k)+n(j.K)n(J.k))
(vi) $\phi$ coefficient $\phi$=n(J.K).n(j.k)-n(j.K).n(J.k)/
$(n(J).n(K).n(j).n(k))^{\frac{1}{2}}$

For discussion of the theorems underlying these measures see Sokal and Sneath (1963). A flowchart for the calculation of the indices is provided in Fig 6.7.

In biogeography it is often necessary to be able to measure the degree of floristic similarity between any two or more vegetation units. By measuring the similarities between units it is possible to group objectively the most similar units into broad vegetation types, as advocated by workers such as Goodall (1953) and Kershaw (1964). The data for this work are often derived from quadrat counts in which a small fixed area (the quadrat, usually 1m$^2$) is studied exhaustively to record all the plant species which occur within it. The quadrat locations are randomly selected.

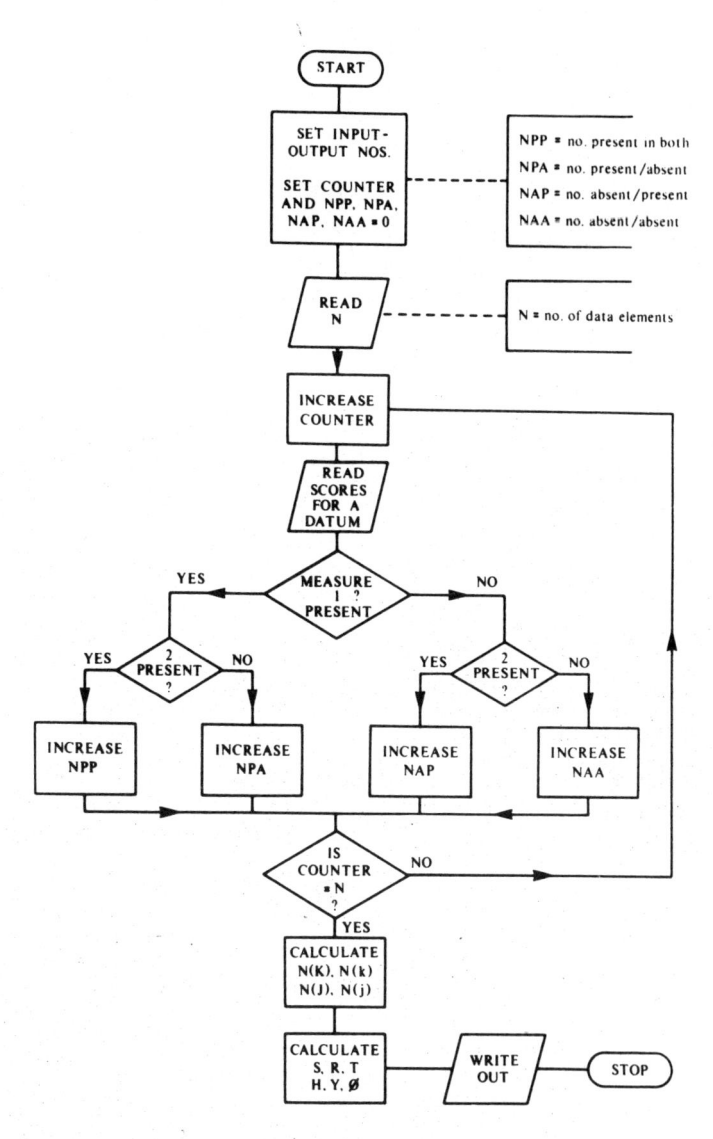

*Fig 6.7* Flowchart for program to calculate several taxonomic measures using incidence data

On Cannock Chase, Staffordshire, quadrat samples were taken in two major vegetation types:

(i) A *Calluna* heath formed on peaty soils after the degeneration of a primitive woodland cover.
(ii) A grass heath in which *Calluna* is replaced by grasses and on which grazing is heavy.

Species which occurred frequently within the quadrat were recorded as *present* (state 1), those which did not as *absent* (state 0), with the following results:

| Plant species | *Calluna* heath | Grass heath |
|---|---|---|
| Birds-foot Trefoil (*Lotus corniculatus*) | 0 | 1 |
| Gorse (*Ulex europaeus*) | 1 | 0 |
| Tormentil (*Potentilla erecta*) | 0 | 1 |
| Sheep's Sorrel (*Rumex acetosella*) | 1 | 1 |
| Heath Bedstraw (*Galium saxatile*) | 0 | 1 |
| Bilberry (*Vaccinium myrtillus*) | 1 | 1 |
| Ling (*Calluna vulgaris*) | 1 | 1 |
| Cat's-ear (*Hypochoerus radicata*) | 1 | 0 |
| Wall Cress (*Arabidopsis thaliana*) | 1 | 0 |
| Bell Heather (*Erica cinerea*) | 1 | 1 |
| Cowberry (*Vaccinium vitis-idaea*) | 1 | 0 |
| Shepherd's Cress (*Teesdalia nudicaulis*) | 1 | 0 |
| Wavy Hair Grass (*Deschampsia flexuosa*) | 1 | 1 |
| Bent Grass (*Agrostis tenuis*) | 0 | 1 |
| Poa (*Poa trivialis*) | 0 | 1 |
| White Clover (*Trifoloum repens*) | 0 | 1 |

Check that you understand the logic of the flowchart in Fig 6.7 and then write the program to calculate the six measures of association. Compare and contrast the different calculated index values.

3 Data collected for geographical analysis are sometimes closely related to a reference scale, for example a time scale or a one-dimensional space scale. The data then occur in a strict sequence. Regular meteorological readings or any economic time-series are of this data form; similarly, certain types of data collected on a landscape transect are of this type. Population density along a line from the centre of a city is another example of data collected by reference to a one-dimensional space-scale. Often it is useful to study the general and long-term trends in such

data. The inor fluctuations in the data can be smoothed and short-term trends partially removed by use of a running or *moving mean*. It is most usual for smoothing of the data to occur over a 3-, 5- or 7-unit period, but any smoothing factor could be used. The general form of the 5-year moving mean would be

$$\sum_{J}^{J+5} X_J/5$$

for J=1 to N-5 where N is the number of ordered data elements and $X_J$ is data element J. The mean of values 1 to 5 is calculated followed by the mean of values 2 to 6, etc. The method has been most widely used with time series (Hoskins, 1964) and is less common with spatial series but an adaptation of the concept to series in two-dimensional space is found in Cole and King (1969). Further explanation of the concept of the moving mean and its relationship to the regression model is discussed in Yamane (1967).

In the study of local rural markets it is difficult to obtain accurate assessments of the total amount of market sales from week to week. Furthermore there occur many small fluctuations in trade depending for example on weather conditions or occurrence of competing events. These problems of assessing a market's importance are often greatest in developing countries. The data below shows the value (in Rp) of the tickets sold to market traders for the most important market day in each week at the main market in Modjokuto, Central Java. The cost of a ticket, which is necessary to trade in the market, depends on the amount and kind of goods brought to the market and thus gives a rough approximation of the market's importance. The Javanese week has only 5 days thus there are usually 6 full weeks per month. Data are for 1952. For a detailed study of this market see Dewey (1962).

| Month | Week | | | | | | |
|---|---|---|---|---|---|---|---|
| | 1 | 2 | 3 | 4 | 5 | 6 | 7 |
| March | 758 | 776 | 792 | 841 | 821 | 882 | 869 |
| April | 858 | 884 | 855 | 843 | 846 | 817 | |
| May | 841 | 731 | 850 | 860 | 730 | 711 | |
| June | 819 | 963 | 845 | 726 | 270 | 445 | |
| July | 642 | 693 | 651 | 641 | 659 | 672 | |
| August | 676 | 689 | 702 | 642 | 746 | 760 | |

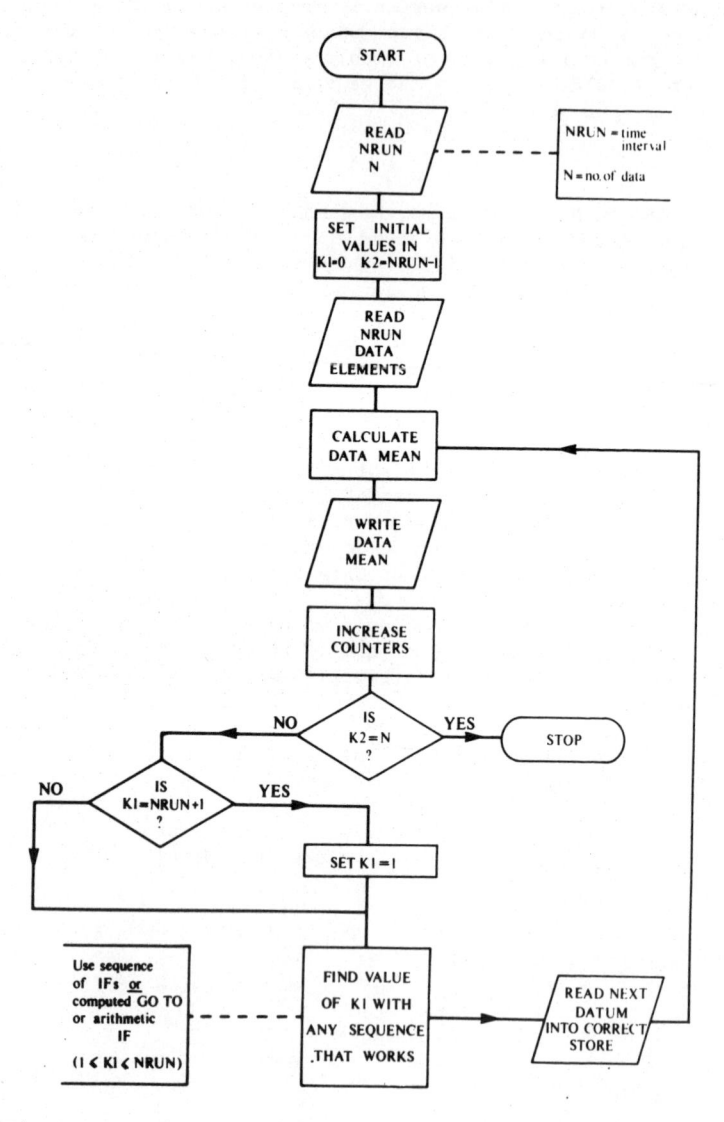

*Fig 6.8* Flowchart for program to calculate moving means

Check that you understand the logic of the flowchart presented in Fig 6.8 then calculate a 5 week moving mean for ticket sales over the six month period. To what extent does the Javanese New Year, at the end of June, affect the general trend of trade?

4 A classification of seasonal climates consisting of 11 types has been devised (Linton, 1973). Nine types are classified according to seasonal temperature characteristics and two additional types are distinguished by their aridity. Summers are classified according to the mean temperature of the warmest month and designated 0,1,2 or 3.

| Type | | Mean temperature of warmest month (degrees Centigrade) |
|---|---|---|
| 0 | no summer | 6 and under |
| 1 | very cool summer | 6-10 |
| 2 | cool summer | 10-20 |
| 3 | full summer | higher than 20 |

Winters are classified according to the mean temperature of the coldest month and designated 0,1 or 2.

| Type | | Mean temperature of coldest month (degrees Centigrade) |
|---|---|---|
| 0 | no winter | higher than 13 |
| 1 | mild winter | 2-13 |
| 2 | cold winter | lower than 2 |

The nine basic types are designated 02 (no summer, cold winter), 12 (very cool summer, cold winter), 11 (very cool summer, mild winter), 22 (cool summer, cold winter), 21 (cool summer, mild winter, 20 (cool summer, no winter), 32 (full summer, cold winter), 31 (full summer, mild winter) and 30 (full summer, no winter). In addition to the basic types there are two arid climates. Firstly there are arid climates, in the middle and low latitudes, in which no month receives as much as 50mm precipitation, and secondly there are extremely arid climates which never have more than 2.5mm precipitation per month for at least 10 months of the year. Implementation of this classification requires the following data at each station:

$X1$ = mean temperature of warmest month,
$X2$ = mean temperature of coldest month,
$X3$ = precipitation in the wettest month,
$X4$ = number of months receiving less than 2.5mm precipitation.

Sample data for 10 stations in the Americas are given below:

| Station | X1 | X2 | X3 | X4 |
|---|---|---|---|---|
| Albany, N.Y. | 23.3 | -3.5 | 109 | 0 |
| Charlestown, S.C. | 26.7 | 10.0 | 196 | 0 |
| Chesterfield Inlet, North-West Territory | 8.8 | -31.6 | 46 | 0 |
| Barrow Point, Alaska | 4.4 | -28.3 | 28 | 0 |
| Winnipeg, Manitoba | 18.9 | -20.0 | 79 | 0 |
| Acapulco, Mexico | 28.8 | 26.3 | 414 | 3 |
| Quixeramobin, Brazil | 29.0 | 26.5 | 170 | 0 |
| Uaupes, Brazil | 16.5 | 7.9 | 452 | 0 |
| Lima, Peru | 22.2 | 15.1 | 5 | 10 |
| San Diego, California | 20.0 | 12.2 | 48 | 4 |

Design a suitable flowchart and write and run a computer program to input these data and to assign each station to its correct category in the scheme.

5 Use the program provided above in section 6.6 to calculate location quotients for female workers in Dublin industries. The data are given in the table below.

Female employment by industry - Republic of Ireland 1966

| | Dublin | Eire |
|---|---|---|
| Agriculture | 81 | 33145 |
| Fishing | 21 | 34 |
| Mining, quarrying and turf production | 94 | 226 |
| Manufacturing industries | 27936 | 63089 |
| Building and construction | 684 | 1301 |
| Electricity, gas and water supply | 388 | 729 |
| Commerce | 15611 | 49719 |
| Insurance, banking, finance | 3074 | 5586 |
| Transport, communication and storage | 3547 | 8327 |
| Public administration and defence | 5439 | 9426 |
| Professions | 15813 | 56194 |
| Personal service | 12189 | 47304 |
| Entertainment and sport | 2773 | 4192 |
| Total | 87873 | 280791 |

Comment on the differences and similarities between male and female employment structure in Dublin as revealed by your answers and those provided as results in section 6.6.

6 Adapt the flowchart and program provided in section 6.6 to calculate location quotients for one industry over

several districts. Data on manufacturing employment for a sample of Irish towns are provided below.

Male employment in manufacturing in some towns in Republic of Ireland in 1966

| District | Manufacturing industry | All industry |
|---|---|---|
| Dublin | 48778 | 160611 |
| Dun Laoghire | 4172 | 2020 |
| Cork | 9865 | 29811 |
| Limerick | 3955 | 13321 |
| Galway | 1217 | 6156 |
| Waterford | 3017 | 7434 |
| Mullingar | 150 | 1761 |
| Tipperary | 270 | 998 |
| Donegal | 55 | 360 |
| Shannon Airport | 291 | 519 |
| Republic of Ireland | 135288 | 785196 |

7 The table below shows the number of males and females in the UK for each census year from 1801-1971. Write a program to calculate (i) total population in each year, and (ii) absolute and percentage increase in total, male and female population between successive census years.

Population (in thousands) in the UK 1801-1971

| Year | Females | Males |
|---|---|---|
| 1801 | 6252 | 5692 |
| 1811 | 7000 | 6368 |
| 1821 | 7974 | 7498 |
| 1831 | 9188 | 8647 |
| 1841 | 10364 | 9819 |
| 1851 | 11404 | 10855 |
| 1861 | 12631 | 11894 |
| 1871 | 14122 | 13309 |
| 1881 | 15955 | 15060 |
| 1891 | 17671 | 16593 |
| 1901 | 19745 | 18492 |
| 1911 | 21725 | 20357 |
| 1921 | 22994 | 21033 |
| 1931 | 23979 | 22060 |
| 1951 | 26107 | 24118 |
| 1961 | 27198 | 25478 |
| 1971 | 28583 | 26982 |

Comment briefly on the trends revealed.

## 6.8 Further Reading

Britton, J.N.H. *Regional analysis and economic geography* (1967)

Cole, J.P. and King, C.A.M. *Quantitative geography* (1969)

Central Statistics Office *Census of population of Ireland - Volume III, industries* (1968)

Dewey, A. *Peasant marketing in Java* (1962)

Doornkamp, J.C. and King, C.A.M. *Numerical analysis in geomorphology* (1971)

Florence, P.S. *Investment, location and size of plant* (1948)

Goodall, D.W. 'Objective methods for the classification of vegetation', *Australian Journal of Botany*, 2, 39-63 (1953)

Hoskins, W.G. 'Harvest fluctuations in British economic history', *Agricultural History Review*, 12, 28-46 (1964)

Isard, W. *Methods of regional analysis* (1960)

Kershaw, K.A. *Quantitative and dynamic ecology* (1964)

Linton, D.L. 'Seasonal climates', *Oxford World Atlas* (1973)

Martin, J.E. *Greater London - an industrial geography* (1966)

Miller, H.B. *Sedimentary petrography* (1962)

Sokal, R.R. and Sneath, P.H.A. *Principles of numerical taxonomy* (1963)

Yamane, D.L. *Statistics* (1967)

## Chapter 7
# The DO Statement

7.1 The Need for Loops
One of the most powerful features of the computer is its ability to repeat some process or calculation a great many times using different data at the start of each cycle. In Chapter 6 a case study program to calculate location quotients was able to repeat its calculations using a conditional transfer of control. The machine was instructed to loop back and repeat some instructions until a critical value of a variable was exceeded, at which point control passed out of the loop and on to the succeeding instructions. The FORTRAN necessary to do this can be written:

```
         READ ( NIN,1ØØ ) N
         .
         .
         .
         KOUNT = Ø
     3Ø  KOUNT = KOUNT + 1
         .
         .
         .
         some calculations
         .
         .
         .
         IF ( N .GT. KOUNT ) GO TO 3Ø
         STOP
         END
```

A flowchart for this operation can be represented by the single loop illustrated in Fig 7.1. The integer variable KOUNT acts as a counting location. When the loop is entered for the first time it has the value 0 + 1 = 1 and this is then increased by 1 each time the loop is executed. At the end of each loop is the conditional statement
            IF ( N .GT. KOUNT ) GO TO 3Ø
While KOUNT remains less than N this statement will be

121

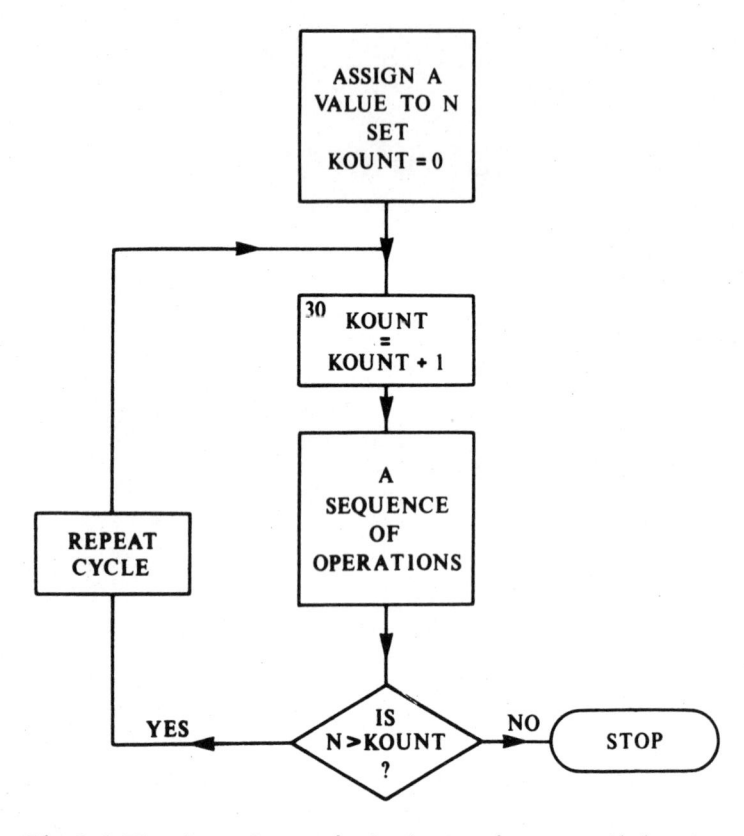

*Fig 7.1* Flowchart for a simple loop using a conditional statement

true and control will be looped back to statement 3∅, but when it is eventually increased by successive additions of 1 to a value equal to N control passes on to the next statement in sequence which in this case will stop the execution and end the program. This type of loop is so frequently required in practical computing that FORTRAN provides a single facility, the DO statement, to enable the programmer to set it up, and to consolidate into one the many statements used in a conditional IF.

## 7.2 The DO Loop

The DO statement is one of the most powerful and frequently used features of FORTRAN, making it possible to repeat a program section with automatic changes in the value of an integer variable between repetitions. When the sequence of statements has been repeated a specified number of times execution automatically continues with the next statement in sequence. Schematically, a DO can be written in one of two possible forms:

$$DO \quad n \quad j = m_1 , m_2 , m_3$$
$$\text{or}$$
$$DO \quad n \quad j = m_1 , m_2$$

In which (i) n is the *statement label* of an executable statement called the *terminal statement* of the DO. This terminal statement must follow the DO statement in the program and may not be a GO TO of any form, an arithmetic or logical IF, a STOP or a further DO. Often it is the dummy CONTINUE statement.

(ii) j is a single *integer* variable such as J, K, KOUNT or L written without sign and called the *control variable*. The values taken by this variable must always be greater than zero.

(iii) $m_1$, $m_2$ and $m_3$ are any *integer constants or variables* such as 1, 9, NSTOP, MS, and NGO which specify the parameters of the DO loop. $m_1$ is called the *initial parameter* and gives the value for j on the first time that the loop is to be executed. $m_2$ is the *terminal parameter* which gives the largest value of j to be used and $m_3$ is the optional *incrementation parameter* which gives the amount by which j is to be increased after each passage through the loop. If $m_3$ is omitted as in the second form of the statement it is assumed to be one.

A possible DO statement might be
$$DO \ 3\emptyset \ J = 1, 21, 2$$
3∅ is the statement label given to the terminal statement of the loop and all statements between this DO up to and

including that labelled 3∅ are referred to as the *range* of this DO.  J is the control variable.  The initial value it will take is 1, the final value 21 and the increase at each step is 2, so that J will have the values 1,3,5,7,9...19,21.  Any or all of the parameters can themselves be integer variables rather than constants so that we might have written

             DO 3∅  J = NSTART, NSTOP, INC

and assigned appropriate values to NSTART, NSTOP and INC earlier in the program.

The effect of these DO statements is as follows.  All the statements following the DO up to and including the one which has the label n (in this case 3∅) are executed repeatedly first with the control variable j (in this case J) set equal to $m_1$ (1 or NSTART in the examples).  Before each succeeding repetition j is increased by an amount $m_3$ (2 or INC in the examples) and execution continues until the statements have been executed with j equal to the largest value that does not *exceed* $m_2$ (21 or NSTOP in the examples).  To illustrate how DO statements work consider the simple program to sum 20 real numbers given as Fig 7.2.

```
C        SAMPLE PROGRAM TO SUM 20 REAL NUMBERS
         NIN = 5
         NOUT = 6
         TOTAL = 0.0
         DO 10 I = 1,20
         READ (NIN,90) X
    10   TOTAL = TOTAL + X
         WRITE(NOUT,91) TOTAL
    90   FORMAT( F7.3)
    91   FORMAT ( 9H TOTAL = ,F12.3)
         STOP
         END
```

*Fig 7.2* Sample program to sum 20 numbers

The statement

             DO 1∅ I = 1, 2∅

causes the READ and the arithmetic assignment labelled 1∅

to be executed 20 times with the control variable I taking on the values 1,2,3...20. No incrementation parameter was written down so it is assumed to be 1 but we might also have written
$$DO\ 10\ I = 1,\ 20,\ 1$$
which would have had precisely the same effect. Each time the loop is executed the computer is instructed to read a real number into location X and then add this to the contents of a location named TOTAL. When all 20 data cards have been read in this way the control variable I will have reached the value 20 which is the largest number that does not exceed the terminal parameter we have specified, and control passes on from the loop to the WRITE which prints out the required summation.

In this simple program we have used the control variable simply as a counter to keep track of the number of times the loop is executed but it can also be used as a variable in an expression, as for example in a program segment to sum the squares of the odd numbers 1,3,5...25,27:
$$NSUM = 0$$
$$DO\ 100\ JJJ = 1,\ 27,\ 2$$
$$100\ NSUM = NSUM + JJJ * JJJ$$
A third and exceedingly powerful use of the control variable as a marker for storage locations in the computer will be dealt with in the next chapter.

DO loops are very flexible in use but there are a number of rules and restrictions which *must* be obeyed but which will be seen to be logically necessary. These are:

(i) None of the DO loop parameters $m_1$, $m_2$, $m_3$ or the control variable j may be altered within the loop although as seen in the last program segment they may be referenced and used. To see why this rule must be enforced consider the simple but exceedingly dangerous program segment
$$DO\ 123\ I = 1,\ NSTOP$$
$$\vdots$$
$$NSTOP = I + 2$$
$$\vdots$$
123 terminal statement

Whatever the instructions involving other variables and constants within the range of this loop an attempt is made to alter the value of the terminal parameter as the loop is being executed. NSTOP will always be set at a value which is just 2 greater than the current value of the control variable I with the result that it can never reach NSTOP. In effect we have constructed a loop without end!

(ii) A DO loop may have any number of executable statements within its range, and it is permissible for this to include other DOs so that the loops become *nested* within each other. All statements in the range of the inner of a pair of such nested loops must also be within that of the outer loop as for example in the sequence:

```
          ┌─── DO 3Ø I = 1, 1ØØ
          │     .
          │     .
          │     .
          │   ┌─ DO 2Ø J = 1, 5Ø
range of  │   │  .
inner DO  │   │  .
          │   └─2Ø terminal statement for inner DO
range of  │     .
outer DO  │     .
          └───3Ø terminal statement for outer DO
```

Within every cycle involving the outer DO the inner loop uses its control variable J to control the execution of all the statements between it and the statement labelled 2Ø fifty times. As there are also one hundred repetitions of the outer loop involving I these instructions in the inner loop will be executed 100 x 50 = 5,000 times.

What is not allowed is for the range of an inner loop to extend beyond that of an outer as might be represented by the following sequence:

```
       ┌──── DO 3Ø L = 1, 56
       │      .
       │      .
       │      .
       │  ┌── DO 2Ø KKK = 1, 56
       │  │   .
       │  │   .
       │  └─3Ø terminal statement for outer DO
       │      .
       │      .
       └────2Ø terminal statement for inner DO
```

Notice that the nested loops may not use the same control variable since this would contravene rule (i) above but that the control variable for the outer loop may enter *as a parameter* of the inner as for example in the sequence:

126

```
          ┌──── DO 11Ø III = 1, 4Ø
          │      ⋮
          │ ┌── DO 12Ø K = III, 1ØØ
          │ │    ⋮
          │ └─ 12Ø terminal statement for inner DO
          │      ⋮
          └── 11Ø terminal statement for outer DO
```
In addition it is also perfectly acceptable if two or more nested loops share the same terminal statement, giving a sequence which might read:
```
          ┌──── DO 9Ø I = 1, 111, 1Ø
          │      ⋮
          │ ┌── DO 9Ø J = 1, 1Ø
          │ │    ⋮
          └─┴─ 9Ø terminal statement for both loops
```
DO loops may be nested to a considerable depth with many more than the two loops shown in these examples but most FORTRAN implementations eventually put some limit on this. Common sense will show that nesting loops can quickly use up a great deal of computer time. Four loops nested together and each specifying 100 operations will ensure that the innermost instructions are executed 100 x 100 x 100 x 100 = 100,000,000 times!

(iii) In general it is not allowed to jump into a DO loop from a point elsewhere in the program but transfers of control out of a loop are permitted. The reason for this rule is simple. Were control to pass to a statement within the range of a DO from outside without 'seeing' the initial DO statement the computer would not have values for the j, $m_1$, $m_2$ and $m_3$ and would be unable to execute any instructions in a sensible manner. The *normal exit* from a loop will occur when that loop is fully satisfied, that is when the control variable has reached a value equal to or in excess of the terminal parameter $m_2$, but it is perfectly possible for control to pass out from a loop before this condition occurs. In the latter situation the DO loop is said to have *extended range*, and there is an important difference between these types of exit from a loop. If the loop terminates normally, the final value of the control variable j is undefined and may differ from installation to installation, depending upon exactly how the particular computer executes a DO, so that it cannot be reliably used in further arithmetic that depends upon its value. After exit from a DO by means of a GO TO or other transfer of control before the loop has been fully satisfied the value of j is reliable and may be used in further arithmetic.

(iv) The final statement within the range of a DO must be executable; it cannot be a GO TO, conditional or arithmetic IF, STOP or other DO. We have met this restriction before in our consideration of the terminal statement but rules i-iii make it obvious that this restriction must be made. At face value this puts a severe limit on what we can use DO loops for, but in practice the CONTINUE statement (*see Chapter 6*) enables the programmer to circumvent this problem.

## 7.3 Standard Functions

A great many calculations, such as finding square roots and evaluating trigonometric ratios such as the sine, cosine and tangent could be programmed by every individual computer user in his own arithmetic expressions but many are so standard that they are supplied by the FORTRAN compiler. Each of these standard or supplied functions has been given a name consisting of 1 to 6 characters and has a type (real or integer) implied by the first letter of the name in the same way that variables are implicitly given a type. These standard functions are brought into action, or *called*, simply by writing the name together with an *argument* (or arguments) in parenthesis after it, all as part of an arithmetic assignment. Figure 7.3 gives a short list of the functions that are of most use to the geographer; a complete list of all the functions specified in ANSI FORTRAN is given in Appendix A.

As an example of the use of standard functions a short program to print out the square roots of the first 20 integer numbers might read:

```
      C THE USE OF SUPPLIED FUNCTIONS
        NOUT = 6
        DO 3Ø I = 1, 2Ø
        W = FLOAT(I)
```

Fig 7.3 Commonly-used standard functions of ANSI FORTRAN

| Function and definition | Number of arguments | Symbolic name | Type of argument | Type of result |
|---|---|---|---|---|
| Absolute value, a | 1 | ABS | Real | Real |
| Truncation, largest integer less than a | 1 | INT | Real | Integer |
| Conversion integer to real | 1 | FLOAT | Integer | Real |
| Exponential, $e^a$ | 1 | EXP | Real | Real |
| Natural logarithm, $\log_e a$ | 1 | ALOG | Real | Real |
| Common logarithm, $\log_{10} a$ | 1 | ALOG10 | Real | Real |
| Trigonometric sine, sin(a) | 1 | SIN | Real | Real |
| Trigonometric cosine, cos(a) | 1 | COS | Real | Real |
| Square root, $a^{\frac{1}{2}}$ | 1 | SQRT | Real | Real |
| Arctangent, arctan(a) | 1 | ATAN | Real | Real |
| Conversion real to integer | 1 | IFIX | Real | Integer |

```
          ROOTW = SQRT(W)
          WRITE(NOUT,9Ø) I, ROOTW
    3Ø CONTINUE
    9Ø FORMAT( I4,F1Ø.4)
          STOP
          END
```
Here the required sequence of numbers is generated using a simple DO loop and the square roots are found using the function SQRT. Reference to Fig 7.3 will show that this function has a single argument, the number whose square root is required, and that this must be of type real. The value of the function is also real. It is therefore necessary to convert the integer values of the control variable I into real numbers, a process which can be accomplished using the function FLOAT before calling SQRT. FLOAT has as its argument an integer and returns a result which is real.

An often-met expression in geographic programs is that for the distance between two points given their spatial co-ordinates X1,Y1 and X2, Y2 which can be programmed as

```
          DIST = SQRT(( X1-X2)**2 + ( Y1-Y2)**2)
```
which makes use of a SQRT call with an argument that is in itself an arithmetic expression. Another example might be the mathematical expression $(1 - \sin^2(x))^{\frac{1}{2}}$ which could be written as

```
          SQRT( 1.Ø - SIN(X)**2)
```

Standard functions are obviously easier to write than the equivalent arithmetic expressions and are likely to be more efficient in their use of computer time so that where possible they should always be used but at the same time great care should be taken to ensure that the programmer understands exactly what the function does and supplies arguments that are correct. It is very easy to supply arguments that ask the computer to evaluate indeterminate or even nonsensical functions. The functions for logarithms and for trigonometric ratios are particularly likely to cause difficulty. Suppose that a part of a program to normalise a skewed data distribution using a logarithmic transform uses an assignment

```
     X = ALOG1Ø (X)  (Common Logarithm, as in tables)
```
or
```
     X = ALOG(X)     (Natural Logarithm)
```
As long as the argument X has values greater than zero all will be well and the results will be as intended, but what happens when X is zero or even negative? For a zero value of X we are asking the computer to evaluate the

logarithm of zero which you will not find given in log tables because it is minus infinity and the computer will almost certainly return the lowest number that it is capable of representing. At the same time it is likely to record a 'run error' and terminate the job. Negative values of X will lead to similarly negative values for the logarithms which may not be quite what the programmer intends.

A second difficulty arises in the use of the supplied functions for the trigonometric ratios whose arguments must be supplied not in the familiar degrees of a circle (°) but in the circular measure called radians. There are $2\pi$ radians in each $360°$, so that 1 radian is equivalent to $57.296°$ and to convert from degrees to radians all that need be done is to divide the angle expressed in degrees by this conversion factor before calling the required function:

```
      C PROGRAM SEGMENT TO FIND COS(72 DEG)
        XDEG = 72.0
        XRAD = XDEG/ 57.296
        COSX = COS(XRAD)
```

## 7.4 Summary

Frequently in computer programs it is necessary to repeat a series of instructions and FORTRAN provides a particular statement for this instruction sequence. This is the DO loop which in generalised form is

DO n j = $m_1$, $m_2$, $m_3$

when n is a labelled statement

    j is a control variable

and  $m_1$, $m_2$, $m_3$ are initial, terminal, and incremental parameters.

With this instruction all statements between the DO and the labelled statement are repeated with the control variable incremented each time by $m_3$ from an initial value of $m_1$ to a terminal value of $m_2$. The second statement introduced in this chapter is the standard function which allows functions such as logarithms or square roots to be calculated in a single statement.

## 7.5 Worked Examples

(i) *Problem.* The survival of men in cold climates largely depends upon the cooling power of the wind as expressed by its speed and the air temperature. To estimate this cooling power it is possible to calculate a number of indices of wind chill. In this case our problem is to

produce a table of values of one particular index for a variety of combinations of air speed and temperature.
*Algorithm and flowchart.* According to Gates (1972) a simple wind chill factor, H, can be calculated from the formula

$$H = ((100 \cdot W)^{\frac{1}{2}} + 10.45 - W) \cdot (33 - T_a)$$

in which the rate of heat loss (H, kcal cm$^{-2}$ hr$^{-1}$) from skin at 33°C is related to the wind velocity (W m.sec$^{-1}$) and air temperature ($T_a$, °C). These values range from about 50 which is considered hot to 2500 which is intolerably cold and frostbite can be expected to occur at values between 1400 and 2100. We wish to write a program to evaluate wind chill factors for all wind speeds at 5mph intervals from 0 to 50mph and for all temperatures at 10° intervals from 40 to -40°F. The required value of H can be found by successive substitution into the formula of values for the wind speed and air temperature, taking care to ensure that both are expressed in the correct units. We can best accomplish these substitutions using two nested DO loops, the outer of which controls the wind speed, the inner the temperature. A suitable flowchart and program for these operations are shown as Figs 7.4 and 7.5. No data cards are required.

*Program and run.* In the program the first two executable statements set the output channel NOUT at 6 and write out a column heading before entry into the nested DOs. With no data no input channel is required. The loops can be shown schematically as:

```
      DO 1Ø I = 1,51,5
         .
         .
         .
      set up wind speed values and calculate first
      half of formula (1ØØW)^½ + 1Ø.45 - W
         .
         .
         .
      T = 4Ø
      DO 2Ø J = 1,9
         .
         .
         .
      set up temperature values and apply rest of
      formula
  2Ø  T = T - 1Ø
  1Ø  CONTINUE
```

The outer loop causes the program to pass through all the instructions as far as 1Ø CONTINUE with the control variable I set first at 1 and increasing by steps of 5 until the value 51 is reached. Remembering that we require values for wind speeds at 5mph intervals from 0 to 50 it

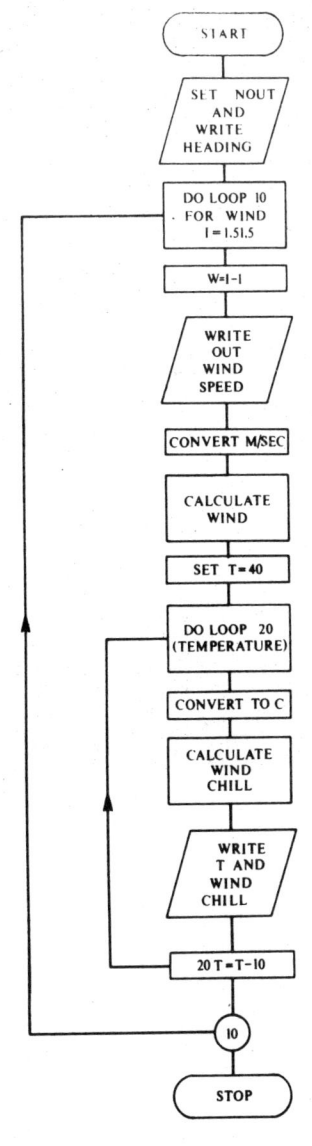

```
                    ┌─────────┐
                    │  START  │
                    └────┬────┘
                         │
                   ╱─────┴─────╲
                  ╱   SET NOUT  ╲
                 ╱     AND       ╲
                 ╲    WRITE      ╱
                  ╲  HEADING    ╱
                   ╲───────────╱
                         │
              ┌──────────┴──────────┐
              │   DO LOOP 10        │
    ┌────────▶│   FOR WIND          │
    │         │   I = 1,51,5        │
    │         └──────────┬──────────┘
    │                    │
    │              ┌─────┴─────┐
    │              │  W = I-1  │
    │              └─────┬─────┘
    │                    │
    │              ╱─────┴─────╲
    │             ╱   WRITE     ╲
    │            ╱    OUT        ╲
    │            ╲   WIND        ╱
    │             ╲  SPEED      ╱
    │              ╲───────────╱
    │                    │
    │           ┌────────┴────────┐
    │           │  CONVERT M/SEC  │
    │           └────────┬────────┘
    │                    │
    │             ┌──────┴──────┐
    │             │  CALCULATE  │
    │             │    WIND     │
    │             └──────┬──────┘
    │                    │
    │              ┌─────┴─────┐
    │              │ SET T=40  │
    │              └─────┬─────┘
    │                    │
    │         ┌──────────┴──────────┐
    │         │   DO LOOP 20        │
    │    ┌───▶│  (TEMPERATURE)      │
    │    │    └──────────┬──────────┘
    │    │               │
    │    │       ┌───────┴────────┐
    │    │       │  CONVERT TO C  │
    │    │       └───────┬────────┘
    │    │               │
    │    │         ┌─────┴─────┐
    │    │         │ CALCULATE │
    │    │         │   WIND    │
    │    │         │   CHILL   │
    │    │         └─────┬─────┘
    │    │               │
    │    │          ╱────┴────╲
    │    │         ╱   WRITE   ╲
    │    │        ╱   T AND     ╲
    │    │        ╲   WIND      ╱
    │    │         ╲   CHILL   ╱
    │    │          ╲─────────╱
    │    │               │
    │    │       ┌───────┴────────┐
    │    └───────┤  20 T = T-10   │
    │            └───────┬────────┘
    │                    │
    │                  ╱─┴─╲
    └─────────────────(  10 )
                       ╲─┬─╱
                         │
                    ┌────┴────┐
                    │  STOP   │
                    └─────────┘
```

*Fig 7.4* Flowchart for a program to calculate wind chill factors

```
C PROGRAM TO CALCULATE WIND CHILL (KCAL./SQ.M./MIN) FOR VARIOUS
C COMBINATIONS OF WIND SPEED AND DRY BULB TEMPERATURE
C ILLUSTRATING THE USE OF A NESTED PAIR OF DO LOOPS AND SQRT(X)
C
C REFERENCE = GATES,D.M.(1972) MAN AND ENVIRONMENT - CLIMATE ,HY
C WRITE TABLE HEADING
      NOUT= 6
      WRITE (NOUT,91)
      DO 10 I = 1 ,51,5
      W = I - 1
      WRITE(NOUT,90)  W
C CONVERT TO METRES PER SECOND
      W = W*0.44704
C CALCULATE WIND FACTOR ONCE IN EACH LOOP
      WIND = (SQRT(100*W) + 10.45 -W )
      T = 40.0
      DO 20 J = 1,9
C CONVERT TO DEGREES CENTIGRADE RETAINING THE F VALUE IN LOCATION T
      TC = 0.5555 * ( T- 32.0)
C WIND CHILL FORMULA
      H = WIND * (33.0 - TC)
C WRITE OUT ANSWER FOR THIS LOOP
      WRITE (NOUT,92) T,H
   20 T = T - 10.0
   10 CONTINUE
      STOP
   90 FORMAT(27H VALUES FOR WIND(M.P.H.) = ,F5.0)
   91 FORMAT ( 21H TEMP(F)    WIND CHILL)
   92 FORMAT ( F7.0,F12.3)
      END
```

*Fig 7.5* Program to calculate wind chill factors

can be seen that the program can generate this sequence using the assignment W = I - 1 in which the location W holds successive wind speed values. Notice too that this arithmetic assignment also changes the number type from an integer (I) to a real (W), a step that ensures that we do not mix types in the succeeding arithmetic. After writing out this current value of the wind speed in mph it is converted to metres/second by a conversion factor of 0.44704 and the first part of the formula which is constant for any given wind speed is evaluated using the expression

$$WIND = SQRT(100 * W) + 10.45 - W$$

All that remains is to generate a sequence of temperature values from 40 to -40° for each value of WIND and complete the calculation for each. The inner DO enables this sequence to be produced. It causes values of J to increase from 1 to 9 by an implied increment of 1 at each step. All the statements down to

$$20 \ T = T - 10.0$$

are within its range and because they are also within the range of the outer DO will be executed 9 times for each and every one of the 11 cycles specified by the outer loop

134

giving a total of 99 applications of the wind chill formula. How does this second loop specify the required temperature values? Immediately before it is entered a location T is set equal to the maximum temperature required, 40, and at the end of each cycle the current value of T is reduced by 10. When J is 1 then T is 40 but as J increases in steps of 1 to 9 so T decreases by successive subtractions of 10 to reach the required minimum value, -40 when J is 9. It can be seen that unlike the control variable I used in the outer loop which is converted directly into a wind speed, the J in the inner loop serves simply as a counter to keep track of the number of times the loop is passed through. Within its range each T value is converted to its Centigrade equivalent and the remainder of the formula evaluated. The results for each pair of wind and temperature values are written out and the program terminates when the outer DO is exhausted. The resulting values of H have been graphed to produce Fig 7.6 which can be used to read off values of the index for any combination of wind and temperature.

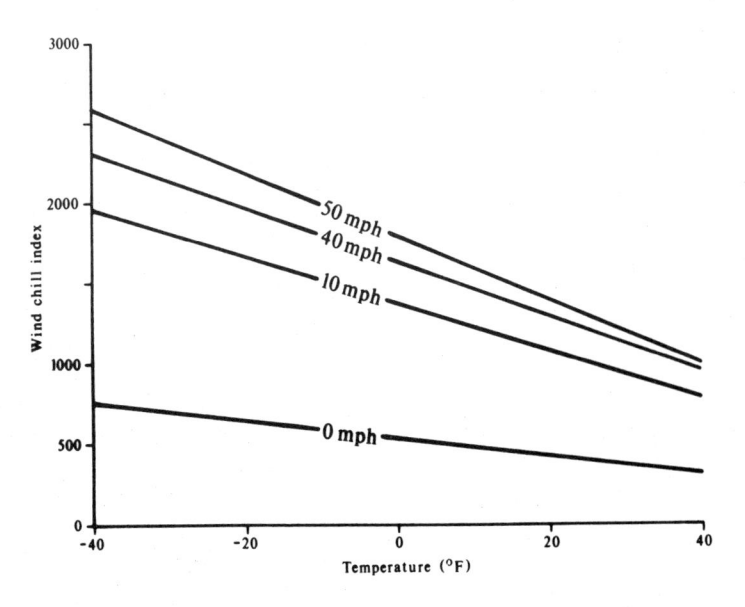

*Fig 7.6* Graph of wind chill factors

(ii) *Problem.* As the summary by Collins (1973) shows, economists and economic geographers have often analysed the functions of manufacturing industries according to various measures of their size at both national and regional level. As might be expected from theoretical considerations (Steindl, 1956; McGuire and Archer, 1965) it has been shown that firms with different sizes tend to have different operational characteristics, but there have been relatively few attempts (Douglas, 1962; McClelland, 1967; Hall, 1971; Retail Business, 1972; National Commission on Food Marketing, 1966) to look at the size characteristics of retail establishments and none of these is concerned with shop size distributions within towns. Our current example ia associated with an attempt to remedy this deficit by considering the operational characteristics associated with groups of different sized shops in two urban areas in Northern Ireland. The procedure adopted and the computer program are of course applicable to other towns and cities.
*Algorithm and flowchart.* The algorithm is extremely simple. It involves the input of employment numbers for different sizes of shops and subsequent calculation of several simple ratios. Both the number of towns and the number of size categories may be varied within the program's data.

For each size class the indices used and the locations in which they are stored are as follows:

   (i) Percentage of total sales, A
  (ii) Percentage of total shop numbers, B
 (iii) Percentage of total employees, calculated in full-time equivalents (FTE) by weighting each part-time employee equal to one half of a full-time employee, C
  (iv) Average sales per shop, D
   (v) Average sales per employee (FTE), E
  (vi) Average employees (FTE) per shop, H
 (vii) Average part-time employees per shop, O
(viii) Percentage of employment on a part-time basis, F
  (ix) Percentage of females employed, G

The flowchart for these operations is shown in Fig 7.7. In the previous worked example a pair of DO loops were used to control the actual sequence of numerical values of the variables that entered into a calculation whereas this second study illustrates the use of nested DOs simply to control the number of times operations are to be performed.

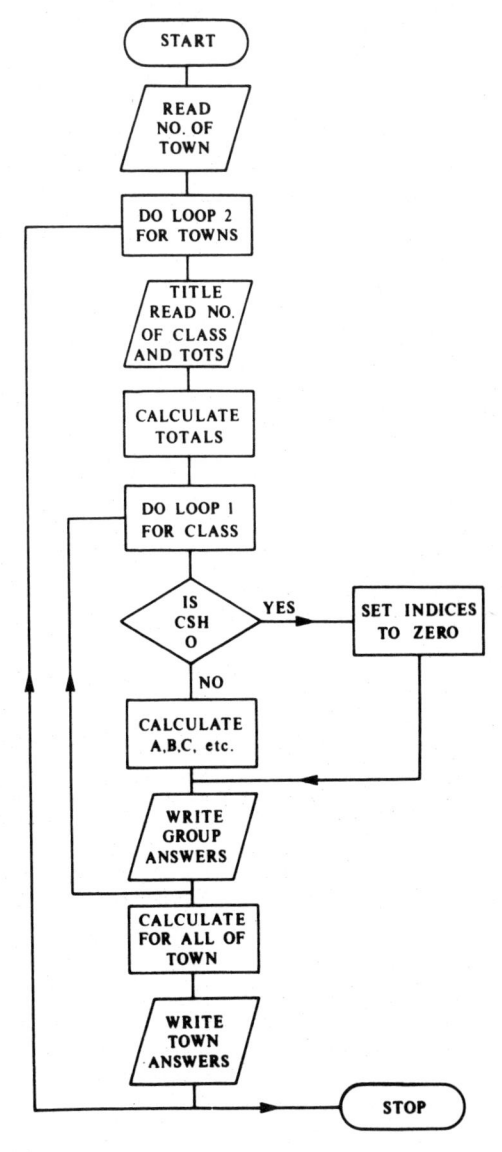

*Fig 7.7* Flowchart for program to calculate shop size indices

```
C           SHOP SIZE DISTRIBUTIONS PROGRAM
C     SET INPUT AND OUTPUT CHANNELS
            NIN = 5
            NOUT = 6
C     READ NUMBER OF TOWNS TO BE STUDIED
            READ(NIN,120) M
C     SET UP DO LOOP TO ALLOW REPEAT CALCULATIONS FOR EACH TOWN
            DO 2 J=1,M
            WRITE(NOUT,130) J
C     SET UP TABLE HEADINGS
            WRITE(NOUT,100)
            WRITE(NOUT,105)
C     READ NUMBER OF SIZE CLASSES
            READ (NIN,120) N
C     READ TOTALS FOR ALL SIZE CLASSES
            READ(NIN,115) TSLS, TSH, TFTM, TPTM,TFTF, TPTF
C     CALCULATE FTE TOTALS AND TOTAL EMPLOYEES
            TFTE = TFTM + TFTF + ((TPTM + TPTF)/2.0)
            TEMP = TFTM + TFTF + TPTM + TPTF
            TPT = TPTM + TPTF
            TFEM = TFTF + TPTF
C     SET DO LOOP TO READ EACH SIZE CLASS, CALCULATE MEASURES, WRITE ANSWERS
            DO 1 I=1,N
            READ(NIN,115) CSLS,CSH,CFTM,CPTM,CFTF,CPTF
C     CHECK IF ANY SHOPS IN THIS SIZE CLASS
            NCSH = IFIX(CSH)
            IF ( NCSH .NE. 0 ) GO TO 3
            A = 0
            B = 0
            C = 0
            D = 0
            E = 0
            F = 0
            G = 0
            H = 0
            O = 0
            GO TO 4
          3 A = CSLS * 100.0/TSLS
            B = CSH * 100.0/TSH
            CFTE = CFTM + CFTF + ((CPTM + CPTF)/2.0)
            C = CFTE *100.0/TFTE
            D = CSLS/CSH
            E = CSLS/CFTE
            CEMP = CFTM + CFTF + CPTM +CPTF
            F = (CPTM + CPTF) * 100.0/CEMP
            G = (CPTF +CFTF) * 100.0/CEMP
            H = CFTE/CSH
            O = (CPTF + CPTM)/(CSH * 2.0)
C           THIS IS AN EXAMPLE OF CONTINUE USED AS A DUMMY STATEMENT
          4 CONTINUE
            WRITE(NOUT,110) I, A, B, C, D, E, F, G,H, O
          1 CONTINUE
C     CALCULATE MEASURES FOR ALL SHOPS
            A = 100.0
            B = A
            C = A
            D = TSLS/TSH
            E = TSLS/TFTE
            F = TPT * 100.0/TEMP
            G = TFEM * 100.0/TEMP
            H = TFTE/TSH
            O = TPT/2.0
            O = O/TSH
            WRITE(NOUT,125) A,B,C,D,E,F,G,H,O
          2 CONTINUE
            STOP
        100 FORMAT ( 90H     SIZE           PERCENTAGE OF        SALES       SALES       PER000
           *C PART     PERC     F,T,E,     PART TIME )
        105 FORMAT ( 90H CLASS     SALES     SHOPS  F,T,E,  PER SHOP   PER FTE  TIMEMP
           *E EMP    FEMALE    PER SHOP    FTE/SHOP )
        110 FORMAT ( 1I6,1F9.1,2F7.1,1F11.2,2F9.2,3F10.2)
        115 FORMAT ( 6F10.0)
        120 FORMAT ( 1I3)
        125 FORMAT ( 6H TOTAL,1F9.1,2F7.1,1F11.2,2F9.2,3F10.2)
        130 FORMAT ( 19H *****     TOWN NO.,1I5,8H *****)
            END
```

*Fig 7.8* Program to calculate indices for shop size distribution

Two loops are incorporated. The first or outer loop cycles the program round M times, where M is the number of towns for which data exist and is read in as data. The inner DO repeats the index calculations detailed above for each of the N size categories of shops recognised within each town.

*Program.* A program-listing together with the data used for two towns are given in Figs 7.8 and 7.9. After the input output channels have been set at 5 and 6 the first datum read specifies M, the number of towns for which analysis is required and the outer DO is entered with this as its terminal parameter. Within this loop, numbered 2, each town number is printed out and a table heading made before data relating to each *town* is read in as follows:

    N = no of size classes (first card)
    TSLS = total sales
    TSH = total number of shops
    TFTM = total full-time employees (male)
    TPTM = total part-time employees (male)
    TFTF = total full-time employees (female)
    TPTF = total part-time employees (female)

A series of simple arithmetic assignments follows these read operations to calculate the full-time equivalent total (TFTE), total employment (TEMP), total part-time (TPT) and total female employment (TFEM). The inner DO, numbered 1, controls the number of size classes for which indices are to be calculated and has as its terminal parameter N, the number of classes. For each such class the following data are read in:

    CSLS = total sales
    CSH = number of shops
    CFTM = total full-time employees (male)
    CPTM = total part-time employees (male)
    CFTF = total full-time employees (female)
    CPTF = total part-time employees (female)

If there are no shops in the size group under consideration then we must set the indices identically equal to zero. In this case the real location CSH will be zero, but before testing this condition using an IF statement we prefer to convert this value to its integer equivalent called NCSH using the standard function IFIX. The reason for doing this is that it is very unlikely that CSH will be stored *exactly* as zero by the machine, yet before action can be taken and control transferred we ask that this be so. Nine times out of ten the programmer can get away with

```
002
010
94238172,     5287,      8514,     1168,     10724,    3135,
  170074,      395,       123,       48,       165,     132,
  352955,      248,       136,       28,       140,      24,
 3415280,      983,       520,      111,       816,     146,
10287269,     1422,      1059,      135,      1501,     295,
18301830,     1339,      1751,      407,      1872,     509,
18401064,      639,      1582,      162,      1568,     193,
11284125,      166,       919,       46,      1067,      54,
 8832526,       63,       737,       58,       786,      58,
 5480258,       19,       492,       66,       584,      65,
17712791,       13,      1185,        7,      2235,       6,
010
10913037,      631,      1220,      151,      1212,     256,
   13319,       31,        20,        2,        12,       9,
   50846,       34,        14,        1,        21,       3,
  456148,      135,        71,       14,       123,      25,
 1025237,      146,       131,       29,       140,      51,
 1906885,      142,       209,       34,       186,      48,
 3213671,      109,       347,       41,       272,      41,
 1218951,       19,       155,       10,       117,      13,
 1266291,        9,       152,       20,       101,       6,
 1240701,        5,       103,        0,       164,      60,
  520988,        1,        18,        0,        76,       0,
```

*Fig 7.9* Data on shop size for Belfast and Londonderry

```
***** TOWN NO.   1 *****
SIZE      PERCENTAGE OF      SALES      SALES   PERC PART   PERC     F.T.E.    PART TIE
CLASS  SALES SHOPS F.T.E. PER SHOP    PER FTE   TIME EMP  FEMALE   PER SHOP   FTE/SHOP
  1     2.2   7.5   1.8      430.57    449.93    38.40    65.40      0.96       0.23
  2     0.4   4.7   1.4     1423.21   1108.73    15.85    50.00      1.22       0.17
  3     3.6  18.6   6.8     3474.34   2332.85    16.13    65.39      1.49       0.15
  4    10.9  26.9  13.0     7234.37   3787.12    14.58    61.07      1.95       0.15
  5    19.4  25.3  14.1    13668.28   4484.64    24.18    52.46      3.05       0.34
  6    19.5  12.1  15.6    28796.66   5530.80    10.13    54.24      5.21       0.29
  7    12.4   3.1   9.5    67976.66   5542.36     4.79    53.74     12.27       1.33
  8     9.4   1.2   7.4   148198.83   5586.67     7.08    51.49     25.10       4.92
  9     5.8   0.4   5.3   288434.63   4836.93    18.85    53.77     64.08       3.45
 10    18.8   0.2  16.0  1302522.38   5169.35     0.38    65.28    263.58       0.50
TOTAL 100.0 100.0 100.0    17824.51   4405.81    18.28    54.87      4.55       0.41
***** TOWN NO.   2 *****
SIZE      PERCENTAGE OF      SALES      SALES   PERC PART   PERC     F.T.E.    PART TIE
CLASS  SALES SHOPS F.T.E. PER SHOP    PER FTE   TIME EMP  FEMALE   PER SHOP   FTE/SHOP
  1     0.1   4.9   1.4      429.05    355.17    25.58    48.84      1.21       0.18
  2     0.6   5.4   1.4     1495.47   1374.22    10.26    61.54      1.29       0.16
  3     4.2  21.4   8.1     3378.87   2136.52    16.74    63.52      1.58       0.14
  4     9.4  23.1  11.8     7022.17   3296.58    22.79    54.42      2.13       0.27
  5    17.5  22.5  16.5    13428.77   4373.59    17.19    49.06      3.07       0.29
  6    29.4  17.3  25.8    29485.22   4869.20    11.78    44.65      6.06       0.38
  7    11.2   3.8  10.8    64155.32   4299.65     7.80    44.07     14.92       0.61
  8    11.5   1.4  10.1   140099.00   4768.49     9.32    38.35     29.56       1.44
  9    11.4   0.6  11.3   248144.20   4177.44    18.35    88.58     59.44       6.00
 10     4.6   0.2   3.6   520988.00   5542.43     0.00    80.85     94.00       0.00
TOTAL 100.0 100.0 100.0    17294.83   4140.78    14.34    51.71      4.18       0.32
```

*Fig 7.10* Results of shop size analysis for Belfast and Londonderry

a statement such as
IF (CSH .NE. ∅.)
but it is not good practice to rely on real numbers being exactly represented in this way. If there are no shops in the size group control passes to a series of statements setting the indices at zero and passing control on to the dummy statement
2 CONTINUE
Usually there will be shops in a size group, control passes to the statement labelled 3, and the various indices are evaluated and printed out. Finally, outside the range of the inner DO, similar index values are calculated for the whole town and printed out.
*Data and run.* The data given in Fig 7.9 are for the towns of Belfast and Londonderry and were collected in connection with the Census of Distribution for Northern Ireland (Ministry of Commerce, 1969). The first line of data is the town totals and then follow the ten size groups related to the volume of sales. The size groups are:

| Class | Sales (£1,000s) |
|---|---|
| 1 | less than 1 |
| 2 | 1 - 1.9 |
| 3 | 2 - 4.9 |
| 4 | 5 - 9.9 |
| 5 | 10 - 19.9 |
| 6 | 20 - 49.9 |
| 7 | 50 - 99.9 |
| 8 | 100 - 199.9 |
| 9 | 200 - 499.9 |
| 10 | 500 and above |

The results are given on Fig 7.10 in a form suitable for graphing. They show several features that have been recognised previously. Although a goodness of fit test would be needed to confirm this, the overall size distributions in Londonderry appear to be very nearly the log-normal said to be characteristic of firms in general (Aitchison and Brown, 1957). The distribution for Belfast probably deviates slightly from this pattern. Productivity, as measured by sales per employee, increases to a peak and then levels off or declines slightly, a feature that seems to be true for most towns (Skillcorn, 1956; McClelland, 1957). Similarly, the use of part-time workers seems to decrease at first as size increases then to increase again. The use of female employees related to sales size is variable but generally tends to decrease as size increases but not every town shows this pattern.

## 7.6 Worksheet

1 Write FORTRAN program segments that use standard functions to evaluate the following mathematical relationships:

(i) Gorczynski's index of the continentality of a climate
$K = 1.7\ A/\sin\theta - 20.4$
 $A$ = mean annual temperature range °C
 $\theta$ = station latitude °
 $1° = 0.0175$ radians

(ii) Student's t-index

$$t = \frac{|\bar{x} - \bar{y}|}{(SE_x^2 + SE_x^2)^{\frac{1}{2}}}$$

$\bar{x}$ = mean of x
$\bar{y}$ = mean of y
$SE_x$ = standard error of x
$SE_y$ = standard error of y

(iii) The distance between two points on the earth's surface measured as the cosine of an angle along a Great Circle is given by
$\cos(d) = \sin(lat_1).\sin(lat_2) + \cos(lat_1).\cos(lat_2).\cos(long_1 - long_2)$
$d$ = angle along Great Circle
$lat_1$, $lat_2$ are the latitudes
$long_1$, $long_2$ are the longitudes measured in the same direction east or west of the Greenwich Meridian.

(iv) The downstream decrease in calibre of a river bedload, averaged across the flow, is usually assumed to be according to a relationship of the form:
$D = d_0\ e^{-kL}$
 $D$ = mean particle diameter at distance L along the channel
 $d_0$ = mean particle diameter at $L = 0$
 $e$ = base of natural logarithms
 $k$ = a constant for the particular river

2 What would be the result of a call of the standard functions ALOG10 and ALOG with the argument X set at 1, 5, 10, 1,000,000 and 2.71828?

3 Convert the location quotient program introduced as worked example in Chapter 6 to use a DO loop whose terminal parameter is read in as data.

4 Adapt the program shown in the worked example in Chapter 5 to calculate demographic rates for the major European countries in 1968. What conclusions can be drawn from the calculated values and to what extent do they tend to cluster around the overall mean values for all countries? Raw data are provided in Table 7.1.

5 The United States Weather Bureau calculates a temperature-humidity index (THI) that provides an index of the combined effects of temperature and humidity on human beings according to the empirical relation:

$$\text{THI} = t_d - (0.55 - 0.55 \text{ rh}) (t_d - 58)$$

$t_d$ = dry bulb temperature (°F)
rh = relative humidity, expressed as a fraction in the range 0 to 1

Write a program using nested DO loops to calculate the information necessary to produce a graph of THI values over the range 60-110°F and from 20-100%rh. According to Gates (1972) 10% of the population feel uncomfortable at THI values of 70, 50% at 75 and almost everyone is uncomfortable at 80. Mark these values on your graph and note how the combined effects of temperature and humidity produce them.

6 Write a program which performs the calculations needed to place countries directly in a table relating the current level (percentage) of tertiary employment (graded high, average and low) to the growth of tertiary employment (graded high, average and low) between 1958 and 1968. Use totals for western Europe to determine average levels with a range for 'high' and 'low' 4% above and below this average. Within the same program calculate the two percentages for the EEC. The necessary data on employment numbers in thousands are shown in Table 7.2.

143

Table 7.1 Data for exercise 4

| Country | Males | <15 | Females 15-49 | >49 | Live Births | Deaths <1 | Deaths >1 | Marriages |
|---|---|---|---|---|---|---|---|---|
| Italy | 26188500 | 6394000 | 13202000 | 7707000 | 930641 | 31511 | 478611 | 375074 |
| Spain | 14726222 | 4067027 | 7731618 | 3850897 | 664948 | 23008 | 256458 | 231546 |
| France | 24249000 | 6119040 | 11519280 | 7868460 | 837100 | 14350 | 525596 | 356615 |
| Portugal | 4580700 | 1329300 | 2323500 | 1263300 | 194962 | 23314 | 73347 | 76553 |
| Switzerland | 2963950 | 700500 | 1494200 | 916800 | 105130 | 1690 | 55684 | 45711 |
| Austria | 3442602 | 868412 | 1692063 | 1346420 | 126115 | 3219 | 92795 | 56001 |
| West Germany | 27477915 | 6569387 | 13571373 | 10079760 | 946325 | 22658 | 624887 | 425112 |
| Belgium | 4697826 | 1116769 | 2183972 | 1582468 | 142494 | 3355 | 111857 | 70019 |
| Netherlands | 6349993 | 1717891 | 3010207 | 1651630 | 237112 | 3224 | 101765 | 117534 |
| United Kingdom | 26983700 | 8389300 | 10962400 | 9198600 | 947241 | 17748 | 638250 | 462758 |
| Denmark | 2402197 | 557422 | 1132368 | 729547 | 81410 | 1497 | 47847 | 40902 |
| Norway | 1885545 | 602518 | 698850 | 597349 | 67350 | 985 | 35231 | 29441 |
| Sweden | 3930696 | 801658 | 1823491 | 1312086 | 113193 | 1560 | 78223 | 52534 |
| Ireland | 1466100 | 446200 | 612800 | 395500 | 61004 | 1280 | 31877 | 18993 |

Table 7.2 Data for exercise 6

| | 1958 Total Civilian Employment | 1958 Service Employment | 1968 Total Civilian Employment | 1968 Service Employment |
|---|---|---|---|---|
| Western Europe | 125685 | 46469 | 129348 | 54974 |
| Austria | 3216 | 1078 | 3145 | 1276 |
| Belgium | 3464 | 1495 | 3615 | 1792 |
| Luxemburg | 134 | 52 | 139 | 59 |
| Denmark | 2045 | 830 | 2350 | 1165 |
| Finland | 1994 | 595 | 2090 | 836 |
| France | 18823 | 7022 | 19741 | 8640 |
| Germany | 25357 | 9296 | 25865 | 10765 |
| Ireland | 1060 | 412 | 1057 | 441 |
| Italy | 20000 | 5948 | 18874 | 6737 |
| Netherlands | 3933 | 1807 | 4428 | 2246 |
| Norway | 1383 | 573 | 1465 | 702 |
| Portugal | 3102 | 857 | 3087 | 973 |
| Spain | 11643 | 3084 | 12113 | 3861 |
| Sweden | 3476 | 1466 | 3782 | 1880 |
| Switzerland | 2399 | 927 | 2713 | 1097 |
| United Kingdom | 23656 | 11029 | 24884 | 12504 |

## 7.7 Further Reading

Aitchison, J. and Brown, J. *The lognormal distribution* (1957)

Collins, L. 'Industrial size distributions and stochastic processes', *Progress in Geography*, 5, 119-165 (1973)

Douglas, E. 'The size of firm and structure of costs in retailing', *Journal of Business*, 35, 158-190 (1962)

Gates, D.M. *Man and his environment: climate* (1972)

Hall, M. 'The small unit in the distributive trades', *Committee of Inquiry on Small Firms, Research Report 8* (1971)

McClelleand, W.G. 'Sales per person and size in retailing: some fallacies', *Journal of Industrial Economics*, 6, 221-229 (1957)

McClelleand, W.G. *Costs and competition in retailing* (1967)

McGuire, J.W. and Archer, S.H. 'Size distributions in industry and the growth of the firm', *Quarterly Review of Economics and Business*, 5, 21-32 (1965)

Ministry of Commerce *Report on the census of distribution and other services in Northern Ireland, 1965* (1969)

National Commission on Food Marketing 'Organisation and competition in food retailing', *Technical Study 7* (1966)

Retail Business 'Economics of scale in retailing', *Retail Business*, 14-25 (September 1972)

Skillcorn, M.D. 'The measurement of efficiency in co-operative retailing', *Cooperative College Papers*, 5 (1956)

Steindl, J. *Random processes and the growth of the firm* (1956)

# Chapter 8
# The Use of Arrays

8.1 The Concept of Arrays
Up to this point in this book all the exercises and examples have used storage locations within the computer that have been referred to by individual unique names, both real and integer, such as ANS, A, NIN and so on. In simple problems with limited data this has been perfectly adequate, but when many data are used this restriction of one name for every storage location can lead to cumbersome and ultimately totally unmanageable programs. Each item must be given its own name and this must be used in all arithmetic, read and write statements. Furthermore, it is often necessary to perform the same operations on many data elements, and if these are contained in uniquely named locations each operation requires a separate statement. For example, a simple task such as reading 20 numbers and expressing each as a percentage of their total would need 20 separate assignments, each calculating the percentage for one individual datum. Fortunately, FORTRAN provides a very powerful set of facilities to enable such strings of numbers to be manipulated. These facilities make use of the idea of an *array*.

An array is a series of storage locations which are all referenced by the same name but are differentiated one from another by the use of subscripts. Each individual storage location in an array is referred to by a name and a number, for example, the third location in array ABC. There may be many memory locations all in array ABC, but each is a distinct location having its own subscript. The subscript is written in brackets after the array name so ABC(3) references the third location in array ABC. Each location is called an *element* of the array. A variable name therefore can be used to signify a set of elements each of which is a memory location. Variable names can now be split into two types:

(i) Those used in this book up to this point and referring to a single location. Strictly these are called scalar variables, for example UNWIN, INK, DAWSON.
(ii) Those which refer to multiple locations and are called array variables.

Array variables may be used almost anywhere that scalar variables may be used, the only difference between the two lying in the presence or absence of a subscript which points to a particular element of the array. The bracketed subscript value follows the array name and must be an integer. Usually it is an integer constant or scalar integer variable, for example the 4th element of A, called A(4), the 99th element of IZ, called IZ(99) or the Ith element of GEOG, called GEOG(I). At least nine out of every ten times an array is used the subscript will be either directly specified by a constant (1, 5, 99, etc) or be an integer variable name (I, NUMBER, J, etc) but occasionally it is useful to have a subscript as an integer expression. This is allowed, but only with a restricted set of seven types of expression. The seven permissible types are of the form shown below where C and K stand for integer constants and V is a scalar integer variable.

| General form of subscript | Example |
|---|---|
| V | X(J) |
| K | X(2) |
| V + K | X(J + 2) |
| V - K | X(J - 2) |
| C * V | X(4 * J) |
| C * V + K | X(4 * J + 2) |
| C * V - K | X(4 * J - 2) |

The first two are special types of expression and the form has been discussed above. In the other five cases, constants follow the operators + and - but precede multiplication. Division is not permitted nor is the use of more than one variable name in the subscript expression. In the last two cases multiplication precedes addition and subtraction in the evaluation of the expression (as in the general rule of evaluation). More complex expressions than these seven can be used, but these must have a separate assignment, as for example

$$K = I * (I - 1)/12 + J$$
$$D(K) = \text{expression}$$
in which the subscript K is first calculated and then used.

## 8.2 The DIMENSION Statement

A variable name can now signify one of two types of variable, scalar or array, but how does the computer decide which type of variable is referred to? Every **array** introduced into a FORTRAN program must be listed in a DIMENSION statement (or related statement - *see Chapters 11 and 12*). This statement must appear *before* the first occurrence of the named array. It is both usual and sensible to place the DIMENSION statement at the beginning of the program, but this is not mandatory. The DIMENSION statement may mention any number of array names and there may be any number of DIMENSION statements. The purpose of the statement is to indicate to the computer which names are arrays and to inform the computer of the number of elements there are to each array so that a suitable block of memory may be reserved for each array.

The DIMENSION statement consists of the word DIMENSION, starting in column 7 of a computer card, followed by the name and size of each array to be used. For example the statement

DIMENSION A (3), I (100), X (15)

will reserve memory space for three arrays, one named A with three elements, one named I with 100 elements and of type integer, and one named X with 15 elements. The size of array is contained in brackets and the information for each array is separated by commas. Because it only provides information the DIMENSION statement is said to be a non-executable statement; it in no way affects evaluation of expressions within the program. The size of an array must be defined by integer constants except under very special circumstances when integer variables are permissible (*see Chapter 11*). The cumulative size of arrays declared by the DIMENSION statement must not exceed the memory store of the computer. Clearly three arrays each of length 10,000 locations cannot be defined in a computer with only 24,000 memory locations, but apart from this physical constraint there is no limit to the size of an array. Having declared an intention to use an array of a certain size then the required memory space is automatically set on one side, but it is not necessary always to fill the whole array if this is not needed in the

program. For example, if after a
                DIMENSION A(5∅), B(5∅)
only 25 locations in both A and B are used this would not
matter; the remaining 25 spaces in each of the two arrays
stay named A and B respectively and cannot be used for any
other storage purpose. Conversely if an attempt is made
to use 55 locations in each of arrays A and B the program
would fail to work satisfactorily. The size of arrays
may be overestimated in the DIMENSION statement but never
underestimated.

8.3 Use of Arrays

How, the perceptive reader will now be asking, does the
use of arrays help in dealing with strings of numbers when
subscripts appear to do nothing more than precisely
specify a unique location name? How is X(42) or X(J)
when J = 12 any different from specifying ATTILA or ANSWER
as scalar variable names? The advantage of arrays lies
in their association with other types of statement rather
than anything in their intrinsic nature.

The example program listed in Fig 8.1, of the calculation
of percentages, shows several aspects of the utility of
arrays. Initially, an array A containing 21 locations is
defined in the DIMENSION statement. The 20 raw data
scores are then read into this array, a DO loop is used to
calculate the total which is placed in the 21st location
(A(21)), and a second loop is used to calculate the percentages which are placed back into array A and overwrite
the raw data. The contents of the whole array A are then
printed.

The input and output statements need more explanation for
they allow either the reading or writing of a string of
numbers. The first parts of both statements are already
familiar and state the numbers of the input-output channels
and the numbers of the FORMAT statements. The remainder
of the READ statement is contained in a separate set of
brackets and consists of the name of the array being dealt
with followed by an integer variable subscript which is
then referenced by an incrementing command as in the DO
loop. Thus
                (A (I), I = 1, 2∅, 1)
means deal with array A the subscripts of which are contained in integer location I which has an initial value of
1 and a terminal value of 20 and increments automatically
from 1 to 20 in steps of 1. As in the normal DO loop,
the incrementation parameter is optional and the above

```
C SAMPLE PROGRAM TO CALCULATE PERCENTAGES
C
      DIMENSION A(21)
C
C SET INPUT-OUTPUT CHANNELS
C
      NIN = 5
      NOUT = 6
C
C READ IN THE DATA INTO THE ARRAY A USING THE FIRST 20 ELEMENTS
C
      READ(NIN,10) ( A(I), I = 1, 20 , 1)
C
C CALCULATE TOTAL AND STORE IN 21ST ELEMENT OF A
C
      A(21) = 0.0
      DO 100 I = 1, 20
  100 A(21) = A(21) + A(I)
C
C CALCULATE PERCENTAGES AND OVERWRITE BACK INTO ELEMENTS OF A
C
      DO 105 I = 1, 20
  105 A(I) = A(I) * 100.0 / A(21)
C
C WRITE OUT THE RESULTS
C
      WRITE(NOUT,15) ( A(I), I = 1,20)
      STOP
   10 FORMAT(10F8.2)
   15 FORMAT( 20F6.1)
      END
```

*Fig 8.1* A sample program to calculate a series of percentages

statement could be shortened to
$$(A (I), I = 1, 2\emptyset)$$
without any change in its effect. The same pattern occurs in the WRITE statement used to output the contents of array B. This form of input-output of an array is called an *implied DO loop*.

If specific values of an array are input or output, or when we want to change the 'natural' order of elements, then the array name and specific subscript may be treated as if they are scalar rather than array variables. Commands such as
  READ (NIN, 1∅) X(15), Y(7), X(2)
  WRITE (NOUT, 15) ABC(32), I(5), Z(1), Z(2), Z(3), Z(4)
may be used in association with suitable FORMAT and DIMENSION statements. If suitable FORMAT statements are used mixtures of scalar and array variables are allowed in input and output statements. For example, all the following would be permissible if provided with DIMENSION and FORMAT statements:
  READ (NIN, 2∅) I(4), ADA

```
    READ (NIN, 25) (A(I), I = 1, 2∅), D, E,
    READ (NIN, 3∅) (A(I), I = 1, 2∅), (B(I), I = 1, 2∅)
    READ (NIN, 35) (A(I), I = 1, N)
    READ (NIN, 4∅) (A(J), J = K, K2)
    WRITE (NOUT, 15) ABC(32), I(5), (Z(K), K = 1, 4)
```
The last example is identical to an earlier example above. Notice that in all these statements the brackets and commas must be correctly positioned; a symbol missing, or in the wrong place, will result in an error. It is clear from these examples, together with others to be introduced later, that arrays are very valuable in the input and output of information.

Returning to Fig 8.1, the remaining use of arrays in the percentage calculation involves their manipulation in arithmetic expressions within DO loops. In both DO loops the control variable I is used to refer successively to the required array elements. First, the total of the new data is accumulated into location A(21) by the statements
$$A(21) = \emptyset.\emptyset$$
and
$$A(21) = A(21) + A(I)$$
where I is incremented within a loop. The location A(21) is treated as any scalar variable with the subscript precisely specified. The remainder of array A is treated as an array variable and is subscripted in a general form by the use of integer variable I which is the DO loop control variable. The percentages are calculated within loop number 105 in the program again using I as its control variable to reference in succession all 20 elements of the array. Here a simple arithmetic statement of the form
$$\text{variable} = \text{expression}$$
is used. The variable is again array A subscripted with I to refer to a DO loop and the expression is
$$A(I) * 1\emptyset\emptyset.\emptyset/A(21)$$
which calculates sequentially (I = 1, 2∅) the percentage for each datum in array A and then replaces it with the calculated percentage. Instead of having a program using only scalar variables and having several dozen statements the program is reduced to one with barely ten executable statements. When arrays are used in place of scalar variables there is a considerable saving in time, programmer's and computer's, coupled with a more elegant programming style.

8.4 Two-dimensional Arrays

Up to this point arrays have been considered as strings of locations. This is analogous to a column of numbers

|   A   |       B       |       C       |
|:-----:|:-------------:|:-------------:|

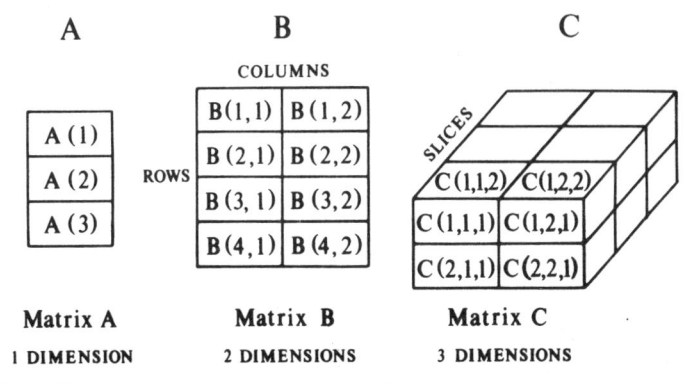

*Fig 8.2* Diagrammatic representation of the structure of arrays

|  Matrix A  | Matrix B    | Matrix C     |
|:----------:|:-----------:|:------------:|
| 1 DIMENSION | 2 DIMENSIONS | 3 DIMENSIONS |

in a matrix and each subscript refers to a row. Thus three rows for a single column may be represented as in Fig 8.2 (A) or A(I), I = 1,3 when the subscript I refers to the row numbers. The matrix and consequently the FORTRAN array has one dimension only. FORTRAN allows an extension to the dimensions of the array to allow manipulation of two- or even three-dimensional arrays. A two-dimensional array has more than one column as well as several rows and a three-dimensional array has slices as well as rows and columns, as in Fig 8.2 (B) and (C). The rules involved in using two- and three-dimensional arrays are basically the same as for a one-dimensional array but with a few additions to deal with the extra dimensions.

(i) The DIMENSION statement must take account of the increased size of the array requiring storage. Instead of the single subscript carried by simple arrays, two subscripts are needed on two-dimensional arrays and three subscripts for three dimensions. With two-dimensional arrays the first subscript refers to ROWS and the second to COLUMNS; with three-dimensional arrays the third subscript refers to SLICES. Three-dimensional arrays are only rarely used and for the present we can ignore them or mention them only in passing. The DIMENSION statement declares the size of the arrays, for example

153

           DIMENSION X (19,4), A(15)
declares two arrays to be used, X has two dimensions with
19 rows and 4 columns and A has one dimension of 15 rows.
Notice that in the array X we have reserved 19 x 4 = 76
locations.   Commas are used between dimension integers
within brackets, and also between each array specification.
Again the only limit on array size is the storage capacity
of the computer but when two- and three-dimensional arrays
are used the area of storage reserved increases very
rapidly indeed.   An array of 50 rows and 50 columns
occupies 2,500 locations while an array of 50 rows, 50
columns and 50 slices occupies 125,000 locations and would
quickly occupy the total available computer store.
(ii) Any two-dimensional array has two subscripts referring
to rows and columns and specific elements in the array may
be identified for the use in operational statements by
giving values to these as for example
                       X (15,20)
or                     JK (4,1)
The first refer to the storage location in array X positioned
at row 15 column 20.   The second is row 4 column 1 of array
JK.   These examples refer to individual locations for use
in assignment, logical or input-output statements and must
not be confused with the overall dimensions given in the
DIMENSION statement.   Thus we might have a sequence
            DIMENSION X(20,25), A(19,4)
            .
            .
            .
            section of program
            .
            .
            IF (X(15,20).GT. 10.0) A(4,1) = 1.0
            .
            .
            .
            section of program
The specified array locations in the conditional statement
are within the total dimension size of the arrays.
(iii) In input-output statements using implied DO loops
then two subscripts are necessary for two-dimensional arrays.
An example would be:
         READ (NIN, 10)((A(I,J), I = 1,15), J = 1,20)
The use of brackets and commas should be carefully noted.
The effect of this read statement would be to fill a two-
dimensional array with numbers.   First all 15 rows (I) of
column 1 (J) are filled then proceeding to column 2, this
will be filled and so on until column 20 has been filled.
The first 15 data elements would thus be placed in the

first column of the array. This process is activated because inner brackets are worked through first and it is the implied DO loop equivalent of normal nested loops. The inner bracket, dealing with subscript I (ie rows) passes through scores 1 to 15 for each value of the outer implied DO (subscript J referring to columns). Supposing, however, that instead of filling the array down each column in turn, it was required to fill in data across each row. Then
       READ (NIN, 1∅)((A(I,J), J = 1,2∅), I = 1,15)
would be required. Now, for each I there are 20 Js (ie for each row there are 20 columns) so the array is processed row by row. This is particularly useful in writing tables where by the nature of the lineprinter, which works down a page, it is usually necessary to print out a table row by row. More complicated input-output arrangements involving arrays are dealt with in Chapter 10.

(iv) A combination of a normal DO loop and an implied DO loop is often used for input and output commands incorporating a two-dimensional array. One advantage of this method relates to the feature of a READ or WRITE whereby either a new card is read or a new line written each time the input-output statement is processed. An example serves to show this feature. With the use of implied DO loops as in:
       WRITE (NIN, 1∅)((A(I,J), I = 1,4), J = 1,5)
it would be possible to output the whole array of 20 numbers in a single line. There is effectively only one WRITE command. Alternatively with the use of an ordinary DO loop and an implied loop as in:
       DO 1∅∅ I = 1,4
       1∅∅ WRITE (NOUT, 15)(A(I,J), J = 1,5)
the WRITE statement is processed four times and each time a new line is begun automatically. Output is thus in the form of a table of four rows each containing five columns. Although this layout could have been achieved using implied loops and a relatively complicated FORMAT (*see Chapter 10*) it is often easier to use the combination of an implied loop within a normal DO loop to produce results in a tabular form.

When using a two-dimensional array in DO loops or implied DO loops the two subscripts must be different integer scalar variables. In the example above, I and J were used but any other integer variable names are permissible and it will be remembered from earlier in the chapter that a limited range of arithmetic expressions are allowable as

**subscripts.** A statement
        READ (NIN, 15)((A(I,I), I = 1,1∅), I = 1,15)
would make little sense. The use of two subscripts in
implied DO loops is basically similar to their use in
normal DO loops.  Consider the statement sequence:
                DO 1∅ I = 1,15
                DO 2∅ J = 1,2∅
                A(I,J) = B(I,J)/C(I,J)
             2∅ CONTINUE
             1∅ CONTINUE
The sequence consists of a nested DO loop with the inner
(2∅) of the two loops passing through the columns of the
three arrays, A, B and C, while the outer loop (1∅) passes
through the rows.  The result of the sequence is a simple
division of the elements of array B by the corresponding
elements in array C with the results being placed into the
corresponding location in array A.

This sequence could be shortened by renumbering the loops
and taking away the CONTINUE statements.  Thus the
commands:
                DO 1∅ I = 1,15
                DO 1∅ J = 1,2∅
             1∅ A(I,J) = B(I,J)/C(I,J)
would have the same effect as the earlier sequence. The
loop involving J is considered as the inner loop and is
dealt with first.

## 8.5 Summary

Although this is a short chapter it has dealt with a very
powerful concept in FORTRAN programming.  Use of an array
allows the programmer considerable freedom in the manipulation of large blocks of data and frequently the geographer is in the position of requiring relatively simple
arithmetic calculations within such a large data block.
Analyses of census material, climatic data or questionnaire
data are usually of this form.  Data are copious, sometimes embarrassingly so, and the problems are ones of
reworking, often very simply, some variables within the
total data block.  While trivial mathematically, calculations of ratios or of percentages can be of immense
value to geographers, but usually they are required by the
hundred rather than the handful.  The use of a computer
program incorporating arrays and DO loops makes light work
of such repetitive calculations.

8.6 Worked Examples
(i) *Problem.* Upland slope deposits exist in at least three forms. Some slopes have a mantle of fine-grained *head* deposits, others a coarser talus or *scree* while others have a cover of large blocks forming a *block-scree*, and it can be suggested that the type of deposit found depends strongly upon the rocks from which it has been derived (Watson, 1961). We can use *contingency analysis* to test this idea (Moroney, 1951). Observations on nominally or ordinally scaled data such as rock type and deposit type are analysed for association between pairs of these attributes by assembling a *contingency table* which is simply a cross-tabulation of the categories for one attribute against those for another. Each cell of the table contains the number of items that have the pair of characteristics corresponding to that cell.

In order to test whether or not parent rock type and slope deposit type are associated together, a sample of 218 deposits in Snowdonia, North Wales, was examined and items were classified into the categories head, scree and block-scree. At the same time the dominant rock making up the deposit was recorded and classified into three groups according to its jointing and cleavage characteristics as gritstones and dolerite, shales and slate and igneous. These observations were then used to draw up the following table:

|  | grits and dolerite | shales and slate | igneous | row sum |
|---|---|---|---|---|
| head | 1 | 28 | 7 | 36 |
| scree | 9 | 53 | 68 | 130 |
| block-scree | 28 | 3 | 21 | 52 |
| column sum | 38 | 84 | 96 | 218 |

Visual inspection suggests that head deposits are relatively more common on the friable shales and slates (28 sites) than on other rocks, that scree is relatively common on all rock types and that block-scree is confined to the harder grits and dolerites (28 sites) and other igneous rocks (21 sites). To test this association statistically it must be established how likely this particular configuration of numbers in the table is, given that the row and column sums are fixed by the sampling strategy chosen. This probability can only have meaning in relation to some idea or hypothesis about these data so that we must also specify an hypothesis about the suspected association. The simplest hypothesis that we can have is the null hypothesis ($H_o$) which proposes that there is *no* association between the attributes. If, in the face of the data collected, this is unlikely to be true then it can be rejected in favour of the alternative which suggests that the attributes are associated.

*Algorithm and flowchart.* The test statistic used is the $\chi^2$ defined as

$$\chi^2 = \sum_{i=1}^{nr} \sum_{j=1}^{nc} (O_{ij} - E_{ij})^2 / E_{ij} \qquad (1)$$

in which $O_{ij}$ is the observed frequency in the cell ij of the table, $E_{ij}$ the expected frequency in that cell assuming that $H_o$ is true and

$$\sum_{i=1}^{nr} \sum_{j=1}^{nc}$$

is the double summation denoting that the total $\chi^2$ for the table is the sum of all the individual cell values, for all nr rows and nc columns. In this particular case the table is of size 3 by 3 so that there are 9 cells. For $\chi^2$ value derived in this way with fixed row and column totals there are

$$df = (nr-1).(nc-1) \qquad (2)$$

degrees of freedom and critical values at various percentage points can be found from published tables (Lindley and Miller, 1962).

How are the expected values calculated? If the null hypothesis is assumed to be true then the expected value $E_{ij}$ for any cell will simply be the proportion of the total got simply by a consideration of the row and column totals for that cell:

$$E_{ij} = (\text{row sum}_i \times \text{column sum}_j)/\text{total} \qquad (3)$$

Because $\chi^2$ values are sensitive to very small expected frequencies most authorities recommend that tables with expected frequencies less than 5 should not be analysed.

A flowchart and computer program to perform contingency analysis are presented as Figs 8.3 and 8.4. The program makes use of both integer and real arrays. There are two major phases in the algorithm, indicated by the two parts of the flowchart. In the first phase the integer array NOS is filled with the observed data and the row, column and overall sums are assembled in the elements of NROW, NCOL and the location NSAMP. The second phase uses these values to calculate and print out the $\chi^2$ values.

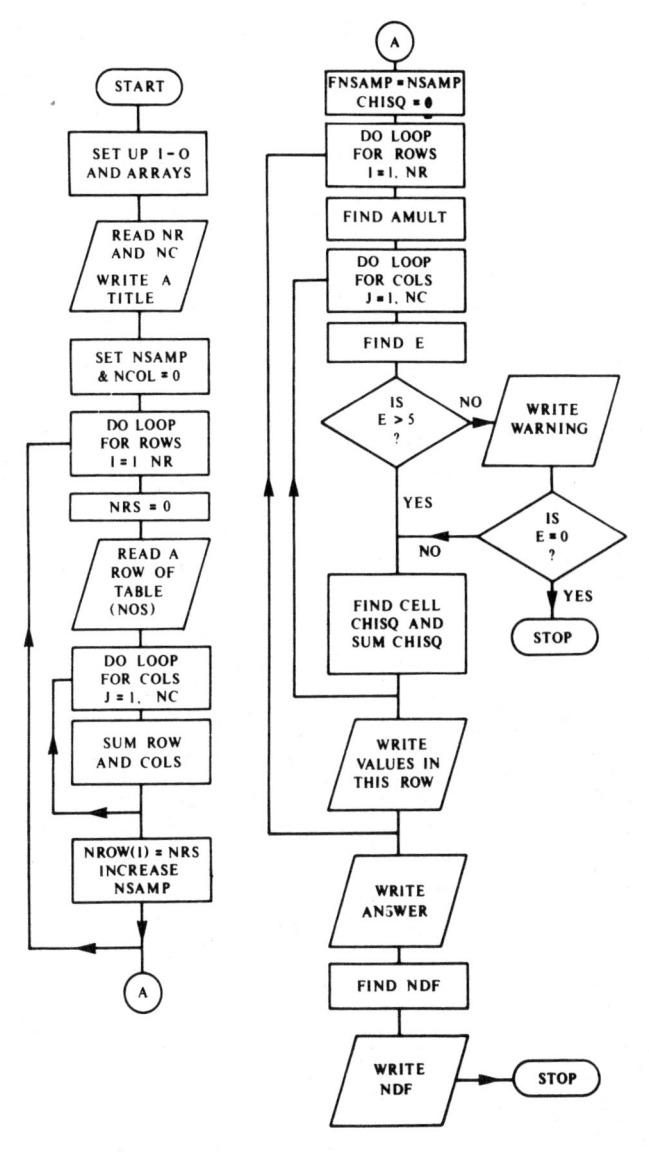

*Fig 8.3* Flowchart for sample program for contingency analysis

```
C PROGRAM TO CALCULATE CHISQUARE FOR A NULL HYPOTHESIS OF NO              10
C ASSOCIATION IN AN NR(ROWS) BY NC(COLS) CONTINGENCY TABLE                20
C DEMONSTRATING THE USE OF ARRAYS                                         30
C AS DIMENSIONED WILL ANALYSE UP TO 10 BY 10 TABLES                       40
C                                                                         50
      DIMENSION NOS(10,10),NCOL(10),NROW(10),CHIS(10)                     60
      NIN = 5                                                             70
      NOUT = 6                                                            80
C READ IN THE NUMBER OF ROWS AND COLUMNS                                  90
      READ(NIN,90)  NR , NC                                              100
C WRITE HEADING                                                          110
      WRITE(NOUT,91)                                                     120
C READ DATA FORMING ROW AND COLUMN SUMS AS YOU GO ALONG                  130
      NSAMP = 0                                                          140
      DO 5 J = 1,NC                                                      150
    5 NCOL(J) = 0                                                        160
      DO 10 I = 1,NR                                                     170
      NRS = 0                                                            180
      READ(NIN,92) (NOS(I,J),J = 1,NC )                                  190
      DO 20 J = 1,NC                                                     200
      NRS = NRS + NOS(I,J)                                               210
   20 NCOL(J) = NCOL(J) + NOS(I,J)                                       220
      NROW (I) = NRS                                                     230
   10 NSAMP = NSAMP + NRS                                                240
      FNSAMP = NSAMP                                                     250
C CALCULATE AND PRINT OUT CHISQUARE VALUES ROW BY ROW                    260
      CHISQ = 0.0                                                        270
      DO 30 I = 1,NR                                                     280
      TOP = NROW(I)                                                      290
      AMULT = TOP/FNSAMP                                                 300
      DO 40 J = 1, NC                                                    310
      TOP = NCOL(J)                                                      320
      EXP = TOP*AMULT                                                    330
C TRAP AND PRINT OUT LOW EXPECTED FREQUENCIES, IF A ZERO THEN STOP.      340
      IF (EXP .GT. 5.0 ) GO TO 50                                        350
      WRITE(NOUT,93) EXP,I,J                                             360
      IF (EXP .EQ. 0.0 ) STOP                                            370
   50 CHIS(J) = (NOS(I,J) - EXP )**2 /EXP                                380
   40 CHISQ = CHISQ + CHIS(J)                                            390
      WRITE(NOUT,94)(NOS(I,J),CHIS(J),J = 1, NC)                         400
   30 CONTINUE                                                           410
      WRITE(NOUT,95) NSAMP,CHISQ                                         420
C CALCULATE AND PRINT OUT DEGREES OF FREEDOM FOR THIS TABLE              430
      NDF = (NR -1 ) * ( NC- 1)                                          440
      WRITE (NOUT,96) NDF                                                450
      STOP                                                               460
   90 FORMAT(2I3)                                                        470
   91 FORMAT( 18H CONTINGENCY TABLE)                                     480
   92 FORMAT (10I3)                                                      490
   93 FORMAT(26H **WARNING**EXPECTED ONLY ,F10.4, 9H IN CELL ,2I5)       500
   94 FORMAT ( I5,F11.4,I5,F11.4,I5,F11.4)                               510
   95 FORMAT(13H TOTAL CASES=,I5,13H TOTAL CHISQ=,F10.4)                 520
   96 FORMAT(19H DEGREES FREEDOM = ,I5)                                  530
      END                                                                540
 3  3
 1 28  7
 9 53 68
28  3 21
```

*Fig 8.4* Program listing for sample program for contingency analysis

160

In the first phase the input and output channel numbers
are set and the number of rows and columns of the table
(NR and NC) read from the first data card.  As dimensioned
the program will accept tables of any size up to 10 by 10.
The next step is to input the table into the integer array
NOS using the DO loop
DO 1∅ I = 1,NR
which at the same time sums each row and column as well as
the overall total.  These summations are a little complicated, involving careful use of the control variables
I and J as array subscripts, and can best be appreciated by
careful study of the appropriate section of the flowchart.

The second phase has a similar structure to the first,
making use of nested DO loops to control array subscripts
in NOS, NROW and NCOL.  The outer DO loop, involving I and
terminating at the statement labelled 30, cycles through
rows of the array NOS.  Within it the quantity (row sum/
grand total) required in the calculation of expected frequencies is found and stored as AMULT.  The inner loop
involves the control variable J and terminates at the
statement labelled 40.  Within it expected frequencies
for each cell of a row of the table are calculated using
the assignments
TOP = NROW (J)
EXP = TOP * AMULT
and then tested to ensure first that they are above 5 and
secondly that they are not zero, indicating that either
a complete row or a complete column of the table has no
recorded occurrences.  Expected frequencies less than 5
but above zero will lead to a warning message whereas
a zero will terminate the program.  The $\chi^2$ values are
calculated using the assignment labelled 50 and the values
for each row are stored in the *real* array CHIS with the
cumulative total in the location CHISQ.  After an entire
row of values has been assembled in this way the values of
the observed frequency and the cell $\chi^2$ for it are printed
and control passes back to the outer loop.  When this
outer loop is satisfied and all the rows have been analysed
the total sample size and total $\chi^2$ for the table are written
together with the appropriate degrees of freedom.
*Data and run.* The test problem is that introduced at the
start of this case study involving a 3 by 3 table showing
frequencies of types of slope deposit cross tabulated
against their dominant lithology for a sample of sites in
Snowdonia.  The input data cards are laid out according

to formats 90 and 92 as

```
  3   3                  ..... FORMAT (2I3) giving NR and NC
  1  28   7
  9  53  68              .....FORMAT (10I3) each card specifying
 28   3  21                    a row of the contingency table
```

```
CONTINGENCY TABLE
    1      4.4346     28     14.3901      7      4.9441
    9      8.2352     53      0.1688     68      2.0195
   28     39.5581      3     14.4859     21      0.1575
 TOTAL CASES=   218  TOTAL CHISQ=      88.3936
 DEGREES FREEDOM =          4
```

*Fig 8.5* Results of contingency analysis program

The results are shown in Fig 8.5, which indicates a total $\chi^2$ associated with the null hypothesis of no association between these attributes of 88.394 with 4 degrees of freedom. If the null hypothesis is true the probability of this value occurring by chance is very low, much less than 0.1% (the critical value is 18.467) so that we are justified in rejecting this hypothesis and inferring that this value is not the result of chance. There is therefore a definite association between deposit type and rock type in Snowdonia.

Although this program has been formulated and used with a geomorphological example it will be found to have widespread applicability in a range of geographical problems for which nominal and ordinal data are all that can be collected.

(ii) *Problem*. In the study of the economic structure of regions the analysis of the growth or otherwise of employment in different parts of the region is often an important part of the study. A number of techniques have been developed to attempt analyses of the changes in regional economic patterns. One is shift-share analysis which in recent years has received considerable attention from geographers and economists (O'Farrell, 1972; Stilwell, 1969). A number of variations on the basic theme have been developed (Ashby, 1970; Chalmers, 1971). The rationale of the method is first to identify the economic sectors which are responsible for higher or lower than average growth rates of different parts of the country or region and secondly to determine whether there is an overall trend to the pattern of growth of the districts over time.

The problem considered in this example is that of isolating the changes in regional employment that took place in Denmark during the 1920s and 30s. While this was a period of slow growth at the national level some parts of Denmark were industrialising quite quickly while others were becoming severely depressed. The questions asked are first which areas were growing and which declining and which industries were responsible for this growth or decline and secondly to what extent was growth associated with the larger urban areas. The example considers the employment change in each of the 26 counties and secondly it considers the change in urban and rural areas. Shift-share analysis reveals the precise way employment patterns changed and whether or not the change was due to national growth processes, regional influences or due to activities in particular industries.

*Algorithm and flowchart.* Growth in a region is considered as being divided into three types:

(i) *A regional share factor.* This is the amount all employment in the region would have grown by had it expanded at the same rate as in the nation as a whole.

(ii) *An industrial shift factor.* This is the extra amount by which employment in the region has grown due to regional specialisations in certain industries. If growth industries are well represented in the region the shift is positive; if declining industries are over-represented then the factor is negative.

(iii) *A competitive factor.* This is the extra employment growth in a particular industry due to that industry growing in the region at a different rate from the change in that industry within the country as a whole. Thus given a particular industrial mix a buoyant region will have a positive competitive factor while a stagnating region will have a negative index.

The three types of growth together account for total employment growth. Although the method has been outlined in terms of growth it is equally applicable if employment is declining and the calculated shifts then become negative. A study of the three components of change reveals the precise way employment patterns are changing and whether the change is due to national growth processes, regional influences or to activities in particular industries.

To calculate the shift and share components it is necessary to have data over at least one time-period relating to the employment composition of a region or series of regions. Three equations are solved, each one giving an arithmetic value for each component. The equations may be simply written using symbols:

let $e_{ij}$ be the employment in industry i in region j (in all the symbols the subscript i refers to industry type and j to regions). If $e_{ij}$ is summed over all industries for a particular region then

$$E_j = \sum_i e_{ij} \quad \text{(ie, total employment in } j^{th} \text{ region)}$$

similarly $E_i = \sum_j e_{ij}$ (ie, total national employment in $i^{th}$ sector)

and $G = \sum_{ij} e_{ij}$ (ie, total national employment)

Superscripts 0 and 1 may be used to refer to the initial date and the final date of the time period with which the analysis is concerned. For example, total employment growth in the region is

$$T_j = E_j^0 - E_j^1$$

Using these symbols the three components can be put into the following equations:

*Regional share factor*

$$R = E_j^0 - E_j^0 \left( \frac{G^1}{G^0} \right)$$

*Industrial shift factor*

$$P = \sum_i e_{ij}^0 \left( \frac{E_i^1}{E_i^0} - \frac{G^1}{G^0} \right)$$

*Competitive factor*

$$D = \sum_i \left[ e_{ij}^1 - e_{ij}^0 \left( \frac{E_i^1}{E_i^0} \right) \right]$$

The regional competitive effect is the difference between the regional share and the industrial shift factor. The competitive factor also may be calculated for each industry in a region to indicate relative growth or decline among the industries.

The flowchart for the shift-share program is shown in Fig 8.6. Initially the full data block for each year is input. Each of the three factors is calculated in turn for all regions and the results for all the regions are written out. Finally, for each region the shift for each industry is calculated and written out.

*Program.* The program listing is shown in Fig 8.7. Arrays BASE and END are defined and contain the employment figures by industry and by region for the base year and the end year of the period under consideration. BREG and EREG contain the regional employment totals for base and end years while BSEC and ESEC contain industrial sector totals for base and end years. The three factors are stored in SHIFT, STRUCT, and COMP respectively. Throughout the program JREG refers to the number of regions and ISEC to the number of industry sectors. Also subscript J refers to regions and I to sectors. Base and end employment data matrices are read using an implied DO loop for the sectors and a normal DO loop to take each region in turn. Statements in nested DO loops then calculate the regional and sectoral totals and the sector totals are added together to calculate national totals. Generally in shift-share analyses there are more regions than industrial sectors so summing over industry totals, rather than over regional totals, is fractionally faster. The ratio of ENAT to BNAT is then calculated as this figure is used extensively in DO loops later in the program, and avoids repeating the division at several points later in the program.

The program then proceeds to the calculation of the factors, and tests are made that employment exists in industry groups. If some groups contain no employment and no checks are carried out then divisions by zero are made and computers have difficulty evaluating such expressions! The WRITE statement incorporates both scalar and array variables within the same statement.

The final block of program consists of a large loop numbered 165 (indexed J) which deals with each region in turn and inside which the shift for each industry is calculated using loop 170 indexed with an I. Again tests are made for zero entries in the data matrix and precautions taken against unnecessary calculations.

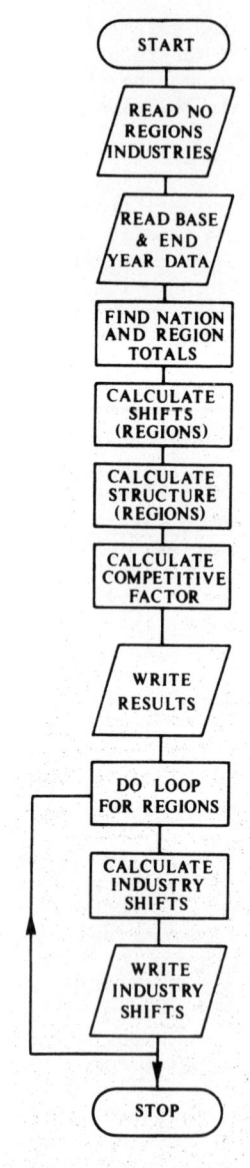

*Fig 8.6* Flowchart for shift-share analysis

```
C PROGRAM TO CALCULATE SHIFT SHARE COEFFICIENTS ON EMPLOYMENT DATA      10
      DIMENSION BASE(20,50),END(20,50),BREG(50),EREG(50),BSEC(20),FSEC(220
     *0),SHIFT(50),STRUCT(50),COMP(50)                                   30
      NIN = 5                                                            40
      NOUT = 6                                                           50
C NUMBER OF REGIONS = JREG                                               60
C NUMBER OF SECTORS = ISEC                                               70
      READ (NIN,10) JREG,ISEC                                            80
C READ BASE AND END YEAR EMPLOYMENT BY SECTOR FOR EACH REGION            90
      DO 100 J=1,JREG                                                   100
  100 READ (NIN,15) (BASE(I,J),I=1,ISEC)                                110
      DO 105 J=1,JREG                                                   120
  105 READ (NIN,15) (END(I,J),I=1,ISEC)                                 130
C CALCULATE REGIONAL TOTALS                                             140
      DO 110 J=1,JREG                                                   150
      BREG(J) = 0.0                                                     160
      EREG(J) = 0.0                                                     170
      DO 115 I=1,ISEC                                                   180
      BREG(J) = BREG(J) + BASE(I,J)                                     190
  115 EREG(J)= EREG(J) + END(I,J)                                       200
  110 CONTINUE                                                          210
C CALCULATE SECTOR TOTALS                                               220
      DO 120 I=1,ISEC                                                   230
      BSEC(I) = 0.0                                                     240
      ESEC(I) = 0.0                                                     250
      DO 125 J=1,JREG                                                   260
      BSEC(I) = BSEC(I) + BASE(I,J)                                     270
  125 ESEC(I) = ESEC(I) + END(I,J)                                      280
  120 CONTINUE                                                          290
C CALCULATE NATIONAL EMPLOYMENT                                         300
      BNAT = 0.0                                                        310
      ENAT = 0.0                                                        320
      DO 130 I=1,ISEC                                                   330
      BNAT = BNAT + BSEC(I)                                             340
  130 ENAT = ENAT + ESEC(I)                                             350
      RNT = ENAT / BNAT                                                 360
C CALCULATE TOTAL SHIFT BY REGION                                       370
      DO 135 J =1,JREG                                                  380
      SHIFT(J) = 0.0                                                    390
  135 SHIFT(J) = EREG(J) -(BREG(J)* RNT)                                400
C CALCULATE STRUCTURAL EFFECT BY REGION                                 410
      DO 140 J=1,JREG                                                   420
      STRUCT(J) = 0.0                                                   430
      TOT = 0.0                                                         440
      DO 145 I=1,ISEC                                                   450
      IF (BSEC(I)) 146,146,147                                          460
  146 ABSEC = 0.0                                                       470
      GO TO 145                                                         480
  147 ABSEC = BASE(I,J) * FSEC(I) / BSEC(I)                             490
  145 TOT = TOT + ABSEC                                                 500
  140 STRUCT(J) = TOT -(BREG(J) * RNT)                                  510
C CALCULATE COMPETITIVE EFFECT BY REGION                                520
      DO 150 J=1,JREG                                                   530
  150 COMP(J) = SHIFT(J) - STRUCT(J)                                    540
C WRITE HEADINGS                                                        550
      WRITE (NOUT,25)                                                   560
      WRITE (NOUT,20)                                                   570
      WRITE (NOUT,21)                                                   580
C CALCULATE EFFECTS AS PERCENTAGES AND WRITE RESULTS                    590
      DO 155 J=1,JREG                                                   600
      PSH = SHIFT(J) * 100.0 / (BREG(J) * RNT)                          610
      PST = STRUCT(J) * 100.0 / (BREG(J) * RNT)                         620
      PCO = PSH - PST                                                   630
  155 WRITE (NOUT,30) J,SHIFT(J),PSH,STRUCT(J),PST,COMP(J),PCO          640
      WRITE (NOUT,31)                                                   650
C CALCULATE SHIFT FOR EACH INDUSTRY IN EACH REGION                      660
      DO 165 J=1,JREG                                                   670
      EXPNAT = 0.0                                                      680
      WRITE (NOUT,35) J                                                 690
      WRITE (NOUT,40)                                                   700
      WRITE (NOUT,41)                                                   710
      DO 170 I=1,ISEC                                                   720
      IF (BSEC(I)) 171,171,172                                          730
  171 EXP = 0.0                                                         740
      GO TO 173                                                         750
```

167

```
    172 EXP = BASE(I,J) * ESEC(I) / BSEC(I)                              760
        EXPNAT = EXPNAT + EXP                                            770
    173 DIF = END(I,J) - EXP                                             780
        PBASE = BASE(I,J) * 100.0 / BREG(J)                              790
        PEND = END(I,J) *100.0 / EREG(J)                                 800
        IF (BASE(I,J)) 174,174,175                                       810
    174 PDIF = 0.0                                                       820
        GO TO 170                                                        830
    175 PDIF = DIF * 100.0 / EXP                                         840
    170 WRITE (NOUT,45) I,BASE(I,J),PBASE,END(I,J),PEND,EXP,DIF,PDIF     850
    165 WRITE (NOUT,50) BREG(J),EREG(J),EXPNAT,COMP(J)                   860
        STOP                                                             870
     10 FORMAT ( 2I3)                                                    88B
     15 FORMAT ( 8F10.0)                                                 890
     25 FORMAT ( 21H SHIFT SHARE ANANYSIS)                               900
     20 FORMAT ( 55H REGION        SHIFT         STRUCTURE    CUMPETITIO910
       *N)                                                               920
     21 FORMAT ( 56H            TOTAL PERCENT TOTAL PERCENT TOTAL PERCE930
       *NT)                                                              940
     30 FORMAT ( I6,F9.1,F8.2,F9.1,F7.2,F9.1,F7.2)                       950
     31 FORMAT ( 59H NOTE -- PERCENTAGES ARE OF EXPECTED EMPLOYMENT IN END960
       * YEAR)                                                           970
     35 FORMAT ( 8H REGION ,1I3)                                         980
     40 FORMAT ( 74H INDUSTRY  EMPLOYMENT BASE    EMPLOYMENT END  EXPECTED 990
       * COMPETITIVE SHIFT  )                                            1000
     41 FORMAT ( 71H            TOTAL    PERCENT   TOTAL   PERCENT  TOTAL 1010
       *  TOTAL  PERCENT)                                                1020
     45 FORMAT ( I7,F10.1,F8.2,F10.1,F8.2,F10.1,F9.2,F7.2)               1030
     50 FORMAT ( 7H TOTALS,1F10.1,1F18.1,1F18.1,1F8.1)                   1040
        END                                                              1050
        FINISH
```

*Fig 8.7* Program for shift-share analysis

The FORMAT statements are collected together at the end
of the program. Some of them appear rather cumbersome
due to the table layout etc. (See Chapter 10 for methods
to reduce the length of such formats.) FORMAT 30 and 40
show how the initial figure in the specification may be
omitted and a 1 is assumed. FORMAT 50 includes the initial
figure. Either version is perfectly permissible.

*Data.* The data for the first run consists of employment
figures for 16 employment groups in each of 26 counties
for 1925 and 1935. These data are shown in Figs 8.8 and
8.9. For the second run the data have been sorted on
a different areal base and 13 groups of communities are
identified. The data are shown in Fig 8.10. The population size groups refer to 1925 so that the same towns
occur in each group in both years. This means that in
1935 some towns, for example, in the 7,000-9,999 group in
fact had a population over 10,000 but an identical set of
towns is retained in each group for both 1925 and 1935.
Unless this is done the results have little meaning.

*Results.* Sample tables from the full print out are shown
in Fig 8.11 for the first run and Fig 8.12 for the second
run. From Fig 8.11 the positive shifts for the regions
are seen to be highest in total for the Copenhagen city
area and highest in percentage terms for the district
immediately around Copenhagen. Employment in and around
Copenhagen grew considerably faster than the national rate.

Fig 8.8 Danish county employment totals in 1925

| County | Food | Textiles | Clothing | Construction | Wood | Leather | Ceramics | Metals | Chemicals | Paper | Books | Personal service | Wholesale | Retail Food | Retail Non-food | Catering |
|---|---|---|---|---|---|---|---|---|---|---|---|---|---|---|---|---|
| Hovedstaden | 23295 | 4036 | 22685 | 16286 | 7274 | 6864 | 2440 | 36303 | 9190 | 1602 | 7465 | 5482 | 27096 | 20083 | 18461 | 8494 |
| Kobenhavns | 1171 | 1251 | 373 | 1379 | 303 | 321 | 581 | 1074 | 706 | 31 | 31 | 306 | 84 | 1442 | 493 | 1280 |
| Roskilde | 1112 | 143 | 471 | 1247 | 715 | 419 | 229 | 858 | 361 | 25 | 118 | 156 | 362 | 1193 | 721 | 389 |
| Frederiksborg | 1726 | 626 | 580 | 2485 | 986 | 322 | 1332 | 2860 | 314 | 8 | 149 | 297 | 365 | 2305 | 1293 | 1356 |
| Holbaek | 1785 | 38 | 577 | 2183 | 984 | 408 | 788 | 1752 | 247 | 0 | 166 | 251 | 532 | 2205 | 1216 | 531 |
| Soro | 2538 | 90 | 678 | 2149 | 1136 | 351 | 464 | 1703 | 368 | 34 | 226 | 311 | 562 | 2095 | 1299 | 689 |
| Praesto | 2011 | 43 | 733 | 2126 | 935 | 411 | 1198 | 1529 | 162 | 622 | 165 | 248 | 521 | 2175 | 1158 | 613 |
| Bornholms | 1354 | 15 | 238 | 743 | 445 | 174 | 1064 | 566 | 87 | 0 | 95 | 123 | 144 | 852 | 473 | 469 |
| Maribo | 3306 | 57 | 882 | 2509 | 1252 | 453 | 410 | 2746 | 274 | 0 | 251 | 327 | 693 | 2475 | 1414 | 653 |
| Svendborg | 2626 | 199 | 1055 | 2583 | 1739 | 585 | 685 | 2621 | 320 | 74 | 288 | 319 | 724 | 2704 | 1570 | 736 |
| Odense | 3469 | 955 | 1860 | 3243 | 1624 | 545 | 647 | 4701 | 636 | 612 | 538 | 427 | 2701 | 2520 | 1870 | 879 |
| Assens | 860 | 45 | 404 | 833 | 628 | 232 | 188 | 1329 | 117 | 0 | 7 | 107 | 201 | 962 | 521 | 227 |
| Vejle | 2841 | 1508 | 1608 | 2960 | 1444 | 758 | 656 | 3318 | 947 | 108 | 418 | 464 | 1170 | 2766 | 1856 | 1102 |
| Skanderborg | 2345 | 927 | 941 | 1839 | 1400 | 315 | 506 | 2282 | 601 | 327 | 288 | 315 | 742 | 2001 | 1549 | 740 |
| Aarhus | 2924 | 531 | 1596 | 3332 | 1946 | 701 | 528 | 4444 | 1760 | 224 | 684 | 511 | 2196 | 2118 | 2292 | 1129 |
| Randers | 2009 | 414 | 963 | 2186 | 1391 | 491 | 958 | 2332 | 823 | 26 | 248 | 340 | 1013 | 2460 | 1737 | 659 |
| Aalborg | 3156 | 571 | 1473 | 3062 | 1425 | 381 | 2707 | 2318 | 1000 | 196 | 390 | 435 | 1441 | 2796 | 2072 | 1036 |
| Hjorring | 1482 | 145 | 724 | 1742 | 860 | 355 | 700 | 2302 | 368 | 0 | 149 | 229 | 440 | 2229 | 1353 | 819 |
| Thisted | 981 | 190 | 452 | 1151 | 480 | 204 | 496 | 998 | 154 | 0 | 109 | 141 | 269 | 1198 | 791 | 372 |
| Viborg | 1708 | 176 | 570 | 1936 | 1054 | 375 | 777 | 1397 | 554 | 62 | 165 | 239 | 325 | 1888 | 1104 | 444 |
| Ringkobing | 1790 | 922 | 769 | 2155 | 901 | 319 | 657 | 1449 | 532 | 2 | 181 | 293 | 499 | 2123 | 1510 | 716 |
| Ribe | 2196 | 179 | 766 | 2228 | 936 | 423 | 804 | 1665 | 557 | 17 | 255 | 376 | 1084 | 2161 | 1465 | 1037 |
| Haderslev | 830 | 97 | 246 | 1191 | 443 | 202 | 156 | 733 | 136 | 7 | 73 | 109 | 439 | 901 | 664 | 356 |
| Aabenraa | 564 | 12 | 167 | 1103 | 292 | 142 | 407 | 526 | 113 | 0 | 56 | 62 | 285 | 646 | 414 | 331 |
| Sonderborg | 997 | 290 | 233 | 1014 | 440 | 160 | 541 | 600 | 136 | 0 | 56 | 96 | 328 | 679 | 486 | 427 |
| Tonder | 445 | 28 | 129 | 633 | 293 | 134 | 62 | 370 | 73 | 0 | 39 | 47 | 108 | 566 | 384 | 276 |

Fig 8.9 Danish county employment totals in 1935

| County | Food | Textiles | Clothing | Construction | Wood | Leather | Ceramics | Metals | Chemicals | Paper | Books | Personal service | Wholesale | Retail Food | Retail Non-food | Catering |
|---|---|---|---|---|---|---|---|---|---|---|---|---|---|---|---|---|
| Hovedstaden | 23933 | 6623 | 28284 | 21792 | 7602 | 8479 | 3210 | 36303 | 11633 | 2855 | 9136 | 9022 | 37119 | 28987 | 23880 | 13788 |
| Kobenhavns | 1204 | 1249 | 1066 | 2975 | 577 | 262 | 869 | 1659 | 1065 | 57 | 75 | 740 | 275 | 2973 | 1120 | 1377 |
| Roskilde | 1138 | 114 | 501 | 1504 | 793 | 367 | 327 | 864 | 1296 | 41 | 149 | 377 | 520 | 1779 | 1017 | 563 |
| Frederiksborg | 1610 | 777 | 548 | 2966 | 1133 | 274 | 1611 | 3474 | 656 | 10 | 196 | 539 | 609 | 3095 | 1597 | 1790 |
| Holbaek | 2046 | 67 | 519 | 2035 | 927 | 416 | 934 | 1669 | 400 | 14 | 141 | 484 | 783 | 2773 | 1399 | 750 |
| Soro | 2662 | 89 | 632 | 2031 | 1151 | 362 | 687 | 1591 | 435 | 29 | 256 | 602 | 869 | 2800 | 1538 | 964 |
| Praesto | 1866 | 25 | 634 | 2216 | 929 | 475 | 1374 | 1470 | 325 | 690 | 223 | 517 | 665 | 2863 | 1469 | 1036 |
| Bornholms | 1318 | 9 | 290 | 735 | 393 | 157 | 1463 | 491 | 101 | 0 | 0 | 199 | 250 | 1024 | 571 | 815 |
| Maribo | 3268 | 246 | 895 | 2163 | 1195 | 401 | 534 | 3198 | 412 | 70 | 261 | 634 | 1338 | 3135 | 1656 | 1089 |
| Svendborg | 2490 | 178 | 1115 | 2627 | 1683 | 490 | 777 | 2347 | 407 | 108 | 276 | 698 | 1251 | 3316 | 1902 | 1044 |
| Odense | 3691 | 1593 | 2120 | 4073 | 2027 | 613 | 897 | 6670 | 873 | 609 | 743 | 999 | 3236 | 3889 | 2623 | 1454 |
| Assens | 903 | 33 | 383 | 1046 | 688 | 213 | 254 | 1601 | 197 | 0 | 71 | 251 | 348 | 1224 | 631 | 411 |
| Vejle | 3259 | 2519 | 1630 | 3262 | 1376 | 651 | 936 | 3554 | 1220 | 169 | 432 | 1020 | 1945 | 4147 | 2453 | 1696 |
| Skanderborg | 2385 | 753 | 1230 | 2112 | 1184 | 319 | 531 | 2771 | 753 | 398 | 397 | 639 | 1236 | 2814 | 1849 | 905 |
| Aarhus | 3326 | 364 | 2345 | 4309 | 1940 | 596 | 668 | 5359 | 2279 | 460 | 728 | 1226 | 3616 | 3116 | 2868 | 1785 |
| Randers | 2205 | 774 | 1006 | 2602 | 1183 | 490 | 807 | 2524 | 727 | 7 | 334 | 736 | 1610 | 4422 | 2009 | 1017 |
| Aalborg | 4175 | 831 | 1398 | 3362 | 1251 | 428 | 2178 | 2537 | 825 | 85 | 495 | 943 | 2460 | 4329 | 2630 | 1499 |
| Hjorring | 1962 | 172 | 796 | 2256 | 717 | 373 | 761 | 2466 | 546 | 0 | 209 | 548 | 1021 | 2939 | 1652 | 1173 |
| Thisted | 1072 | 175 | 408 | 1172 | 386 | 189 | 508 | 970 | 250 | 0 | 124 | 300 | 605 | 1562 | 921 | 458 |
| Viborg | 1474 | 437 | 564 | 2164 | 1157 | 423 | 829 | 1363 | 581 | 64 | 191 | 436 | 772 | 2485 | 1376 | 686 |
| Ringkobing | 2172 | 2060 | 1177 | 2164 | 757 | 305 | 599 | 1758 | 718 | 4 | 269 | 589 | 995 | 2784 | 1897 | 946 |
| Ribe | 2594 | 240 | 862 | 2370 | 994 | 499 | 540 | 1987 | 803 | 33 | 223 | 723 | 1607 | 2179 | 2535 | 1379 |
| Haderslev | 1023 | 333 | 490 | 1286 | 429 | 165 | 269 | 839 | 155 | 14 | 110 | 224 | 541 | 1086 | 783 | 532 |
| Aabenraa | 621 | 8 | 289 | 924 | 366 | 167 | 321 | 478 | 120 | 0 | 67 | 153 | 444 | 845 | 465 | 526 |
| Sonderborg | 738 | 333 | 265 | 847 | 311 | 148 | 480 | 419 | 135 | 2 | 153 | 186 | 467 | 818 | 529 | 514 |
| Tonder | 542 | 127 | 150 | 530 | 241 | 136 | 81 | 345 | 89 | 0 | 35 | 92 | 167 | 670 | 424 | 431 |

170

Fig 8.10 Employment totals by type of town 1925 and 1935

| 1925 | Food | Textiles | Clothing | Construction | Wood | Leather | Ceramics | Metals | Chemicals | Paper | Books | Personal service | Wholesale | Retail Food | Retail Non-food | Catering |
|---|---|---|---|---|---|---|---|---|---|---|---|---|---|---|---|---|
| København | 17892 | 3224 | 21656 | 11471 | 6167 | 5843 | 1247 | 31525 | 8136 | 1318 | 7030 | 4480 | 25628 | 15479 | 16335 | 7457 |
| Frederiksborg | 3515 | 809 | 918 | 3678 | 908 | 917 | 965 | 4198 | 814 | 282 | 374 | 760 | 1304 | 3528 | 1646 | 701 |
| Gentofte | 1888 | 3 | 111 | 1137 | 199 | 104 | 201 | 580 | 240 | 2 | 61 | 242 | 64 | 1076 | 480 | 336 |
| Aarhus | 2138 | 439 | 1336 | 2368 | 1357 | 550 | 170 | 3191 | 1577 | 197 | 629 | 412 | 2126 | 1885 | 1945 | 971 |
| Odense | 2286 | 924 | 1306 | 1681 | 832 | 290 | 326 | 2936 | 517 | 169 | 514 | 326 | 1783 | 1319 | 1479 | 709 |
| Aalborg | 1958 | 462 | 965 | 1247 | 410 | 107 | 451 | 1333 | 505 | 196 | 294 | 284 | 1179 | 1153 | 1285 | 684 |
| 20,000–50,000 | 3963 | 1527 | 1665 | 2056 | 1551 | 662 | 464 | 3463 | 1109 | 129 | 603 | 571 | 4939 | 2401 | 2787 | 1272 |
| 10,000–19,999 | 7818 | 1859 | 2453 | 5884 | 3029 | 1215 | 1186 | 8642 | 1287 | 1061 | 1391 | 1272 | 3475 | 6310 | 6174 | 2956 |
| 7,000–9,999 | 2374 | 839 | 847 | 2084 | 1077 | 403 | 639 | 3679 | 454 | 36 | 416 | 387 | 1215 | 2341 | 2458 | 1196 |
| 5,000–6,999 | 1995 | 108 | 589 | 1887 | 1023 | 360 | 857 | 1953 | 588 | 0 | 325 | 393 | 964 | 2145 | 2066 | 1098 |
| 3,000–4,999 | 2014 | 349 | 532 | 1268 | 748 | 296 | 125 | 1302 | 289 | 34 | 272 | 307 | 697 | 1779 | 1668 | 804 |
| Less than 3,000 | 1719 | 60 | 650 | 1334 | 871 | 460 | 450 | 1720 | 370 | 0 | 172 | 319 | 455 | 2107 | 1655 | 896 |
| Rural areas | 19961 | 2885 | 8155 | 32874 | 13154 | 4838 | 13472 | 18254 | 4650 | 553 | 592 | 2258 | 2129 | 24428 | 8188 | 6682 |

**1935**

| | Food | Textiles | Clothing | Construction | Wood | Leather | Ceramics | Metals | Chemicals | Paper | Books | Personal service | Wholesale | Retail Food | Retail Non-food | Catering |
|---|---|---|---|---|---|---|---|---|---|---|---|---|---|---|---|---|
| København | 18482 | 5573 | 26284 | 14712 | 6118 | 7423 | 1691 | 35388 | 10202 | 2560 | 8565 | 7268 | 34528 | 22119 | 20482 | 11834 |
| Frederiksborg | 3477 | 872 | 1745 | 4591 | 1209 | 901 | 1276 | 6200 | 901 | 295 | 495 | 1261 | 2284 | 4903 | 2527 | 1297 |
| Gentofte | 1974 | 178 | 255 | 2489 | 275 | 155 | 243 | 1072 | 530 | 0 | 76 | 493 | 307 | 1965 | 871 | 657 |
| Aarhus | 2755 | 356 | 2123 | 3686 | 1554 | 498 | 497 | 4896 | 2169 | 460 | 690 | 1061 | 3464 | 3593 | 2573 | 1586 |
| Odense | 2827 | 1468 | 1763 | 2840 | 1160 | 398 | 592 | 5664 | 739 | 609 | 689 | 729 | 2998 | 2494 | 2205 | 1116 |
| Aalborg | 2782 | 788 | 947 | 1731 | 542 | 209 | 1184 | 1491 | 491 | 85 | 442 | 554 | 1856 | 2158 | 1674 | 1021 |
| 20,000–50,000 | 4386 | 1861 | 1824 | 3067 | 1307 | 746 | 465 | 4575 | 1448 | 211 | 726 | 1303 | 3666 | 4566 | 3594 | 1750 |
| 10,000–19,999 | 8689 | 3336 | 3597 | 7388 | 3042 | 1395 | 2016 | 9773 | 1872 | 1264 | 1507 | 2440 | 5759 | 9458 | 7643 | 4313 |
| 7,000–9,999 | 2491 | 1294 | 1379 | 2462 | 1089 | 436 | 919 | 3391 | 540 | 35 | 452 | 811 | 2127 | 3067 | 2546 | 1557 |
| 5,000–6,999 | 2578 | 219 | 890 | 2464 | 1232 | 497 | 1166 | 2990 | 1959 | 11 | 431 | 964 | 1970 | 3388 | 3040 | 1528 |
| 3,000–4,999 | 1781 | 702 | 581 | 1473 | 644 | 242 | 143 | 1318 | 313 | 59 | 294 | 522 | 1182 | 2085 | 1820 | 1165 |
| Less than 3,000 | 1660 | 111 | 798 | 1326 | 886 | 483 | 743 | 1662 | 396 | 9 | 225 | 609 | 678 | 2418 | 2006 | 1278 |
| Rural areas | 18758 | 2023 | 6623 | 27691 | 11867 | 3823 | 11224 | 14949 | 4357 | 70 | 659 | 4216 | 3639 | 27639 | 9419 | 8220 |

```
SHIFT SHARE ANANYSIS
REGION      SHIFT         STRUCTURE        COMPETITION
         TOTAL   PERCENT   TOTAL   PERCENT   TOTAL   PERCENT
    1    9372.1    4.19    4096.5    1.83    5275.6    2.36
    2    3580.9   11.68     -60.8   -0.20    3641.7   11.88
    3    3398.8   41.75    -130.8   -1.61    3529.6   43.35
    4    6182.5   23.98     155.0    0.60    6027.5   23.38
    5    7227.2   34.31     100.0    0.47    7127.2   33.84
    6    2804.6   18.51     187.4    1.24    2617.3   17.28
    7     186.5    0.53     744.5    2.11    -558.0   -1.58
    8    5674.3    8.37     343.5    0.45    5330.8    7.92
    9    -158.2   -0.64      76.2    0.31    -234.4   -0.95
   10    5529.7   27.93      52.2    0.26    5477.5   27.67
   11    -791.3   -5.23     127.1    0.84    -918.3   -6.08
   12    -740.2   -4.62      15.3    0.10    -755.5   -4.71
   13   -4226/.0 -21.41   -5666.0   -2.87  -36600.9  -18.54
NOTE -- PERCENTAGES ARE OF EXPECTED EMPLOYMENT IN END YEAR




REGION   1
INDUSTRY  EMPLOYMENT BASE   EMPLOYMENT END   EXPECTED    COMPETITIVE SHIFT
          TOTAL   PERCENT   TOTAL   PERCENT   TOTAL      TOTAL    PERCENT
    1    17892.0    9.68   18482.0    7.92   18694.7    -212.71    -1.14
    2     3224.0    1.74    5573.0    2.39    4489.2    1083.83    24.14
    3    21656.0   11.71   26284.0   11.27   25666.1     617.88     2.41
    4    11471.0    6.20   14712.0    6.31   12627.1    2084.90    16.51
    5      616/.0   3.34    6118.0    2.62    6088.1      29.94     0.49
    6     5843.0    3.16    7423.0    3.18    6265.8    1157.21    18.47
    7     1247.0    0.67    1691.4    0.73    1344.4     346.56    25.78
    8    31525.0   17.05   35388.0   15.17   35559.3    -171.31    -0.48
    9     8136.0    4.40   10202.0    4.37   10267.9     -65.86    -0.64
   10     1318.2    0.71    2560.0    1.16    1878.4     681.59    30.29
   11     7030.0    3.80    8565.0    3.67    8460.1     104.93     1.24
   12     4480.0    2.42    7268.0    3.12    8292.0   -1023.97   -12.35
   13    25628.0   13.86   34528.0   14.80   35944.3   -1416.33    -3.94
   14    15479.0    8.37   22119.0    9.48   21088.9    1030.09     4.88
   15    16335.0    8.84   20482.0    8.78   20484.0      -2.03    -0.01
   16     7457.0    4.03   11534.0    5.27   10803.1    1030.97     9.54
TOTALS  184888.0          233229.0           227953.4    5275.6
```

172

REGION 2

| INDUSTRY | EMPLOYMENT BASE | | EMPLOYMENT END | | EXPECTED | COMPETITIVE SHIFT | |
|---|---|---|---|---|---|---|---|
| | TOTAL | PERCENT | TOTAL | PERCENT | TOTAL | TOTAL | PERCENT |
| 1 | 1171.0 | 10.82 | 1204.0 | 6.80 | 1241.0 | -37.00 | -2.98 |
| 2 | 1251.0 | 11.56 | 1249.0 | 7.12 | 1866.9 | -617.95 | -33.10 |
| 3 | 373.0 | 3.45 | 1000.0 | 6.08 | 449.3 | 610.68 | 137.25 |
| 4 | 1379.0 | 12.74 | 2975.0 | 16.96 | 1619.0 | 1356.03 | 83.70 |
| 5 | 303.0 | 2.80 | 577.0 | 3.29 | 303.6 | 273.38 | 90.04 |
| 6 | 321.0 | 2.97 | 262.0 | 1.49 | 348.1 | -86.07 | -24.73 |
| 7 | 581.0 | 5.37 | 869.0 | 4.95 | 652.6 | 216.35 | 33.15 |
| 8 | 1074.0 | 9.92 | 1659.0 | 9.46 | 1234.7 | 424.27 | 34.36 |
| 9 | 706.0 | 6.52 | 1065.0 | 6.07 | 928.3 | 136.74 | 14.73 |
| 10 | 31.0 | 0.29 | 57.0 | 0.32 | 44.6 | 12.42 | 27.86 |
| 11 | 31.0 | 0.29 | 75.0 | 0.43 | 37.6 | 37.35 | 99.22 |
| 12 | 306.0 | 2.83 | 740.0 | 4.22 | 582.8 | 157.17 | 26.97 |
| 13 | 84.0 | 0.78 | 275.0 | 1.57 | 122.7 | 152.29 | 124.11 |
| 14 | 1442.0 | 13.32 | 2973.0 | 16.95 | 2025.3 | 947.74 | 46.80 |
| 15 | 493.0 | 4.55 | 1120.0 | 6.38 | 632.5 | 487.51 | 77.08 |
| 16 | 1280.0 | 11.82 | 1377.0 | 7.85 | 1919.4 | -542.40 | -28.26 |
| TOTALS | 10820.0 | | 17543.0 | | 14006.5 | 3534.5 | |

REGION 25

| INDUSTRY | EMPLOYMENT BASE | | EMPLOYMENT END | | EXPECTED | COMPETITIVE SHIFT | |
|---|---|---|---|---|---|---|---|
| | TOTAL | PERCENT | TOTAL | PERCENT | TOTAL | TOTAL | PERCENT |
| 1 | 997.0 | 15.38 | 738.0 | 11.58 | 1056.6 | -318.60 | -30.15 |
| 2 | 290.0 | 4.47 | 333.0 | 5.22 | 432.8 | -99.79 | -23.06 |
| 3 | 233.0 | 3.59 | 265.0 | 4.16 | 280.7 | -15.67 | -5.58 |
| 4 | 1014.0 | 15.64 | 867.0 | 13.60 | 1190.5 | -323.45 | -27.17 |
| 5 | 440.0 | 6.79 | 311.0 | 4.88 | 440.9 | -129.90 | -29.46 |
| 6 | 160.0 | 2.47 | 148.0 | 2.32 | 173.5 | -25.49 | -14.69 |
| 7 | 541.0 | 8.34 | 480.0 | 7.53 | 607.7 | -127.71 | -21.02 |
| 8 | 000.0 | 9.25 | 519.0 | 8.14 | 689.8 | -170.79 | -24.76 |
| 9 | 136.0 | 2.10 | 135.0 | 2.12 | 178.8 | -43.81 | -24.50 |
| 10 | 0.0 | 0.00 | 2.0 | 0.03 | 0.0 | 2.00 | 0.00 |
| 11 | 56.0 | 0.86 | 62.0 | 0.97 | 68.0 | -6.01 | -8.83 |
| 12 | 96.0 | 1.48 | 186.0 | 2.92 | 182.8 | 3.15 | 1.72 |
| 13 | 328.0 | 5.06 | 467.0 | 7.33 | 479.1 | -12.15 | -2.53 |
| 14 | 679.0 | 10.47 | 818.0 | 12.83 | 953.6 | -135.64 | -14.22 |
| 15 | 486.0 | 7.58 | 529.0 | 8.30 | 623.5 | -94.51 | -15.15 |
| 16 | 427.0 | 6.59 | 514.0 | 8.06 | 640.3 | -126.30 | -19.73 |
| TOTALS | 6483.0 | | 6374.0 | | 7996.7 | -1624.7 | |

Fig 8.11 Sample results for counties from the shift-share program

SHIFT SHARE ANALYSIS

| REGION | SHIFT TOTAL | SHIFT PERCENT | STRUCTURE TOTAL | STRUCTURE PERCENT | COMPETITION TOTAL | COMPETITION PERCENT |
|---|---|---|---|---|---|---|
| 1 | 6061.2 | 2.98 | 3527.6 | 1.30 | 4533.6 | 1.67 |
| 2 | 4029.4 | 29.82 | 494.8 | 3.66 | 3534.5 | 26.16 |
| 3 | 716.1 | 6.73 | -132.5 | -1.25 | 848.6 | 7.98 |
| 4 | -340.4 | -1.60 | -90.7 | -0.43 | -249.7 | -1.18 |
| 5 | -1697.9 | -9.96 | -286.0 | -1.68 | -1411.9 | -8.28 |
| 6 | -1642.7 | -8.96 | -316.3 | -1.72 | -1326.4 | -7.23 |
| 7 | -1510.0 | -8.26 | -225.6 | -1.23 | -1284.3 | -7.02 |
| 8 | -609.6 | -7.14 | -262.6 | -3.08 | -346.9 | -4.06 |
| 9 | -1601.7 | -7.25 | -567.0 | -2.57 | -1034.7 | -4.68 |
| 10 | -2793.2 | -11.88 | -501.6 | -2.13 | -2291.6 | -9.75 |
| 11 | 2123.7 | 6.25 | -102.2 | -0.30 | 2225.8 | 6.55 |
| 12 | -60.6 | -0.73 | -231.9 | -2.79 | 171.2 | 2.06 |
| 13 | 425.7 | 1.36 | 49.8 | 0.17 | 355.9 | 1.19 |
| 14 | -1091.7 | -5.11 | -68.8 | -0.32 | -1022.9 | -4.79 |
| 15 | 2692.9 | 8.02 | -204.0 | -0.61 | 2896.9 | 8.62 |
| 16 | -1384.1 | -6.14 | -154.6 | -0.69 | -1229.5 | -5.46 |
| 17 | -1105.1 | -3.62 | -275.8 | -0.90 | -829.3 | -2.72 |
| 18 | 244.0 | 1.41 | -125.0 | -0.72 | 369.0 | 2.13 |
| 19 | -868.6 | -6.71 | -91.3 | -0.92 | -777.2 | -7.80 |
| 20 | -1003.2 | -6.29 | -289.6 | -1.82 | -713.7 | -4.48 |
| 21 | 697.3 | 3.77 | 87.0 | 0.47 | 610.3 | 3.30 |
| 22 | -590.1 | -2.93 | 24.4 | 0.12 | -614.5 | -3.05 |
| 23 | 61.7 | 0.75 | -39.9 | -0.49 | 101.6 | 1.24 |
| 24 | -597.1 | -9.34 | -85.1 | -1.33 | -512.0 | -8.01 |
| 25 | -1718.5 | -21.24 | -93.8 | -1.16 | -1624.7 | -20.08 |
| 26 | -417.5 | -9.32 | -39.6 | -0.88 | -377.9 | -8.44 |

NOTE -- PERCENTAGES ARE OF EXPECTED EMPLOYMENT IN END YEAR

REGION 1

| INDUSTRY | EMPLOYMENT BASE TOTAL | PERCENT | EMPLOYMENT END TOTAL | PERCENT | EXPECTED TOTAL | COMPETITIVE SHIFT TOTAL | PERCENT |
|---|---|---|---|---|---|---|---|
| 1 | 23295.0 | 10.73 | 23933.0 | 8.58 | 24687.6 | -754.59 | -3.06 |
| 2 | 4036.0 | 1.86 | 6623.0 | 2.37 | 6023.2 | 599.82 | 9.96 |
| 3 | 22685.0 | 10.45 | 28284.0 | 10.14 | 27326.4 | 957.65 | 3.50 |
| 4 | 16286.0 | 7.50 | 21792.0 | 7.81 | 19120.1 | 2671.95 | 13.97 |
| 5 | 7274.0 | 3.35 | 7602.0 | 2.72 | 7288.9 | 313.14 | 4.30 |
| 6 | 6864.0 | 3.16 | 8479.0 | 3.04 | 7442.8 | 1036.19 | 13.92 |
| 7 | 2440.0 | 1.12 | 3210.0 | 1.15 | 2740.9 | 469.11 | 17.12 |
| 8 | 36303.0 | 16.73 | 42660.0 | 15.29 | 41736.0 | 924.01 | 2.21 |
| 9 | 9190.0 | 4.23 | 11633.0 | 4.17 | 12083.1 | -450.13 | -3.73 |
| 10 | 1602.0 | 0.74 | 2855.0 | 1.02 | 2303.7 | 551.29 | 23.93 |
| 11 | 7465.0 | 3.44 | 9136.0 | 3.27 | 9065.7 | 70.26 | 0.77 |
| 12 | 5482.0 | 2.53 | 9022.0 | 3.23 | 10441.4 | -1419.40 | -13.59 |
| 13 | 27096.0 | 12.48 | 37119.0 | 13.30 | 39582.1 | -2463.14 | -6.22 |
| 14 | 20083.0 | 9.25 | 28987.0 | 10.39 | 28206.2 | 780.78 | 2.77 |
| 15 | 18461.0 | 8.51 | 23880.0 | 8.56 | 23684.3 | 195.68 | 0.83 |
| 16 | 8494.0 | 3.91 | 13788.0 | 4.94 | 12737.0 | 1050.90 | 8.25 |
| TOTALS | 217056.0 | | 279003.0 | | 274469.4 | 4533.6 | |

REGION 3

| INDUSTRY | EMPLOYMENT BASE |  | EMPLOYMENT END |  | EXPECTED |  | COMPETITIVE SHIFT |  |
|---|---|---|---|---|---|---|---|---|
|  | TOTAL | PERCENT | TOTAL | PERCENT | TOTAL | TOTAL | PERCENT |
| 1 | 1888.0 | 28.08 | 1974.0 | 17.11 | 1972.7 | 1.30 | 0.07 |
| 2 | 3.0 | 0.04 | 178.0 | 1.54 | 4.2 | 173.824 | 161.16 |
| 3 | 111.0 | 1.65 | 255.0 | 2.21 | 131.6 | 123.45 | 93.84 |
| 4 | 1137.0 | 16.91 | 2489.0 | 21.57 | 1251.6 | 1237.41 | 98.87 |
| 5 | 199.0 | 2.96 | 275.0 | 2.38 | 196.5 | 78.55 | 39.98 |
| 6 | 104.0 | 1.55 | 155.0 | 1.34 | 111.5 | 43.47 | 38.96 |
| 7 | 201.0 | 2.99 | 243.0 | 2.11 | 216.7 | 26.29 | 12.13 |
| 8 | 580.0 | 8.63 | 1072.0 | 9.29 | 654.2 | 417.78 | 63.86 |
| 9 | 240.0 | 3.57 | 530.0 | 4.59 | 302.9 | 227.11 | 74.98 |
| 10 | 2.0 | 0.03 | 0.0 | 0.00 | 2.9 | -2.85 | -100.00 |
| 11 | 61.0 | 0.91 | 76.0 | 0.66 | 73.4 | 2.59 | 3.53 |
| 12 | 242.0 | 3.60 | 493.0 | 4.27 | 447.9 | 45.09 | 10.07 |
| 13 | 64.0 | 0.95 | 307.0 | 2.66 | 89.8 | 217.24 | 242.01 |
| 14 | 1076.0 | 16.00 | 1965.0 | 17.03 | 1466.0 | 499.04 | 34.04 |
| 15 | 480.0 | 7.14 | 871.0 | 7.55 | 601.9 | 269.08 | 44.70 |
| 16 | 336.0 | 5.00 | 657.0 | 5.69 | 486.8 | 170.23 | 34.97 |
| TOTALS | 6724.0 |  | 11540.0 |  | 8010.4 | 3529.6 |  |

REGION 9

| INDUSTRY | EMPLOYMENT BASE |  | EMPLOYMENT END |  | EXPECTED |  | COMPETITIVE SHIFT |  |
|---|---|---|---|---|---|---|---|---|
|  | TOTAL | PERCENT | TOTAL | PERCENT | TOTAL | TOTAL | PERCENT |
| 1 | 2374.0 | 11.61 | 2491.0 | 10.13 | 2480.5 | 10.49 | 0.42 |
| 2 | 839.0 | 4.10 | 1294.0 | 5.26 | 1168.2 | 125.76 | 10.76 |
| 3 | 847.0 | 4.14 | 1379.0 | 5.61 | 1003.8 | 375.16 | 37.37 |
| 4 | 2084.0 | 10.19 | 2462.0 | 10.01 | 2294.0 | 167.97 | 7.32 |
| 5 | 1077.0 | 5.27 | 1089.0 | 4.43 | 1063.2 | 25.79 | 2.43 |
| 6 | 403.0 | 1.97 | 436.0 | 1.77 | 432.2 | 3.84 | 0.89 |
| 7 | 639.0 | 3.13 | 919.0 | 3.74 | 688.9 | 230.07 | 33.40 |
| 8 | 3679.0 | 17.99 | 3391.0 | 13.79 | 4149.8 | -758.81 | -18.29 |
| 9 | 454.0 | 2.22 | 540.0 | 2.20 | 573.0 | -32.96 | -5.75 |
| 10 | 36.0 | 0.18 | 35.0 | 0.14 | 51.3 | -16.31 | -31.78 |
| 11 | 410.0 | 2.03 | 452.0 | 1.84 | 500.6 | -48.62 | -9.71 |
| 12 | 387.0 | 1.89 | 811.0 | 3.30 | 716.3 | 94.71 | 13.22 |
| 13 | 1215.0 | 5.94 | 2127.0 | 8.65 | 1744.1 | 422.91 | 24.82 |
| 14 | 2341.0 | 11.45 | 3067.0 | 12.47 | 3189.4 | -122.43 | -3.84 |
| 15 | 2458.0 | 12.02 | 2546.0 | 10.35 | 3082.3 | -535.32 | -17.40 |
| 16 | 1196.0 | 5.85 | 1557.0 | 6.33 | 1732.7 | -175.67 | -10.14 |
| TOTALS | 20445.0 |  | 24596.0 |  | 24830.4 | -234.4 |  |

Fig 8.12 Sample results for town types from the shift-share program

175

In south Jutland and the southern islands the growth rate was considerably below the national figure. The only counties other than those around Copenhagen to show notable positive shifts were those centred on Odense and Aarhus - the two major regional centres. The structure factor shows a widespread 'poor' structure with very few areas, other than the Copenhagen region, which had any, even slight, concentrations of growth industries. The competition factor again highlights the relatively buoyant nature of employment growth in the Copenhagen district and in Odense and Aarhus counties compared with the very low level of growth over much of the remainder of Denmark and the apparently stagnating districts of Svendborg, Sonderborg and to a lesser extent south Zealand. The growth effects of Copenhagen appear to have had little effect beyond the immediate Copenhagen region. The second analysis tends to confirm this general view with particularly prominent shifts in the large cities but additionally considerable growth in towns in the 5,000 - 6,999 population bracket. The contrast of large towns with small towns and rural areas is particularly vivid.

The second stage of consideration of the results is to look at the competitive shifts of industry for each region to determine which industries in which regions were responsible for the competitive factor figures. A sample output for the first run shows regions 1, 2 and 25 (Fig 8.11). Although the competitive factor had a net positive value in the Copenhagen region (regions 1 and 2), some industries did not grow as expected. In the case of region 1, the capital city, several of the service industries showed a negative competitive shift; in the area around the core, the growing suburbs in Copenhagen county (region 2), almost all employment groups showed a positive shift. These tables can be interpreted as evidence of the suburbanisation process and its particular effect on the service trades. In contrast to the growth area is county 25 (Sonderborg) where all but two groups had a negative shift and the two positive shifts were too small to be of much consequence. Major negative shifts occurred for the food processing industries and in construction industries. The second run of the program provides a more detailed breakdown for the Copenhagen area and Fig 8.12 shows results for the inner city core (region 1) and the inner suburbs (region 3). Suburban growth was particularly evident in the construction industries, food retailing and the metal industries. The inner city showed notably poor growth in wholesaling and personal services both of which were strongly concentrated into central Copenhagen before 1927. Sample output is also shown for region 9 (towns with 7,000 - 9,999 population). Positive and negative shifts are well mixed here with a negative shift in retailing and a positive shift in wholesaling. Metal industry employment failed to grow while the clothing industries and ceramics performed better than expected.

This cursory inspection of the results serves only to indicate the types of conclusion which may be drawn from shift-share analysis. The comments on the sample tables are indicative of analysis rather than substantive. To answer fully the questions posed by the spatial patterns of employment change in Denmark between 1925 and 1935 a full study of all the tables produced by the shift-share program is required.

## 8.7 Worksheet

1 Many of the previous worked examples and answers to exercises could have been reduced in length and made more elegant by the use of arrays. If you have missed out some exercises now is the chance to go back and solve them using your new-found computing power.

2 What is wrong with the following dimension statements:
   (i) DIMENSION (A20), (B-20), (A20)
  (ii) DIMENSION X(3,m). 9m(3)
 (iii) DIMENSION P(3/2,4)
       ODIMENSION PP(3.), P(6)
  (iv) DIMENSION H(6*3)

3 What is wrong with the following input-output statements:
   (i)     READ (NIN,15) AI, I = 1,14
        15 FORMAT 10F8.2
  (ii)    READ (NIN,15) XY(Z(I) (I = 1,12)
        15 FORMAT ( 12I2)
 (iii)   WRITE (NOUT,20) (X(k,m), m = 1,10, Y(L,m),
        L = 1,15, m = 6,2,3)
        20 FORMAT (8F4.1) (10F8.2)
  (iv)   READ (NIN,15) ABC, (Din(A), I = B,C)
        15 FORMAT ( 15I5)
   (v)   WRITE (NOUT,20) (X(4,5), I = 1,4, J = 1,5)
        20 FORMAT (20I20)

4 Given the following highway statistics for Wales in 1971 write a program to calculate the following measures for

each vehicle licensing authority (Table 8.1):
- (i) percentage of Welsh total road mileage accounted for by each authority
- (ii) percentage of Welsh 'other' road mileage accounted for by each authority
- (iii) percentage of all private cars licensed by each authority
- (iv) percentage of all motor vehicles licensed by each authority
- (v) percentage of roads of each type
- (vi) private cars per person
- (vii) total road mileage per person
- (viii) 'other' road mileage per person
- (ix) road mileage per vehicle
- (x) for each type of road and for all roads (a) cost of road maintenance per mile (b) cost of road maintenance per person and (c) cost of road maintenance per vehicle.

Comment on the indices you have calculated with particular reference to the differences between urban and rural Wales.

5 Write a program to read in N pairs of (x,y) co-ordinates and then calculate the matrix of inter-point distances.

6 Write a program to read a list of N real numbers and sort them into descending rank order. Keep associated with each number an identification of its position in the original list. Write out the answers in a table with the following headings:

RANK        VALUE         POSITION IN L ST

7 Given that the area of the pentagon PQRST shown below in terms of its (x,y) co-ordinate positions is

$$\text{AREA} = \tfrac{1}{2}X_1(Y_2-Y_5) + \tfrac{1}{2}X_2(Y_3-Y_1) + \tfrac{1}{2}X_3(Y_4-Y_2) + \tfrac{1}{2}X_4(Y_5-Y_3) + \tfrac{1}{2}X_5(Y_1-Y_4)$$

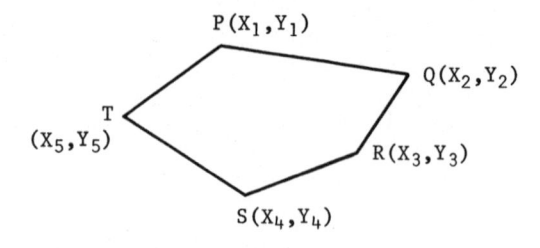

$P(X_1,Y_1)$

$Q(X_2,Y_2)$

$T(X_5,Y_5)$

$R(X_3,Y_3)$

$S(X_4,Y_4)$

Table 8.1 Data for exercise 4 - Highway statistics, 1971

| Authority | Population | Road Mileage trunk | Road Mileage principal | Road Mileage other | Vehicles private cars | Vehicles other | Maintenance Cost (£1000) trunk | Maintenance Cost (£1000) principal | Maintenance Cost (£1000) other |
|---|---|---|---|---|---|---|---|---|---|
| Angelsey | 59705 | 22 | 64 | 637 | 16450 | 4840 | 41 | 41 | 272 |
| Breconshire | 53234 | 108 | 53 | 1103 | 13580 | 3710 | 195 | 73 | 340 |
| Caernarvonshire | 122852 | 78 | 101 | 1128 | 31950 | 7230 | 139 | 140 | 487 |
| Cardiganshire | 54844 | 71 | 99 | 1364 | 16230 | 5020 | 156 | 87 | 489 |
| Carmarthenshire | 162313 | 95 | 150 | 1962 | 39820 | 13280 | 137 | 103 | 585 |
| Denbighshire | 184824 | 80 | 128 | 1855 | 43400 | 11310 | 168 | 126 | 690 |
| Flintshire | 175396 | 43 | 130 | 724 | 43210 | 8720 | 97 | 224 | 511 |
| Glamorgan | 749372 | 73 | 260 | 2229 | 143290 | 27660 | 204 | 712 | 1975 |
| Cardiff | 278221 | 6 | 29 | 370 | 59600 | 18630 | 4 | 188 | 821 |
| Merthyr Tydfil | 55215 | 15 | 9 | 148 | 8480 | 1480 | 31 | 14 | 94 |
| Swansea | 172566 | 0 | 31 | 240 | 40810 | 7580 | 0 | 83 | 317 |
| Merionethshire | 35277 | 104 | 90 | 775 | 9500 | 2450 | 165 | 100 | 260 |
| Monmouthshire | 349411 | 101 | 120 | 1666 | 70230 | 14240 | 222 | 268 | 1328 |
| Newport | 112048 | 7 | 21 | 143 | 23490 | 5250 | 0 | 61 | 276 |
| Montgomeryshire | 42761 | 121 | 43 | 1826 | 12350 | 5670 | 180 | 43 | 687 |
| Pembrokeshire | 97295 | 71 | 95 | 1586 | 27640 | 8400 | 86 | 49 | 570 |
| Radnorshire | 18262 | 45 | 63 | 708 | 5740 | 2610 | 69 | 54 | 227 |

Write a program to read any set of co-ordinates and calculate the area of the polygon formed by them.  Check your program with the following (x,y) co-ordinate points which approximately specify the shore line of Lough Neagh, Northern Ireland.  The points are from a millimetre grid on a map of 8 miles to the inch.  Calculate the area in square kilometres.

Co-ordinate points (in sequence across the page):
60,15;  58,20;  52,18;  49,27;  50,32;  55,45;  51,48;
39,39;  30,37;  31,44;  23,43;  17,35;  15,36;  12,34;
 6,34;   3,25;  10,25;   9,14;   7,11;   8, 5;  14, 0;
20, 2;  29,13;  37,13;  50,10.

8 In climatology, an *anomaly* is the departure of one observation in either a spatial or temporal series from the average of all the observations in that series.  Maps and graphs of anomaly values are useful in the study of both spatial and temporal change.  One interesting time series is that of the Zurich relative sunspot number, R, first introduced by Wolf about 1849 (Chernosky and Hagen, 1958), that expresses the extent to which the sun's surface has sunspots and which is believed by some authors to affect weather conditions on earth (Tucker, 1964).  Given the annual average R values in Table 8.2,
   (i) write a program to input these values into an array, calculating the average yearly value as the series is read in.  Use this average to determine and print out the anomaly for each year.  Is there any suggestion of a cycle or cycles in these data?
 (ii) use the data as input to a program to calculate moving means (*see Chapter 6*) for three-, five-, seven-, nine- and eleven-year time-periods.  Is there still evidence of any trends in solar conditions?

8.8 Further Reading
Ashby, L.D. 'Changes in regional industrial structure : a comment', *Urban Studies*, 6, 162-78 (1970)
Chalmers, J.A. 'Measuring changes in regional industrial structure : a comment on Stillwell and Ashby', *Urban Studies*, 8, 289-92 (1971)
Chernosky, E.J. and Hagen, M.P. 'The Zurich sunspot number and its variations for 1700-1957', *Journal of Geophysical Research*, 63, 775-788 (1958)

Table 8.2 Data for exercise 8

| Decade beginning | 0 | 1 | 2 | 3 | 4 | 5 | 6 | 7 | 8 | 9 |
|---|---|---|---|---|---|---|---|---|---|---|
| 1840 | 63.2 | 36.8 | 24.2 | 10.7 | 15.0 | 40.1 | 61.5 | 98.5 | 124.3 | 95.9 |
| 1850 | 66.5 | 64.5 | 54.2 | 39.0 | 20.6 | 6.7 | 4.3 | 22.8 | 54.8 | 93.8 |
| 1860 | 95.7 | 77.2 | 59.1 | 44.0 | 47.0 | 30.5 | 16.3 | 7.3 | 37.3 | 73.9 |
| 1870 | 139.1 | 111.2 | 101.7 | 66.3 | 44.7 | 17.1 | 11.3 | 12.3 | 3.4 | 6.0 |
| 1880 | 32.3 | 54.3 | 59.7 | 63.7 | 63.5 | 52.2 | 25.4 | 13.1 | 6.8 | 6.3 |
| 1890 | 7.1 | 35.6 | 73.0 | 84.9 | 78.0 | 64.0 | 41.8 | 26.6 | 26.7 | 12.1 |
| 1900 | 9.5 | 2.7 | 5.0 | 24.4 | 42.0 | 63.5 | 53.5 | 62.0 | 48.5 | 43.9 |
| 1910 | 18.6 | 5.7 | 3.6 | 1.4 | 9.6 | 47.4 | 57.1 | 103.9 | 80.6 | 63.6 |
| 1920 | 37.6 | 26.1 | 14.2 | 5.8 | 16.7 | 44.3 | 63.9 | 69.0 | 77.8 | 65.0 |
| 1930 | 35.7 | 21.2 | 11.1 | 5.7 | 8.7 | 36.1 | 79.6 | 114.4 | 109.6 | 88.8 |
| 1940 | 67.8 | 47.5 | 30.6 | 16.3 | 9.6 | 33.2 | 92.6 | 151.6 | 136.2 | 135.1 |

Year of Decade

Lindley, D.V. and Miller, J.C.P. *Cambridge elementary statistical tables* (1962)
Moroney, M.J. *Facts from figures* (1951)
O'Farrell 'A shift and share anslysis of regional employment change in Ireland, 1951-66', *Economic and Social Review*, 4, 59-86 (1972)
Stilwell, F.J.B. 'Regional growth and structural adaptation', *Urban Studies*, 6, 162-78 (1969)
Tucker, G.B. 'Solar influences on the weather', *Weather*, 19, 302-11 (1964)
Watson, E. 'Periglacial action in the uplands of Mid-Wales', *Report of the Welsh Soils Discussion Group*, 2, 11-14 (1961)

## Chapter 9
# Recapitulation

9.1 Introduction
With the use of DO loops and arrays we have reached a point
in FORTRAN programming beyond which not everyone will want
to progress. Using the statements outlined in Chapters
4-8 arithmetic operations can be performed on stored data,
information can be read into the computer store and
answers written from it, logical and arithmetic tests can
be made, tedious data manipulations and calculations can
be carried out with ease using DO loops, and large data
blocks can be handled with speed and accuracy by the use
of arrays. With this battery of FORTRAN techniques the
computer has become a useful tool capable of being programmed to solve a wide variety of common geographical
problems. Some readers may feel this is far enough - so
be it. Others, and hopefully there are many of these,
will want to push ahead to discover the more subtle aspects
of FORTRAN programming dealt with in the remaining chapters,
for example ways of using the lineprinter to produce
diagrams and methods for splitting up a program into
sections so that sections may be incorporated, as a unit,
into other programs.
Because we have now reached a watershed this chapter
differs from earlier ones in providing no new FORTRAN
commands. It has two objectives. The first is to review
the aspects of the FORTRAN that have been dealt with up to
this point and to provide a checklist of the commands
introduced so far; this is achieved in a few worked
examples utilising most of the known commands. The
second is to provide a rather longer list of exercises
than in earlier chapters; these exercises extend some of
the earlier ones and also begin to solve rather more complex geographical problems.
The first example deals with questionnaire analysis, or,

at a more general level, the production of contingency tables (as for example those used as input in the example in Chapter 8). This type of simple classification exercise is frequently used in geography, as in the analysis of questionnaires on shopping habits or recreation patterns, in analyses of land use surveys and in many other circumstances. The data used in the present case are from a survey of operating results of different types of engineering firms within the UK. The program uses integer arrays and rather more complicated subscript methods than have been dealt with in earlier examples. The second example uses several intrinsic functions in a program to calculate vector resultants. The data deal with wind velocity but the concept is equally applicable to a wide set of geographical problems including network analysis and the solution of Weberian-type location problems. The third example is a program to minimise a function subject to certain inequalities using a simple variety of linear programming applicable to many problems.

The exercises all deal with actual problems encountered in the geographical analysis of data and drawn from as wide a variety of data sources as possible so that the reader may become familiar with using different data types, but all are concerned with numerical data. Ways of dealing with non-numerical data are dealt with in the next chapter.

9.2 Questionnaire Analysis

*Problem.* Data are frequently collected by questionnaire methods and before conclusions can be drawn from such data counts and cross-tabulations are required. While large suites of programs to provide such analyses are available from computer manufacturers and software houses frequently they are much larger and more complex than is required for the analysis of relatively small and simple questionnaires which constitute the majority of those used by geographers.
*Algorithm and flowchart.* The algorithm is very simple. Initially the different answers to each question have to be counted and secondly the program must be able to form a cross-tabulation of the responses to any specified question against those of any another.

The flowchart is shown in Fig 9.1. Although the basic problem is simple, considerable data manipulations are necessary, giving a fairly complicated flowchart. Initially a set of controls are read; these are the number

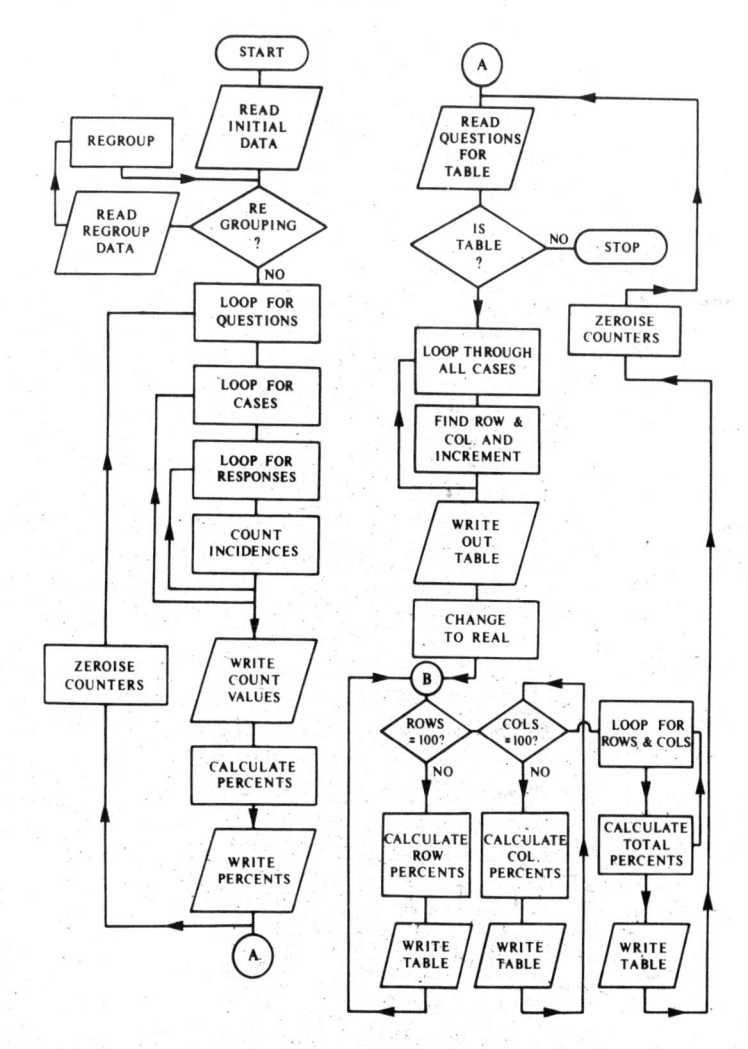

*Fig 9.1* Flowchart for questionnaire analysis

of questionnaires, the number of questions and the maximum scores on each of these questions. The main data block of answers to the questions, which must be integer numbers, is then read into an integer array. At this point it may be necessary to re-group answers to some questions, for example exact distances grouped into broad distance zones. If re-grouping is necessary then the questions are to be re-worked and the new class intervals must be read and the relevant column of the raw data array re-formed using the re-grouped values. Having obtained the data in the correct form for analysis, a simple count procedure is begun. The responses to a particular question on all the questionnaires are sorted into as many classes as there are possible answers to the question, and a count made of the numbers of items that fall into each answer group for each question. The counts are then changed to percentages and these are written out. The storage locations used in the incrementing process are set to zero and the next question taken.

A succeeding major unit of program provides the commands necessary to cross-classify each item and form a two-dimensional contingency table. The numbers of the questions to be cross-tabulated are read and if the program is to be terminated dummy zero question numbers serve to stop execution. If cross-tabulations are required a counting system is set up to increment a row and cell location in an integer storage array. The particular row and cell location for each questionnaire is specified by the responses to the two questions used in the cross-tabulations. Headings are then set up and the item count from the array is written out. Percentages are calculated first across the rows, secondly down the columns, and thirdly as a percentage of the total, and the program employs various checks to bypass zero entries. Control is then returned to the point where the next pair of questions used in cross-tabulations are read.

*Program.* The program listing corresponding to the flowchart is shown in Fig 9.2. Throughout, I refers to the number of questionnaires and J to the number of questions. As the program is listed in Fig 9.2 a maximum of 500 questionnaires each with 20 questions is its limit but it could be expanded by increasing the size of the arrays in the DIMENSION statement up to a limit given by the available computer storage capacity. The data from the questionnaires are read into the integer array IA and then re-grouped as

```
      DIMENSION IC(20),JQ(20),IA(500,20),IS(20),SI(20),ITAB(20,20),       10
     *TAB(20,20)                                                          20
C SET INPUT AND OUTPUT CHANNELS                                           30
      NOUT = 6                                                            40
      NIN = 5                                                             50
C READ NUMBER OF QUESTIONAIRES AND QUESTIONS                              60
      READ (NIN,10)I,J                                                    70
C READ MAXIMUM SCORE FOR EACH QUESTION                                    80
      READ (NIN,25) (JQ(K),K=1,J)                                         90
C READ QUESTIONAIRES                                                     100
      DO 99 L=1,I                                                        110
   99 READ (NIN,20) (IA(L,M),M=1,J)                                      120
C SET UP HEADINGS                                                        130
      WRITE (NOUT,60)                                                    140
      DO 160 K=1,20                                                      150
  160 IS(K) = K+1                                                        160
      WRITE (NOUT,65) (IS(K),K=1,17)                                     170
C READ NUMBER OF QUESTION TO BE REGROUPED AND NUMBER OF CLASSES REQUIRED 180
  115 READ (NIN,10) NR,NC                                                190
      IF (NR.EQ.0) GO TO 100                                             200
      JQ(NR) = NC                                                        210
      NCP = NC+1                                                         220
C READ VALUES OF NEW CLASS DIVIDERS                                      230
      READ (NIN,15) (IC(K),K=1,NCP)                                      240
C REGROUP DATA AS SPECIFIED                                              250
      DO 105 L=1,I                                                       260
      DO 110 IL=1,NC                                                     270
      ILP = IL+1                                                         280
      IF (IA(L,NR).GE.IC(IL).AND.IA(L,NR).LT.IC(ILP)) GO TO 106          290
      GO TO 110                                                          300
  106 IA(L,NR) = IL                                                      310
      GO TO 105                                                          320
  110 CONTINUE                                                           330
  105 CONTINUE                                                           340
      GO TO 115                                                          350
  100 CONTINUE                                                           360
C CALCULATE SIMPLE AND PERCENTAGE COUNTS                                 370
      DO 135 LM = 1,20                                                   380
  135 IS(LM) = 0                                                         390
C LOOP THROUGH QUESTIONS                                                 400
      DO 120 M=1,J                                                       410
      KX = JQ(M)                                                         420
C LOOP THROUGH QUESTIONAIRES                                             430
      DO 125 L=1,I                                                       440
      KXP = KX + 1                                                       450
      KKK = IA(L,M)+1                                                    460
      IS(KKK) = IS(KKK) + 1                                              470
  125 CONTINUE                                                           480
C WRITE SIMPLE COUNT                                                     490
      ITOT = 0                                                           500
      IM = M-1                                                           510
      WRITE (NOUT,50) M,(IS(LN),LN=1,KXP)                                520
      DO 145 LM = 1,KXP                                                  530
  145 ITOT = IS(LM)+ITOT                                                 540
      TOT = FLOAT (ITOT)                                                 550
      DO 150 LM=1,KXP                                                    560
      SI(LM) = FLOAT (IS(LM))                                            570
      IF(TOT.NE.0.0.AND.SI(LM).NE.0.0)SI(LM) = SI(LM)*100.0/TOT          580
  150 CONTINUE                                                           590
C WRITE PERCENTAGE COUNT                                                 600
      WRITE (NOUT,55) M,(SI(LM),LM=1,KXP)                                610
      DO 155 LM=1,KXP                                                    620
  155 IS(LM) = 0                                                         630
  120 CONTINUE                                                           640
  265 CONTINUE                                                           650
C READ QUESTIONS REQIRED IN TWO DIMENSIONAL TABLES                       660
      READ (NIN,10) NVR,NVC                                              670
      IF (NVR.EQ.0) GO TO 270                                            680
      DO 190 K = 1,20                                                    690
      DO 190 KK = 1,20                                                   700
  190 ITAB(K,KK)=0                                                       710
      KB = JQ(NVC)+1                                                     720
      KA = JQ(NVR)+1                                                     730
C LOOP THROUGH QUESTIONAIRES                                             740
      DO 165 L=1,I                                                       750
C IDENTIFY ROW ENTRY TO BE COUNTED                                       760
      K = IA(L,NVR) + 1                                                  770
C IDENTIFY COLUMN ENTRY TO BE COUNTED                                    780
      KK = IA(L,NVC) + 1                                                 790
      DO 180 KKM=1,KB                                                    800
      KKM = KK-1                                                         810
      IF (IA(L,NVC).EQ.KKM) GO TO 185                                    820
  180 CONTINUE                                                           830
C ENTER COUNTED VALUE INTO APPROPRIATE ROW AND COLUMN                    840
  185 ITAB(K,KK) = ITAB(K,KK)+1                                          850
  165 CONTINUE                                                           860
C SET UP TABLE HEADINGS                                                  870
      DO 201 IJK = 1,20                                                  880
  201 IS(IJK) = IJK+1                                                    890
      WRITE (NOUT,75)NVR,NVC                                             900
      WRITE (NOUT,60)                                                    910
```

```
            WRITE (NOUT,80) (IS(M) ,M=1,KB)                          920
      C WRITE COUNT RESULTS                                          930
            DO 195 L=1,KA                                             940
            IL = L-1                                                  950
        195 WRITE (NOUT,70) IL,(ITAB(L,M),M=1,KB)                     960
            KAP = KA+1                                                970
            KBP = KB+1                                                980
            TAB(1,KBP) = 0.0                                          990
            TAB(KAP,1) = 0.0                                         1000
        250 CONTINUE                                                 1010
      C SET UP TABLE HEADINGS                                        1020
            WRITE (NOUT,85)                                          1030
            WRITE (NOUT,80) (IS(M),M=1,KB)                           1040
      C CHANGE INTEGER TO REAL SCORES                                1050
            DO 200 L=1,KA                                            1060
            DO 200 M=1,KB                                            1070
        200 TAB(L,M) = FLOAT (ITAB(L,M))                             1080
            IF (TAB(1,KBP).EQ.100.0) GO TO 225                       1090
      C CALCULATE ROW PERCENTAGES                                    1100
            DO 205 L=1,KA                                            1110
        205 TAB(L,KBP)=0.0                                           1120
            DO 215 L=1,KA                                            1130
            DO 220 M=1,KB                                            1140
        220 TAB(L,KBP) = TAB(L,KBP)+TAB(L, M)                        1150
            DO 221 M=1,KB                                            1160
            IF (TAB(L,KBP).NE.0.0.AND.TAB(L,M).NE.0.0) TAB(L,M) = TAB(L,M)* 1170
           *100.0/TAB(L,KBP)                                         1180
        221 CONTINUE                                                 1190
            ML = L-1                                                 1200
      C WRITE ROW PERCENTAGES                                        1210
            TAB(L,KBP) = 100.0                                       1220
            WRITE (NOUT,90) ML,(TAB(L,M),M=1,KBP)                    1230
        215 CONTINUE                                                 1240
            GO TO 250                                                1250
        225 CONTINUE                                                 1260
            IF (TAB(KAP,1).EQ.100.0) GO TO 245                       1270
      C CALCULATE COLUMN PERCENTAGES                                 1280
            DO 210 M=1,KB                                            1290
        210 TAB(KAP,M) = 0.0                                         1290
            DO 230 M=1,KB                                            1300
            DO 235 L=1,KA                                            1310
        235 TAB(KAP,M) = TAB(KAP,M)+TAB(L,M)                         1320
            DO 236 L=1,KA                                            1330
            IF (TAB(KAP,M).NE.0.0.AND.TAB(L,M).NE.0.0) TAB(L,M) = TAB(L,M)* 1350
           *100.0/TAB(KAP,M)                                         1360
        236 CONTINUE                                                 1370
            TAB(KAP,M) = 100.0                                       1380
        230 CONTINUE                                                 1390
      C WRITE COLUMN PERCENTAGES                                     1400
            DO 240 L=1,KAP                                           1410
            ML = L-1                                                 1420
            WRITE (NOUT,90) ML,(TAB(L,M),M=1,KB)                     1430
        240 CONTINUE                                                 1440
            GO TO 250                                                1450
        245 CONTINUE                                                 1460
      C CALCULATE PERCENTAGE OF THE TOTAL                            1470
            TOTQ = FLOAT (I)                                         1480
            DO 255 L=1,KA                                            1490
            DO 260 M=1,KB                                            1500
            IF (TAB(L,M).NE.0.0) TAB(L,M) = TAB(L,M)*100.0/TOTQ      1510
        260 CONTINUE                                                 1520
            ML = L-1                                                 1530
      C WRITE PERCENTAGES OF TOTAL                                   1540
            WRITE (NOUT,90) ML,(TAB(L,M),M=1,KB)                     1550
        255 CONTINUE                                                 1560
            GO TO 265                                                1570
        270 CONTINUE                                                 1580
      C FORMAT STATEMENTS                                            1590
         10 FORMAT ( 2I4)                                            1600
         15 FORMAT ( 20I4)                                           1610
         20 FORMAT ( 6I2,2I3)                                        1620
         25 FORMAT ( 20I2)                                           1630
         50 FORMAT ( 1I7,8H    ITEM  ,20I6)                          1640
         55 FORMAT ( 1I7,8H    PERC  ,20F6.1)                        1660
         60 FORMAT ( 23H SIMPLE COUNT    SCORE)                      1670
         65 FORMAT ( 9H QUESTION,5H TYPE,1I7,16I6)                   1680
         70 FORMAT ( 1I3,3H ..,20I6)                                 1690
         75 FORMAT ( 22H THE ROWS ARE QUESTION,1I4,29H AND THE COLUMNS ARE QUE 1700
           *STION,1I4)                                               1710
         80 FORMAT ( 6H  ....,20I6)                                  1720
         85 FORMAT ( 12H PERCENTAGES )                               1730
         90 FORMAT ( 1I3,3H ..,20F6.1)                               1740
            STOP                                                     1750
            END                                                      1760
```

*Fig 9.2* Program listing for questionnaire analysis

necessary. These re-grouped data replace the original data which are then lost. The simple count then proceeds with outer loop 120 passing through the J questions and loop 125 passing through the I questionnaires. The value is picked out of the raw data array and used as a subscript value in the counting array. To allow for counts of non-responses (coded as zero) each raw data score is incremented by one before counting occurs. The count values are stored in array IS and the corresponding percentage values calculated and stored in SI.

The remainder of the program calculates the cross-tabulations. The numbers of the two questions to be calculated are held in NVR and NVC and the incremented totals are accumulated in ITAB (statement label 185). After the contents of ITAB are written its integer scores are converted to real values in the array TAB using the supplied function FLOAT. The row scores are totalled and placed in the array element one beyond the last element containing a count score. Row percentages are written within the loop 215 in which they are calculated, but this is not possible with the column percentages so that these are calculated in loop 230 and then written afterwards in loop 240. Percentages of the total are dealt with in loop 255. The FORMAT statements are collected together at the end of the program.

*Data and run.* The data used in this study are a small section of a data block on the operating characteristics of engineering firms in the UK. Eight measures of each of 102 firms have been selected to show typical output from the program. Non-response to any question is coded zero. The eight measures all in integer number form are:

1 Type of firm by SIC order   01 is agricultural machinery manufacture
02 is construction machinery manufacture
03 is mining equipment manufacture
04 is diversified mechanical engineering

2 Sales size in groups   01 M£ less than 1
02 M£1-4
03 M£5-9
04 M£10-24
05 M£25-49

|   |   |
|---|---|
|   | 06 M£50-99 |
|   | 07 M£100-249 |
|   | 08 M£250 and over |
| 3 Percentage of sales exported | - actual value to nearest whole number (to be grouped within the program) |
| 4 Capital employed, in groups | 01 M£ less than 1<br>02 M£1-2.4<br>03 M£2.5-4.9<br>04 M£5.0-9.9<br>05 M£10.0-14.9<br>06 M£15.0-24.9<br>07 M£25.0-49.9<br>08 M£50 and over |
| 5 Profit or loss, in groups | 01 loss of over M£-0.25<br>02 loss of M£0-0.24<br>03 profit of M£0-0.24<br>04 profit of M£0.25-0.49<br>05 profit of M£0.50-0.74<br>06 profit of M£0.75-0.99<br>07 profit of over M£1.00 |
| 6 Sales: capital ratio (x 10.0) | - actual value (to be grouped within the program) |
| 7 Percentage profit on sales | - actual value (to be grouped within the program) |
| 8 Profit as a percentage of capital | - actual value (to be grouped within the program) |

Variables 3,6,7 and 8 require grouping in the following ways:
Variable 3: 1-9, 10-19, 20-29, 30-39, 40-49, 50-59, 60-69, 70-79, 80-89, 90 and over.
Variable 6: 1-4, 5-9, 10-14, 15-19, 20-24, 25-29, 30-34, 35-39, 40-44, 45-49, 50 and over.
Variable 7: (-199)-(-6), (-5)-(-2), (-1)-(+4), 5-9, 10-14, 15-19, 20-24, over 24.
Variable 8: (-99)-(-6), (-5)-(-2), (-1)-(+4), 5-9, 10-14, 15-19, 20-24, over 24.
  Because it consists of over 100 cards, the full data block is not reproduced but Fig 9.3 shows its beginning and end. The first data card in 2I4 format specifies the number of responses and the number of questions (01020008).

```
01020008                          number of questionnaires and questions
14089908/999999                   maximum upper limits to each category
01021202032102025 ⎤
01037604051001001b
01024003040900000b
01020102042500/01b
01022002032120110
01050107011005
01020101030005014
010768080/20003000
010432050/11009010
010100010323000022
010441040/1001102                 data
   •    •    •
   •    •    •
   •    •    •
04072900/13004005
0400150/0/17005009
040519007/1400912
040412050/190005  ⎦
0003001
000100100020005004200, ...    ⎤
000000011                         reprouping
000100050001005002020025...       cards for
0407000B                          variables
-199-005-001000500100015...       3, 6, 7 and 8
000B0000
-099-005-00100050010001500...  ⎦
00000000                       ⎤  zero's card
0001002
0002004                           cross tabulations
0007000B                          control cards
0002003                        ⎦
0000000                           zero's card
```

*Fig 9.3* Section of data block for questionnaire analysis

The second card, format 8I2, specifies the maximum value that the answer to each question may take. The main data block in format 6I2,2I3 then follows. After the responses come the control data to re-group particular questions. First question 3 is re-grouped into 10 classes with the class divisions shown on the subsequent card. Re-grouping of questions 6,7 and 8 follows and the re-grouping control data are terminated by a dummy question number (zero). The remaining data cards specify the questions to be used in each cross-tabulation. The first two-dimensional table deals with questions 1 and 2, the second with 1 and 3, and so on. Zero dummy question numbers terminate the final section of data.

Results are shown in Fig 9.4. Responses to the second question, on the size of firm, show that the dominant size class is class 2 (M£1-4) with a secondary peak in class 4 followed by a long upper tail. The size of profit

| SIMPLE COUNT | | SCORE | | | | | | | | | | | |
|---|---|---|---|---|---|---|---|---|---|---|---|---|---|
| QUESTION | TYPE | 0 | 1 | 2 | 3 | 4 | 5 | 6 | 7 | 8 | 9 | 10 | 11 |
| 1 | ITEM | 0 | 16 | 19 | 16 | 51 | | | | | | | |
| 1 | PERC | 0.0 | 15.7 | 18.6 | 15.7 | 50.0 | | | | | | | |
| 2 | ITEM | 0 | 4 | 39 | 17 | 23 | 9 | 5 | 4 | 1 | | | |
| 2 | PERC | 0.0 | 3.9 | 38.2 | 16.7 | 22.5 | 8.8 | 4.9 | 3.9 | 1.0 | | | |
| 3 | ITEM | 2 | 23 | 21 | 19 | 11 | 8 | 6 | 6 | 5 | 1 | 0 | |
| 3 | PERC | 2.0 | 22.5 | 20.6 | 18.6 | 10.8 | 7.8 | 5.9 | 5.9 | 4.9 | 1.0 | 0.0 | |
| 4 | ITEM | 0 | 12 | 26 | 19 | 16 | 12 | 4 | 8 | 5 | | | |
| 4 | PERC | 0.0 | 11.8 | 25.5 | 18.6 | 15.7 | 11.8 | 3.9 | 7.8 | 4.9 | | | |
| 5 | ITEM | 0 | 2 | 6 | 34 | 13 | 5 | 7 | 35 | | | | |
| 5 | PERC | 0.0 | 2.0 | 5.9 | 33.3 | 12.7 | 4.9 | 6.9 | 34.3 | | | | |
| 6 | ITEM | 0 | 0 | 4 | 32 | 39 | 15 | 5 | 3 | 1 | 1 | 1 | 1 |
| 6 | PERC | 0.0 | 0.0 | 3.9 | 31.4 | 38.2 | 14.7 | 4.9 | 2.9 | 1.0 | 1.0 | 1.0 | 1.0 |
| 7 | ITEM | 0 | 1 | 4 | 21 | 46 | 23 | 5 | 1 | 1 | | | |
| 7 | PERC | 0.0 | 1.0 | 3.9 | 20.6 | 45.1 | 22.5 | 4.9 | 1.0 | 1.0 | | | |
| 8 | ITEM | 0 | 3 | 2 | 7 | 24 | 27 | 18 | 10 | 11 | | | |
| 8 | PERC | 0.0 | 2.9 | 2.0 | 6.9 | 23.5 | 26.5 | 17.6 | 9.8 | 10.8 | | | |

THE ROWS ARE QUESTION 2 AND THE COLUMNS ARE QUESTION 3

SIMPLE COUNT SCORE

| ..... | 0 | 1 | 2 | 3 | 4 | 5 | 6 | 7 | 8 | 9 | 10 |
|---|---|---|---|---|---|---|---|---|---|---|---|
| 0 .. | 0 | 0 | 0 | 0 | 0 | 0 | 0 | 0 | 0 | 0 | 0 |
| 1 .. | 0 | 3 | 0 | 0 | 0 | 1 | 0 | 0 | 0 | 0 | 0 |
| 2 .. | 0 | 12 | 6 | 11 | 3 | 2 | 2 | 0 | 3 | 0 | 0 |
| 3 .. | 0 | 2 | 2 | 3 | 3 | 3 | 2 | 0 | 1 | 1 | 0 |
| 4 .. | 2 | 5 | 4 | 3 | 3 | 2 | 2 | 1 | 1 | 0 | 0 |
| 5 .. | 0 | 1 | 6 | 0 | 0 | 0 | 0 | 2 | 0 | 0 | 0 |
| 6 .. | 0 | 0 | 2 | 0 | 1 | 0 | 0 | 2 | 0 | 0 | 0 |
| 7 .. | 0 | 0 | 0 | 2 | 1 | 0 | 0 | 1 | 0 | 0 | 0 |
| 8 .. | 0 | 0 | 1 | 0 | 0 | 0 | 0 | 0 | 0 | 0 | 0 |

PERCENTAGES

| ..... | 0 | 1 | 2 | 3 | 4 | 5 | 6 | 7 | 8 | 9 | 10 | |
|---|---|---|---|---|---|---|---|---|---|---|---|---|
| 0 .. | 0.0 | 0.0 | 0.0 | 0.0 | 0.0 | 0.0 | 0.0 | 0.0 | 0.0 | 0.0 | 0.0 | 100.0 |
| 1 .. | 0.0 | 75.0 | 0.0 | 0.0 | 0.0 | 25.0 | 0.0 | 0.0 | 0.0 | 0.0 | 0.0 | 100.0 |
| 2 .. | 0.0 | 30.8 | 15.4 | 28.2 | 7.7 | 5.1 | 5.1 | 0.0 | 7.7 | 0.0 | 0.0 | 100.0 |
| 3 .. | 0.0 | 11.8 | 11.8 | 17.6 | 17.6 | 17.6 | 11.8 | 0.0 | 5.9 | 5.9 | 0.0 | 100.0 |
| 4 .. | 8.7 | 21.7 | 17.4 | 13.0 | 13.0 | 8.7 | 8.7 | 4.3 | 4.3 | 0.0 | 0.0 | 100.0 |
| 5 .. | 0.0 | 11.1 | 66.7 | 0.0 | 0.0 | 0.0 | 0.0 | 22.2 | 0.0 | 0.0 | 0.0 | 100.0 |
| 6 .. | 0.0 | 0.0 | 40.0 | 0.0 | 20.0 | 0.0 | 0.0 | 40.0 | 0.0 | 0.0 | 0.0 | 100.0 |
| 7 .. | 0.0 | 0.0 | 0.0 | 50.0 | 25.0 | 0.0 | 0.0 | 25.0 | 0.0 | 0.0 | 0.0 | 100.0 |
| 8 .. | 0.0 | 0.0 | 100.0 | 0.0 | 0.0 | 0.0 | 0.0 | 0.0 | 0.0 | 0.0 | 0.0 | 100.0 |

PERCENTAGES

| ..... | 0 | 1 | 2 | 3 | 4 | 5 | 6 | 7 | 8 | 9 | 10 |
|---|---|---|---|---|---|---|---|---|---|---|---|
| 0 .. | 0.0 | 0.0 | 0.0 | 0.0 | 0.0 | 0.0 | 0.0 | 0.0 | 0.0 | 0.0 | 0.0 |
| 1 .. | 0.0 | 13.0 | 0.0 | 0.0 | 0.0 | 12.5 | 0.0 | 0.0 | 0.0 | 0.0 | 0.0 |
| 2 .. | 0.0 | 52.2 | 28.6 | 57.9 | 27.3 | 25.0 | 33.3 | 0.0 | 60.0 | 0.0 | 0.0 |
| 3 .. | 0.0 | 8.7 | 9.5 | 15.8 | 27.3 | 37.5 | 33.3 | 0.0 | 20.0 | 100.0 | 0.0 |
| 4 .. | 100.0 | 21.7 | 19.0 | 15.8 | 27.3 | 25.0 | 33.3 | 16.7 | 20.0 | 0.0 | 0.0 |
| 5 .. | 0.0 | 4.3 | 28.6 | 0.0 | 0.0 | 0.0 | 0.0 | 33.3 | 0.0 | 0.0 | 0.0 |
| 6 .. | 0.0 | 0.0 | 9.5 | 0.0 | 9.1 | 0.0 | 0.0 | 33.3 | 0.0 | 0.0 | 0.0 |
| 7 .. | 0.0 | 0.0 | 0.0 | 10.5 | 9.1 | 0.0 | 0.0 | 16.7 | 0.0 | 0.0 | 0.0 |
| 8 .. | 0.0 | 0.0 | 4.8 | 0.0 | 0.0 | 0.0 | 0.0 | 0.0 | 0.0 | 0.0 | 0.0 |
| 9 .. | 100.0 | 100.0 | 100.0 | 100.0 | 100.0 | 100.0 | 100.0 | 100.0 | 100.0 | 100.0 | 100.0 |

PERCENTAGES

| ..... | 0 | 1 | 2 | 3 | 4 | 5 | 6 | 7 | 8 | 9 | 10 |
|---|---|---|---|---|---|---|---|---|---|---|---|
| 0 .. | 0.0 | 0.0 | 0.0 | 0.0 | 0.0 | 0.0 | 0.0 | 0.0 | 0.0 | 0.0 | 0.0 |
| 1 .. | 0.0 | 2.9 | 0.0 | 0.0 | 0.0 | 1.0 | 0.0 | 0.0 | 0.0 | 0.0 | 0.0 |
| 2 .. | 0.0 | 11.8 | 5.9 | 10.8 | 2.9 | 2.0 | 2.0 | 0.0 | 2.9 | 0.0 | 0.0 |
| 3 .. | 0.0 | 2.0 | 2.0 | 2.9 | 2.9 | 2.9 | 2.0 | 0.0 | 1.0 | 1.0 | 0.0 |
| 4 .. | 2.0 | 4.9 | 3.9 | 2.9 | 2.9 | 2.0 | 2.0 | 1.0 | 1.0 | 0.0 | 0.0 |
| 5 .. | 0.0 | 1.0 | 5.9 | 0.0 | 0.0 | 0.0 | 0.0 | 2.0 | 0.0 | 0.0 | 0.0 |
| 6 .. | 0.0 | 0.0 | 2.0 | 0.0 | 1.0 | 0.0 | 0.0 | 2.0 | 0.0 | 0.0 | 0.0 |
| 7 .. | 0.0 | 0.0 | 0.0 | 2.0 | 1.0 | 0.0 | 0.0 | 1.0 | 0.0 | 0.0 | 0.0 |
| 8 .. | 0.0 | 0.0 | 1.0 | 0.0 | 0.0 | 0.0 | 0.0 | 0.0 | 0.0 | 0.0 | 0.0 |

*Fig 9.4* Results of sample run of questionnaire analysis

(question 5) shows a similar basic pattern. Question 3 was one of those which were re-grouped and it includes two firms who did not respond to the question. Almost a quarter of firms exported less than 10 per cent of their output measured by sales volume. Question 7, percentage profit on sales, was also re-grouped and the simple count shows that 45 per cent of the sample of firms had a profit of between 5 and 9 per cent. The responses to question 8 show that profit expressed as a percentage of capital was generally much higher.

Typical of the results of the two-dimensional tables is that relating questions 2 and 3. First the item count is presented. For example, twelve firms have a sales size of M£1-4 (class 2 in rows) and export 1-9 per cent of sales (class 1 in columns). These crude totals are then given in percentage form, first as a percentage of the row totals. Thus, of the firms with sales size M£1-4 (class 2 in rows) 30.8 per cent export 1-9 per cent of sales. Secondly the totals are given as a per cent of the column totals. Hence, with the same data for firms exporting 1-9 per cent of sales, 52.2 per cent have a sales size of M£1-4. Finally percentages of the grand table total are presented from which it can be seen that 11.8 per cent of all firms export 1-9 per cent of their sales and also have a sales size of M£1-4. Although Fig 9.4 only shows one such cross-tabulation, the complete output generated by the data in Fig 9.3 has 4 sets of tables of which this is the last.

## 9.3 Determination of a Resultant Vector for Orientation Data

*Problem.* In geography it is often necessary to deal with measurements on data that are *vector* quantities such as the wind, stones in a sediment, and roads that have both a magnitude (speed, dip or road length) and a direction (measured as an orientation in $^oE$ of north). The most representative summary measure for a series of such observations is the *resultant* which is not the same as would be obtained by simply taking an average of the orientations and an average of the magnitudes. Because orientation data are periodic there is a repetition of the values every $360^o$ and this makes the calculation of a normal average difficult. Resultants are calculated using the method of *radius vector summation* (Pincus, 1953) and a specific problem will illustrate their use in the analysis of wind observations.

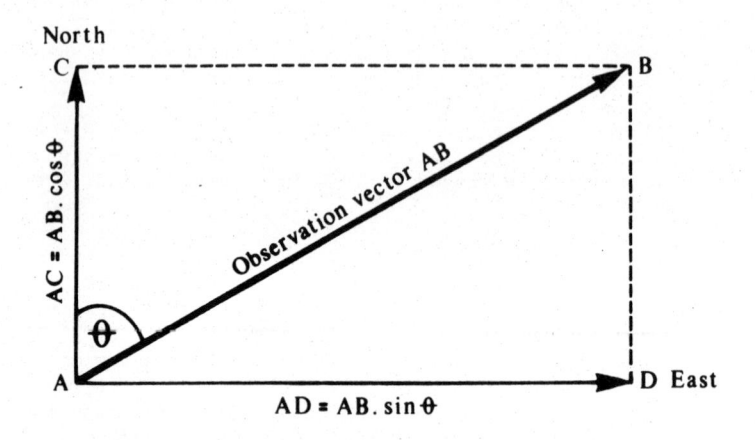

*Fig 9.5* Radius vector summation

*Algorithm and flowchart.* In radius vector summation each observation is broken down, or resolved, into two perpendicular components as shown in Fig 9.5. Each datum can be represented by a line AB of length proportional to its magnitude and with a direction plotted as the angle θ. The components towards the north and east are obtained by simple trigonometry as follows. In the triangle ABC,

$$\cos θ = AC/AB$$

hence

$$AC = AB \cdot \cos(θ) \quad (1)$$

Similarly the component to the east AD is given by

$$AD = AB \cdot \sin(θ) \quad (2)$$

The vector magnitude AB in our example will be wind speed but it could be any measure of importance such as road length, a weighting for drumlin size, a cirque altitude, the dip of cobbles in a boulder clay or even a constant value of unity which would give all the observations the same weight. The individual components are next summed over all the observations and the tangent of the resultant's orientation obtained from

$$\tan(θ) = ΣAD/ΣAC \quad (3)$$

The vector mean magnitude of this resultant is given by the Pythagoras relation

$$V_{mag} = (SQRT( (ΣAD^2) + (ΣAC^2)))/N_{obs} \quad (4)$$

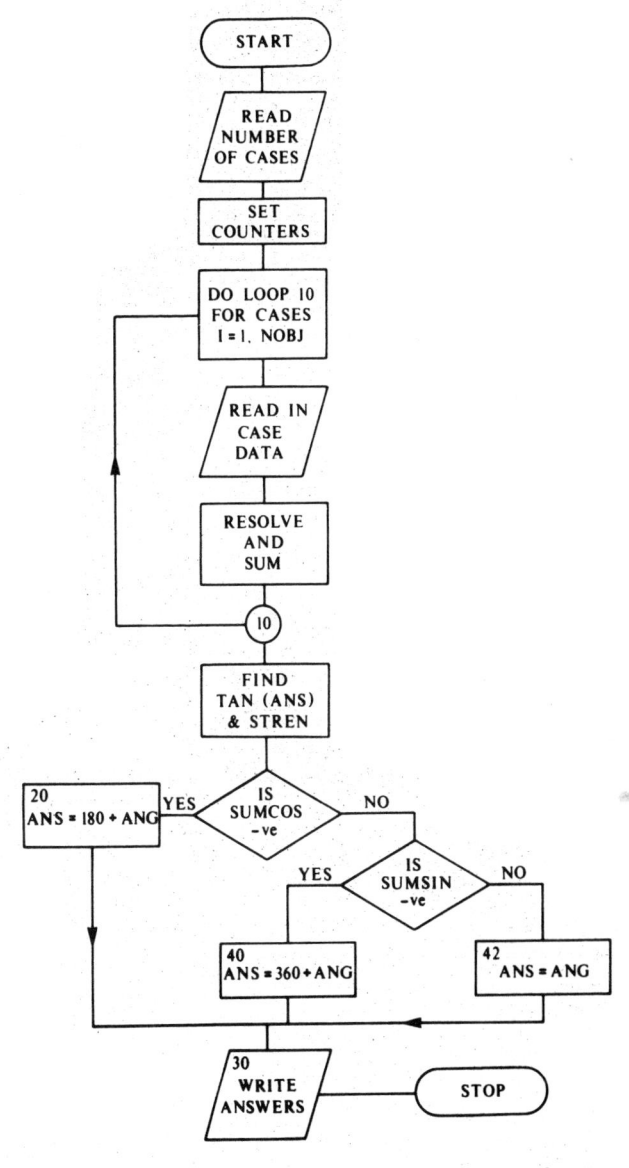

*Fig 9.6* Flowchart for program to calculate resultant vectors

195

```
C   PROGRAM TO CALCULATE THE RESULTANT DIRECTION AND MEAN MAGNITUDE          10
C                   OF DATA THAT ARE                                          20
C           VECTOR QUANTITIES USING RADIUS VECTOR METHODS                     30
C                                                                             40
C   REFERENCE    =    PINCUS,H.J.,1953 THE ANALYSIS OF AGGREGATES OF          50
C                     ORIENTATION DATA IN THE EARTH SCIENCES                  60
C                     JOURN.GEOL. 61,482-509                                  70
C                                                                             80
C   INPUT IS A CARD (FORMAT I3) GIVING THE NUMBER OF OBSERVATIONS             90
C        FOLLOWED BY THE DATA, ONE CASE PER CARD ACCORDING TO                100
C        FORMAT 91 EACH WITH THE VECTOR MAGNITUDE AND DIRECTION              110
C        IN DEGREES EAST OF NORTH ROUND THE CIRCLE                           120
C                                                                            130
C   SET INPUT AND OUTPUT CHANNELS                                            140
        NIN = 5                                                              150
        NOUT = 6                                                             160
C   READ IN NUMBER OF OBSERVATIONS AND CLEAR SUMMING LOCATIONS               170
        READ(NIN,90) NOBJ                                                    180
        SUMSIN = 0.0                                                         190
        SUMCOS = 0.0                                                         200
        SUMS2 = 0.0                                                          210
        SUMC2 = 0.0                                                          220
C   READ IN DATA ONE CARD AT A TIME                                          230
        DO 10 I = 1, NOBJ                                                    240
        READ(NIN,91) DIRECT,VMAG                                             250
C   CONVERT TO RADIANS BEFORE CALLING STANDARD TRIG. FUNCTIONS               260
        DIRECT = DIRECT/ 57.296                                              270
        TEMP1 = VMAG * SIN(DIRECT)                                           280
        TEMP2 = VMAG * COS(DIRECT)                                           290
        SUMSIN = SUMSIN + TEMP1                                              300
        SUMCOS = SUMCOS + TEMP2                                              310
        SUMS2 = SUMS2 + TEMP1 * TEMP1                                        320
        SUMC2 = SUMC2 + TEMP2 * TEMP2                                        330
   10   CONTINUE                                                             340
C   CALCULATE TANGENT OF RESULTANT ANGLE                                     350
        TANANG = SUMSIN/SUMCOS                                               360
C   CALCULATE MEAN VECTOR MAGNITUDE                                          370
        ANOBJ = NOBJ                                                         380
        STREN = (SQRT(SUMS2 + SUMC2))/ANOBJ                                  390
C   ALLOCATE ANGLE TO ITS PROPER QUADRANT AS EXPLAINED IN TEXT               400
        ANG = ATAN(TANANG) * 57.296                                          410
        IF ( SUMCOS ) 20,22,22                                               420
C   ANSWER IS IN LOWER HALF OF CIRCLE FROM 90 - 270 DEGREES EAST             430
   20   ANSWER = 180.0 + ANG                                                 440
        GO TO 30                                                             450
C   ANSWER IS IN UPPER HALF OF CIRCLE FROM 270 - 90 DEGREES EAST             460
   22   IF ( SUMSIN ) 40,42,42                                               470
   40   ANSWER = 360. + ANG                                                  480
        GO TO 30                                                             490
   42   ANSWER = ANG                                                         500
C   WRITE OUT ANSWERS AND STOP                                               510
   30   WRITE(NOUT,92) ANSWER, STREN                                         520
        STOP                                                                 530
   90   FORMAT(I3)                                                           540
   91   FORMAT(2F4.0)                                                        550
   92   FORMAT( 30H RESULTANT VECTOR DIRECTION = ,F7.3,                      560
       1        25H VECTOR MEAN MAGNITUDE = ,F7.3)                           570
        END                                                                  580
```

*Fig 9.7* Program listing for calculation of resultant vectors

A flowchart and computer program to perform these calculations are shown in Figs 9.6 and 9.7. Initially the number of observations, NOBJ, is read from the first of the data cards, and the locations SUMSIN, SUMCOS, SUMS2 and SUMC2 which are used to accumulate the sums of the resolved components, are set at zero. This second step will ensure that at the first addition these locations contain zero and not any other spurious numbers that have no relevance to this program and which would otherwise result in serious error. Data for each observation, one card per case, are read under the control of a single DO loop using I as its control variable. On each card is punched the observed direction (DIRECT) and magnitude (VMAG) according to the FORMAT (2F4.∅) and these are used to evaluate the required components using the functions SIN and COS. Notice that immediately before these functions are called the angular values obtained from the data cards are converted into radian measure using the factor 1/57.296. On exit from the DO loop SUMSIN, SUMCOS, SUMS2 and SUMC2 will contain $\Sigma AD$, $\Sigma AC$, $\Sigma AD^2$ and $\Sigma AC^2$ so that equations (3) and (4) can be used directly to evaluate the resultant vector. The vector mean magnitude calculation is straightforward as is that for the tangent of its direction but to convert this tangent back into degrees of a circle is by no means simple and illustrates the care that must be taken when standard functions are used. The tangent value is first converted into circular measure using the function ATAN and then into degrees using the factor 57.296 but the resulting angle will be in the range +90 to -90° and not the full 360° of a circle required. How can we assign this angle into its correct quadrant of the circle? The information necessary to do this is contained in the signs of the values of SUMSIN and SUMCOS as shown below:

| Quadrant | Sign of: SUMSIN | SUMCOS |
|---|---|---|
| 0 - 90 | + | + |
| 90 - 180 | + | - |
| 180 - 270 | - | - |
| 270 - 360 | - | + |

The remainder of the program is a decision network correctly to assign ANG. If SUMCOS is negative then the answer must lie between 90° and 270° and all that need be done is to add or subtract ANG from 180° according to whether it is itself positive or negative (statement 20). If SUMCOS is positive the answer lies between 270 and 90° and a second decision is made according to the sign of SUMSIN. Should this be positive the angle must lie between 0 and 90° (statement 42) but if it is negative then the required angle is obtained from 360 + ANG as in statement 40. The plus sign is to allow for the fact that in this quadrant ANG will already carry a minus sign in the computer store.

*Data and run.* The problem is to find the resultant of a series of daily high-level wind speed and direction observations made at Mombasa over a period of fifteen November days. The data are:

| Direction (°E of N) | Magnitude (knots) |
|---|---|
| 60 | 12 |
| 200 | 16 |
| 70 | 10 |
| 50 | 5 |
| 130 | 12 |
| 120 | 8 |
| 210 | 4 |
| 80 | 9 |
| 100 | 20 |
| 140 | 5 |
| 350 | 23 |
| 240 | 17 |
| 220 | 24 |
| 190 | 22 |
| 340 | 10 |

There are 15 observations. Calculation of the normal arithmetic means gives a statistically 'average' wind direction of 167° with a speed of 13kt, whereas the resultant wind has a direction 150.4° and mean vector magnitude of 3.79kt in this direction. The weakness of this mean vector magnitude, less than any of the individual observations, is caused by the directional inconstancy of the winds which is evident in the data. Resultant winds are in themselves useful climatic parameters but they might be further used in studies of the average atmospheric transport of aerosols and gaseous pollutants.

9.4 Simple Factory-Warehouse Allocation

*Problem.* The problem is one frequently met in simple spatial allocation models and is the minimisation of movement in the allocation of goods from factory to warehouses, as in the following table of distances between factories and warehouse and tables of warehouse requirements and factory output.

|  | Warehouses | | |
|---|---|---|---|
| Distances | w1 | w2 | w3 |
| Factories $\frac{f1}{f2}$ | $\frac{5}{3}$ | $\frac{2}{3}$ | $\frac{1}{4}$ |
| Warehouse requirements | $\frac{w1}{8}$ | $\frac{w2}{10}$ | $\frac{w3}{7}$ |
| Factory outputs | $\frac{f1}{10}$ | $\frac{f2}{15}$ | |

In general terms the problem is to minimise an objective function (in this case movement) subject to a series of linear constraints (in this case factory output and warehouse requirements). As such it is a type of linear program (Cox, 1965; Williams, 1969).

*Algorithm and flowchart.* Let the number of units from f1 to w1 be x and from f1 to w2 be y. An allocation table becomes:

|  | w1 | w2 | w3 |
|---|---|---|---|
| f1 | x | y | 10-(x+y) |
| f2 | 8-x | 10-y | (x+y)-3 |

The entries for w3 are simply what is left over from the total factory production after goods have been allocated to w1 and w2, while the entries for f2 are the total warehouse requirements less f1. Negative flows are not possible so that each entry in this table must be greater than or equal to zero. Six restrictions or constraints on the values that x and y may take can be deduced from this table:

| | | |
|---|---|---|
| x ⩾ 0 | or | 0 ⩽ 1x+0y |
| y ⩾ 0 | or | 0 ⩽ 0x+1y |
| x+y ⩽ 10 | or | 1x+1y ⩽ 10 |
| x ⩽ 8 | or | 1x+0y ⩽ 8 |
| y ⩽ 10 | or | 0x+1y ⩽ 10 |
| x+y ⩾ 3 | or | 3 ⩽ 1x+1y |

The maximum that x and y may take are 8 and 10 and the minimum values are 0 and 0 respectively. By reference to the distance table the total distance in the allocation table is therefore:

Distance = 5x + 2y + 10 - (x+y) + 3(8-x) + 3(10-y)
           + 4((x+y)-3)
         = 5x + 2y + 52

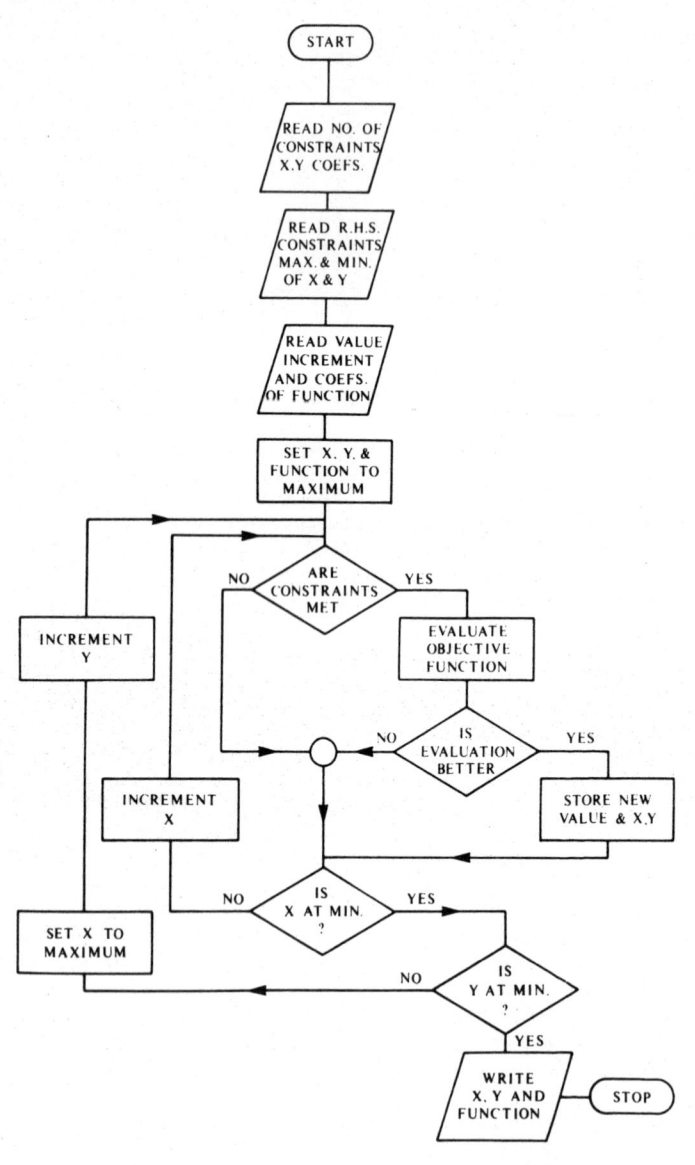

*Fig 9.8* Flowchart for factory-warehouse allocation program

This distance equation is to be minimised subject to the six restraints mentioned above and may be termed the *objective function*. Remembering that not every problem has a minimum solution, a number of mathematical methods are available for such minimisation procedures (Williams, 1969). The method used here is a systematic search procedure suitable for small problems of this type but for larger problems other and more complex methods of solution are more efficient. A maximum value for x and y is assumed and if the constraints are met then the objective function is evaluated. Values of x and y are each successively reduced, the constraints checked, and objective function calculated. The minimum for this function can then be determined and the associated x and y values placed into the allocation table and the remaining elements of this table calculated.

The flowchart for the systematic search is shown in Fig 9.8. The data are read and values of x and y assumed. Tests are made to determine whether the current values of x and y are permissible within the constraints. If so the objective function is evaluated and a check is made to determine whether this value is less than the previously stored value for the function. If it is less the new value is substituted for the old. If the new value is greater than a previously calculated value then the flowchart proceeds as if the constraints had not been satisfied and x is decreased by the incrementing value and control returns to check if the constraints are satisfied with the new x value. This procedure is repeated until x reaches its minimum value. The value of y is then successively reduced, x reset to its maximum value and the whole procedure repeated for each successive y value. When the minimum y value has been reached then the stored values for x and y and the evaluated objective function are written. The program is then terminated. This algorithm is thus a very good example of a problem that is easily stated for computer solution but that would be extremely intractable to hand calculation.

*Program.* The program is shown in Fig 9.9. Data are initially read in a series of statements. Array A holds the coefficients of the x and y variables of the constraints and B holds the constant values in the constraints. Values for x and y are taken and an initial dummy maximum value for the objective function is set. The evaluated value of this function as it is successively minimised is stored

```
C  PROGRAM TO MINIMISE SIMPLE LINEAR FUNCTIONS                              10
C  TO BE USED IN FINDING SOLUTION TO FACTORY-WAREHOUSE ALLOCATION MODEL     20
   DIMENSION A(2,10),B(10),EQ(10)                                           30
   NIN = 5                                                                  40
   NOUT = 6                                                                 50
C  N IS THE NUMBER OF INEQUALITIES                                          60
   READ (NIN,4) N                                                           70
C  READ MATRIX OF INEQUALITY COEFFICIENTS                                   80
   DO 100 I = 1,2                                                           90
   READ (NIN,5) (A(I,J),J=1,N)                                             100
100 CONTINUE                                                                110
   READ (NIN,5) (B(J),J=1,N)                                               120
C  READ MINIMUM AND MAXIMUM VALUES ALLOWED BY THE INEQUALITIES             130
   READ (NIN,5) XMI,YMI,XMA,YMA                                            140
C  READ LEVEL OF INCREMENTATION REQUIRED                                   150
   READ (NIN,10) RNC                                                       160
C  READ COEFFICIENTS OF OBJECTIVE FUNCTION                                 170
   READ (NIN,5) C,D,E                                                      180
   X = XMA                                                                 190
   Y = YMA                                                                 200
   Z = 100000.0                                                            210
130 CONTINUE                                                                220
C  TEST INEQUALITIES                                                       230
   DO 105 J = 1,N                                                          240
   EQ(J) = A(1,J)*X + A(2,J)*Y                                             250
   IF (EQ(J) - B(J)) 105,105,104                                           260
105 CONTINUE                                                                270
C  CALCULATE OBJECTIVE FUNCTION                                            280
   W = C*X +D*Y + E                                                        290
C  TEST FOR MINIMUM                                                        300
   IF (W - Z) 110,110,115                                                  310
110 Z = W                                                                   320
   ZX = X                                                                  330
   ZY = Y                                                                  340
115 CONTINUE                                                                350
104 IF (XMI + X) 120,120,125                                               360
C  DECREASE X BY INCREMENTATION VALUE                                      370
125 X = X - RNC                                                             380
   GO TO 130                                                               390
120 IF (YMI + Y) 135,135,140                                               400
C  DECREASE Y BY INCREMENTATION VALUE                                      410
140 Y = Y - RNC                                                             420
   X = XMA                                                                 430
   GO TO 130                                                               440
135 CONTINUE                                                                450
C  SET HEADINGS AND OUTPUT MINIMUM SOLUTION                                460
   WRITE (NOUT,25)                                                         470
   WRITE ( NOUT,20) ZX, ZY, Z                                              480
   STOP                                                                    490
4  FORMAT ( I3)                                                            500
5  FORMAT ( 6F5.0)                                                         510
10 FORMAT ( F6.2)                                                          520
20 FORMAT ( 3F12.4)                                                        530
25 FORMAT ( 36H SOLUTION FOR X,Y, AND EVALUATION IS)                       540
   END                                                                     550
```

*Fig 9.9* Program listing for factory-warehouse allocation program

in Z and the corresponding x and y values in ZX and ZY. Loop 105 tests if the constraints are met. If any are not then control passes out of the loop (to statement label 104) and incrementation occurs. If all constraints are met then the objective function is calculated and placed into a holding location (W) from where it is tested against the value of Z. If necessary the value of Z is moved across

202

```
 WWb
  -1,    0,    1,    1,    2,   -1,
  0,   -1,    1,    0,    1,   -1,
  0,    0,   10,    8,   10,   -3,
  0, 1
  5,    2,   52,
 SOLUTION FOR X, Y, AND EVALUATION IS
      0.0000        3.0000       58.0000
```

*Fig 9.10* (A) Data and (B) results for factory-warehouse allocation program

to W, otherwise incrementation occurs. Control is returned to the testing of the constraints. When the full search is complete the results are written. The FORMAT statements are collected together at the end of the program.

The data and results of the run are shown in Fig 9.10. The constraints are shown above and the data consists of the coefficients. An incrementation value of 0.1 is taken. The coefficients of the objective function are 5,2 and 52 as shown above. The results for x and y are:

$x = 0$   $y = 3$   and objective function = 58

These values may then be placed into the allocation table and the full matrix written as

|    | w1 | w2 | w3 |
|----|----|----|----|
| f1 | 0  | 3  | 7  |
| f2 | 8  | 7  | 0  |

and the distance travelled to allocate the goods from factories to warehouses is minimised at 58 distance units.

9.5 Worksheet
1 The decay of the last ice sheet to affect the British

Isles had been completed in large part by 12,500 to 13,000 radiocarbon years ago, but a subsequent deterioration in climate led to a re-establishment of glacier ice in much of Highland Britain during pollen zone III about 10,800 to 10,300 radiocarbon years ago. In Snowdonia, North Wales, this led to the formation of a series of small but distinct 'fresh' corrie moraines. According to Seddon (1957), these moraines and thus the ice masses from which they resulted, show a preferred orientation towards the north-east. Using the radius vector program shown earlier and the data for 41 such moraines listed below
  (i) Find the radius vector orientation mean and the mean vector magnitude.
  (ii) What climatic inferences may be drawn from these results?

The orientation and elevation of 41 Zone III moraines in Snowdonia

| Orientation | Height | Orientation | Height |
|---|---|---|---|
| 54  | 1099 | 62  | 1569 |
| 359 | 1131 | 55  | 1497 |
| 359 | 1370 | 70  | 1750 |
| 0   | 1567 | 51  | 1825 |
| 352 | 1426 | 30  | 1250 |
| 1   | 1200 | 36  | 2550 |
| 34  | 1325 | 19  | 1400 |
| 354 | 1512 | 167 | 1756 |
| 70  | 1507 | 122 | 2175 |
| 0   | 1700 | 180 | 1715 |
| 45  | 1273 | 148 | 1965 |
| 44  | 1707 | 132 | 1660 |
| 191 | 1648 | 29  | 1635 |
| 117 | 1389 | 0   | 1329 |
| 65  | 1420 | 49  | 2020 |
| 145 | 1975 | 355 | 1775 |
| 39  | 1551 | 29  | 2400 |
| 352 | 1970 | 79  | 2080 |
| 102 | 1626 | 89  | 1742 |
| 83  | 1140 | 61  | 1600 |
| 80  | 1789 |     |      |

2 Incorporate the results from exercise 8.5 into a program to calculate the population potential of
  (i) total population
  (ii) urban population - defined as population of urban districts (shi)

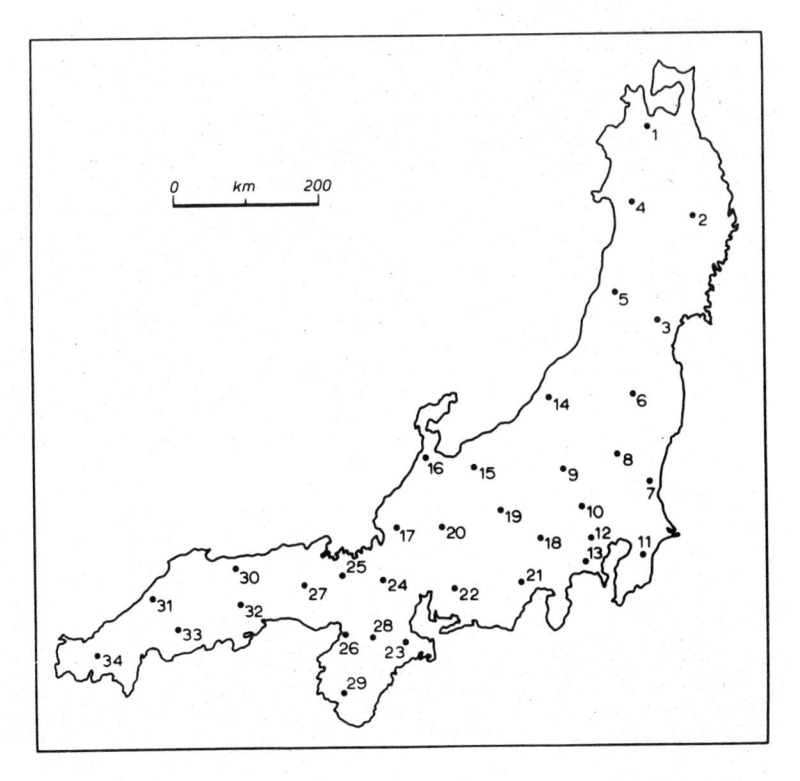

*Fig 9.11* Central points of prefectures of Honshu Island

(iii) rural population
for 1920 and 1965 for Honshu Island, Japan. Data are provided in the table and the control points are shown in Fig 9.11. The control points are the central points of the 34 prefectures of the island. Map your results using isolines. What conclusions do you draw on both the urban-rural dichotomy in Japan and the macro regional changes in population distribution that have taken place since 1920? The formula for the population potential, V, over N points, for point i is

$$V_i = \frac{P_i}{(d_{in}/2.0)} + \sum_{j=2}^{n} \frac{P_j}{d_{ij}}$$

*Table 9.1* Data for exercise 2

| Point | Population 1920 '000 total | urban | Population 1965 '000 total | urban | Co-ordinates x | y |
|---|---|---|---|---|---|---|
| 1 | 756.4 | 81.7 | 1416.6 | 774.0 | 106 | 121 |
| 2 | 845.5 | 42.4 | 1411.1 | 711.1 | 93 | 128 |
| 3 | 961.7 | 119.0 | 1753.1 | 849.4 | 74 | 122 |
| 4 | 898.5 | 36.3 | 1279.8 | 544.0 | 95 | 117 |
| 5 | 968.9 | 91.4 | 1263.1 | 783.6 | 78 | 114 |
| 6 | 1362.8 | 73.3 | 1983.7 | 998.4 | 60 | 118 |
| 7 | 1350.4 | 39.4 | 2056.1 | 932.3 | 45 | 121 |
| 8 | 1046.5 | 63.8 | 1521.6 | 910.7 | 50 | 115 |
| 9 | 1052.6 | 99.1 | 1605.6 | 945.5 | 47 | 104 |
| 10 | 1319.5 | 0 | 3014.9 | 1895.3 | 41 | 108 |
| 11 | 1336.1 | 0 | 2701.8 | 1744.9 | 32 | 119 |
| 12 | 3699.4 | 2212.2 | 10869.2 | 10341.1 | 35 | 110 |
| 13 | 1323.4 | 512.8 | 4430.7 | 4066.6 | 31 | 109 |
| 14 | 1736.5 | 162.1 | 2398.9 | 1362.3 | 60 | 102 |
| 15 | 724.3 | 98.5 | 1025.5 | 669.0 | 48 | 88 |
| 16 | 747.3 | 129.3 | 980.5 | 627.6 | 49 | 79 |
| 17 | 599.2 | 56.6 | 750.6 | 450.6 | 36 | 73 |
| 18 | 583.5 | 56.2 | 763.2 | 371.4 | 35 | 100 |
| 19 | 1562.7 | 113.6 | 1958.0 | 1037.5 | 40 | 93 |
| 20 | 1070.4 | 91.0 | 1700.4 | 977.5 | 32 | 82 |
| 21 | 1550.4 | 138.8 | 2912.5 | 2050.6 | 27 | 97 |
| 22 | 2089.8 | 533.7 | 4798.7 | 3789.4 | 26 | 84 |
| 23 | 1069.3 | 122.2 | 1514.5 | 928.4 | 17 | 75 |
| 24 | 651.1 | 31.5 | 853.4 | 345.7 | 27 | 71 |
| 25 | 1287.1 | 591.3 | 2102.8 | 1713.8 | 28 | 64 |
| 26 | 2587.8 | 1337.9 | 6657.2 | 6301.6 | 18 | 64 |
| 27 | 2301.8 | 726.0 | 4309.9 | 3438.0 | 27 | 56 |
| 28 | 564.6 | 40.3 | 825.9 | 483.6 | 18 | 69 |
| 29 | 750.4 | 83.5 | 1027.0 | 579.9 | 7 | 64 |
| 30 | 454.7 | 29.3 | 579.9 | 291.3 | 30 | 43 |
| 31 | 714.7 | 37.5 | 821.6 | 412.9 | 24 | 28 |
| 32 | 1217.7 | 94.6 | 1645.1 | 958.4 | 23 | 45 |
| 33 | 1541.9 | 347.1 | 2281.1 | 1330.9 | 19 | 33 |
| 34 | 1041.0 | 72.3 | 1543.6 | 1085.6 | 14 | 17 |

The co-ordinates are on a 7.5km grid

where P stands for population
$d_{ij}$ is the distance between the points i and j
$d_{in}$ is the distance between point i and its nearest neighbour.

For further discussion of the method see Anderson (1956).

3 Using the following data relating to agricultural acreages in sample districts of the Irish Republic in June 1970, and the map provided (*Fig 9.12*), write a program to calculate the following:
  (i) The percentage of agricultural land in each area accounted for by pasture (ie total less crops)
  (ii) The arable (ie excluding meadow but including first year hay acreages) crop combination, using Weaver's method, for each district. Map your results and comment on the pattern shown. How does the map differ if the Thomas method of calculation is used?
  (iii) The effect of the inclusion of meadow acreages on the crop combinations for each district.
The methods of calculation of crop combinations is dealt with step by step in Yeates (1968).

4 The best known method for estimating potential evapotranspiration ($E_o$) using air temperatures alone is that developed by Thornthwaite (1948) who gives the following equations:

$E_o = c(1.6(10T_i/I)^a)$ cm/month
$T_i$ = mean monthly screen temperature in °C for month i
$I$ = the station *heat index* = $\sum_{1}^{12} (T_i/5)^{1.514}$
$a = (0.6751\ I^3 - 77.11\ I^2 + 17920\ I + 492390) \times 10^{-6}$
$c$ = a correction factor for day length depending on station latitude and month. For stations in the British Isles north of 50°N, appropriate c values are:

|   | J | F | M | A | M | J |
|---|---|---|---|---|---|---|
| c = | 0.74 | 0.78 | 1.02 | 1.15 | 1.33 | 1.36 |
|   | J | A | S | O | N | D |
|   | 1.37 | 1.25 | 1.06 | 0.92 | 0.76 | 0.70 |

Notice that these formulae are only valid for temperatures in the range 0 - 26.5°C.
  (i) Write a computer program to evaluate annual total and monthly values of potential evapotranspiration.
  (ii) Run the program using the data provided below and produce a map of the total annual amounts for England, Wales and Scotland.

5 In comparative studies of manufacturing activity a useful measure is the *index of industrial diversification*, providing a measure of the extent to which manufacturing

Table 9.2 Data for exercise 3

| | | wheat | oats | barley | potatoes | greens and other roots | hay first year | meadow | TOTAL crops and pasture |
|---|---|---|---|---|---|---|---|---|---|
| | | | | | CROPS | | | | |
| 1 | Balrothery | 7136 | 1144 | 11101 | 5781 | 3338 | 1502 | 6982 | 80144 |
| 2 | Celbridge No 2 | 1256 | 149 | 2076 | 294 | 239 | 603 | 2428 | 21132 |
| 3 | Rest of Dublin Co | 2967 | 232 | 3552 | 1550 | 1767 | 723 | 6819 | 64237 |
| 4 | Rathdown No 2 | 274 | 162 | 777 | 227 | 523 | 495 | 1921 | 22035 |
| 5 | Dunshaughlin | 5718 | 730 | 7523 | 611 | 709 | 2051 | 14858 | 105173 |
| 6 | Trim | 5034 | 1546 | 6450 | 1093 | 1081 | 2333 | 18488 | 133339 |
| 7 | Celbridge No 1 | 2509 | 419 | 5015 | 226 | 420 | 1766 | 6570 | 49730 |
| 8 | Edenderry No 2 | 2327 | 567 | 3582 | 369 | 779 | 1535 | 6830 | 51919 |
| 9 | Naas | 5826 | 1091 | 7632 | 774 | 1557 | 3595 | 15940 | 140324 |
| 10 | Delvin | 898 | 672 | 2387 | 433 | 374 | 577 | 9437 | 65795 |
| 11 | Mullingar | 2450 | 1530 | 5573 | 1419 | 2081 | 2222 | 25849 | 197906 |
| 12 | Ballymore | 108 | 190 | 360 | 216 | 209 | 161 | 4009 | 28552 |
| 13 | Athlone No 1 | 300 | 346 | 690 | 630 | 640 | 652 | 9322 | 58468 |
| 14 | Ballymahon | 19 | 305 | 666 | 322 | 157 | 168 | 9263 | 53600 |
| 15 | Edenderry No 1 | 1098 | 355 | 2338 | 392 | 883 | 716 | 8124 | 52398 |
| 16 | Tullamore | 2258 | 875 | 8849 | 1321 | 4989 | 3444 | 17373 | 126890 |
| 17 | Birr | 3340 | 775 | 5830 | 1739 | 5219 | 3132 | 19188 | 126823 |
| 18 | Mountmellick | 3873 | 1040 | 9729 | 1225 | 5282 | 3132 | 20026 | 119307 |
| 19 | Borrisokane | 3643 | 598 | 6859 | 551 | 6091 | 2903 | 15407 | 89094 |
| 20 | Athlone No 2 | 138 | 1971 | 1321 | 1419 | 1285 | 1527 | 16600 | 89928 |
| 21 | Ballinasloe | 201 | 1884 | 1496 | 1199 | 1971 | 1185 | 16190 | 101140 |
| 22 | Portumna | 430 | 940 | 1414 | 768 | 1753 | 890 | 10946 | 57604 |
| 23 | Loughrea | 170 | 2760 | 1775 | 1604 | 2110 | 1314 | 22522 | 133251 |
| 24 | Gort | 586 | 1336 | 2421 | 1112 | 2758 | 1217 | 11946 | 69787 |
| 25 | Mountbellow | 57 | 2617 | 1110 | 1426 | 767 | 1287 | 13411 | 71465 |
| 26 | Glennamaddy | 117 | 1860 | 747 | 1302 | 791 | 663 | 11835 | 61032 |
| 27 | Tuam | 285 | 6356 | 3189 | 4337 | 3735 | 3240 | 19469 | 163653 |
| 28 | Galway | 323 | 252 | 2396 | 2428 | 3017 | 2396 | 14342 | 100423 |
| 29 | Oughterard | 9 | 499 | 38 | 712 | 109 | 320 | 3412 | 38182 |
| 30 | Clifden | 0 | 252 | 1 | 664 | 88 | 84 | 4661 | 33996 |

208

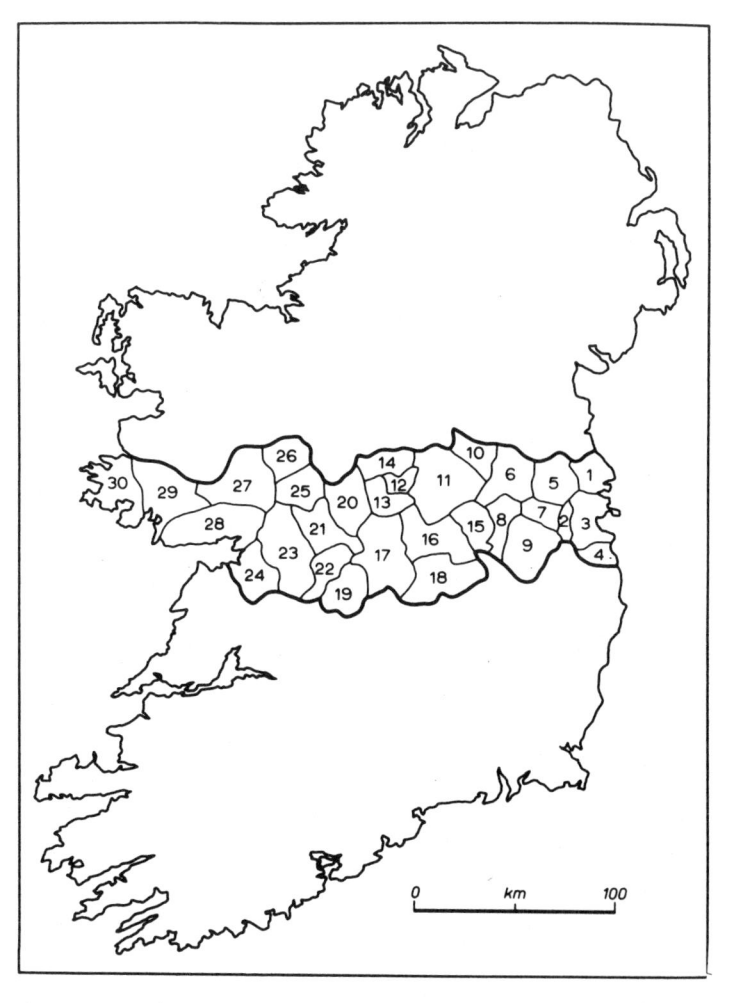

*Fig 9.12* Agricultural districts of the central belt of the Irish Republic

Table 9.3 Data for exercise 4

Mean Monthly Temperature (°C)

| No | Station | J | F | M | A | M | J | J | A | S | O | N | D |
|---|---|---|---|---|---|---|---|---|---|---|---|---|---|
| 1 | Aberdeen | 4 | 4 | 5 | 7 | 9 | 12 | 14 | 14 | 14 | 9 | 6 | 4 |
| 2 | Aberystwyth | 6 | 5 | 7 | 9 | 11 | 14 | 16 | 16 | 14 | 11 | 8 | 6 |
| 3 | Bath | 5 | 5 | 7 | 9 | 12 | 16 | 17 | 17 | 14 | 11 | 7 | 5 |
| 4 | Birmingham | 4 | 4 | 6 | 8 | 11 | 14 | 16 | 16 | 14 | 10 | 6 | 4 |
| 5 | Blackpool | 4 | 4 | 6 | 8 | 11 | 14 | 16 | 16 | 13 | 11 | 7 | 5 |
| 6 | Bournemouth | 6 | 5 | 7 | 9 | 12 | 15 | 17 | 17 | 15 | 11 | 8 | 6 |
| 7 | Brighton | 6 | 5 | 7 | 9 | 12 | 15 | 17 | 17 | 16 | 12 | 8 | 6 |
| 8 | Cambridge | 4 | 4 | 6 | 9 | 12 | 15 | 17 | 17 | 14 | 10 | 7 | 4 |
| 9 | Cardiff | 5 | 5 | 7 | 9 | 12 | 15 | 16 | 16 | 14 | 11 | 7 | 6 |
| 10 | Dover | 5 | 5 | 6 | 9 | 12 | 14 | 17 | 17 | 15 | 12 | 8 | 6 |
| 11 | Dundee | 3 | 4 | 5 | 7 | 9 | 13 | 15 | 14 | 12 | 9 | 6 | 4 |
| 12 | Durham | 3 | 3 | 5 | 7 | 10 | 13 | 16 | 15 | 13 | 9 | 6 | 4 |
| 13 | Edinburgh | 4 | 4 | 5 | 7 | 9 | 13 | 15 | 14 | 12 | 9 | 6 | 4 |
| 14 | Falmouth | 7 | 7 | 8 | 9 | 12 | 15 | 16 | 17 | 15 | 12 | 9 | 7 |
| 15 | Fort William | 4 | 4 | 6 | 8 | 11 | 13 | 14 | 14 | 12 | 9 | 6 | 4 |
| 16 | Glasgow | 3 | 4 | 6 | 7 | 10 | 13 | 15 | 14 | 12 | 9 | 6 | 4 |
| 17 | Ilfracombe | 7 | 6 | 8 | 9 | 12 | 14 | 16 | 17 | 15 | 12 | 9 | 7 |
| 18 | Inverness | 3 | 4 | 5 | 7 | 9 | 12 | 14 | 14 | 12 | 8 | 6 | 4 |
| 19 | Liverpool | 4 | 4 | 6 | 8 | 11 | 14 | 16 | 16 | 13 | 10 | 7 | 5 |
| 20 | Llandudno | 6 | 6 | 7 | 9 | 11 | 14 | 16 | 16 | 14 | 11 | 8 | 7 |
| 21 | London | 4 | 4 | 7 | 9 | 12 | 16 | 18 | 17 | 15 | 11 | 7 | 4 |
| 22 | Luton | 3 | 4 | 6 | 8 | 11 | 14 | 17 | 16 | 14 | 10 | 6 | 4 |
| 23 | Manchester | 4 | 4 | 7 | 9 | 12 | 15 | 17 | 16 | 14 | 11 | 7 | 5 |
| 24 | Newquay | 7 | 6 | 7 | 9 | 11 | 14 | 16 | 16 | 14 | 12 | 9 | 7 |
| 25 | Norwich | 4 | 4 | 6 | 8 | 12 | 15 | 17 | 17 | 14 | 11 | 7 | 4 |
| 26 | Nottingham | 4 | 4 | 6 | 8 | 12 | 14 | 17 | 16 | 14 | 10 | 7 | 4 |
| 27 | Oxford | 4 | 4 | 7 | 9 | 12 | 15 | 17 | 17 | 14 | 11 | 5 | 4 |
| 28 | Penzance | 8 | 7 | 8 | 10 | 12 | 15 | 17 | 17 | 15 | 12 | 9 | 8 |
| 29 | Plymouth | 7 | 6 | 8 | 9 | 12 | 15 | 17 | 17 | 15 | 12 | 9 | 7 |
| 30 | Portsmouth | 6 | 6 | 7 | 9 | 13 | 16 | 18 | 18 | 16 | 12 | 8 | 6 |
| 31 | Ryde (IoW) | 6 | 6 | 7 | 9 | 12 | 16 | 17 | 17 | 16 | 12 | 8 | 6 |
| 32 | Scarborough | 4 | 5 | 6 | 8 | 11 | 14 | 16 | 16 | 14 | 11 | 7 | 5 |
| 33 | Sheffield | 4 | 4 | 6 | 8 | 11 | 14 | 17 | 16 | 14 | 10 | 7 | 4 |
| 34 | Shrewsbury | 3 | 4 | 6 | 9 | 11 | 15 | 17 | 16 | 14 | 10 | 7 | 4 |
| 35 | Skegness | 4 | 4 | 6 | 8 | 11 | 14 | 16 | 16 | 14 | 11 | 7 | 4 |
| 36 | Tenby | 6 | 6 | 7 | 9 | 12 | 14 | 16 | 16 | 14 | 11 | 8 | 7 |
| 37 | Weymouth | 6 | 6 | 7 | 9 | 12 | 15 | 17 | 17 | 16 | 12 | 8 | 7 |
| 38 | York | 4 | 4 | 6 | 8 | 11 | 14 | 17 | 16 | 14 | 10 | 6 | 4 |

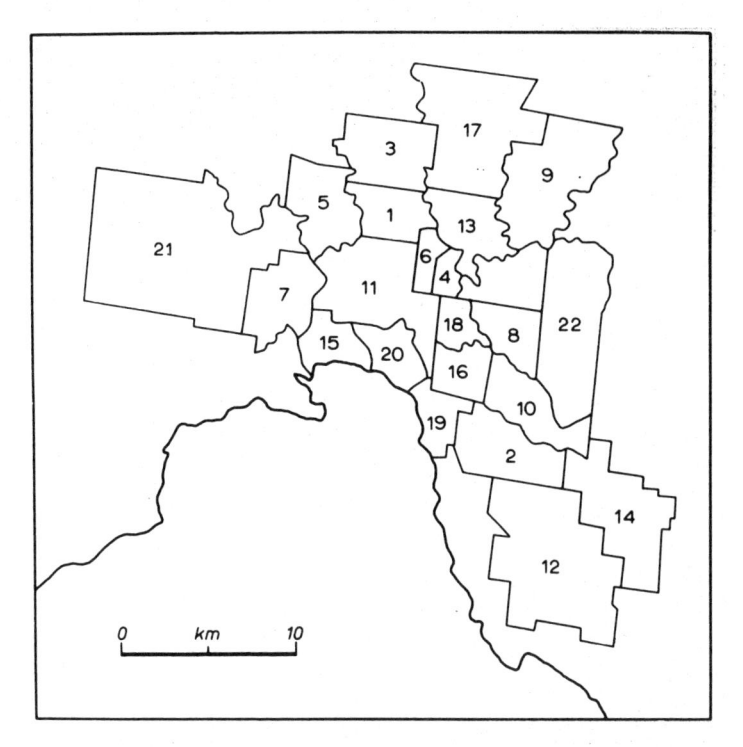

*Fig 9.13* Local government areas in greater Melbourne

activity, usually measured by employment numbers, is concentrated into a few industrial groups or is diversified through all groups. The index is usually calculated from a graph on which are plotted accumulated ranked employment percentages of the industry groups on the y axis against the appropriate industry group on the x axis (Dawson and Thomas, 1974). If total diversification is present the graph is the diagonal passing through the origin and any concentration into groups will show a concentrated industrial structure. Programs to rank data and to calculate the area of a polygon have been the subjects of earlier exercises. Combine these earlier results to calculate indices of diversification on the data given below relating to the main manufacturing districts of Greater Melbourne in 1968 (*Fig 9.13*). Is the pattern of

211

Table 9.4 Data for exercise 5

| District | Map code | food | textiles | clothing | wood | paper | chemicals | non-metallic minerals | basic metals | fabricated metals | transport equipment | other machinery | miscellaneous manufacturing |
|---|---|---|---|---|---|---|---|---|---|---|---|---|---|
| a) *Male employment* | | | | | | | | | | | | | |
| Brunswick | 1 | 428 | 1091 | 1751 | 323 | 431 | 68 | 388 | 385 | 1194 | 161 | 965 | 422 |
| Camberwell | 2 | 94 | 24 | 229 | 100 | 77 | 34 | 5 | 20 | 56 | 35 | 175 | 95 |
| Caulfield | 3 | 125 | 33 | 83 | 135 | 135 | 24 | 4 | 2 | 204 | 50 | 218 | 93 |
| Coburg | 4 | 270 | 458 | 450 | 237 | 321 | 160 | 231 | 768 | 1683 | 175 | 1830 | 112 |
| Collingwood | 5 | 1347 | 1182 | 2381 | 675 | 650 | 191 | 43 | 96 | 757 | 317 | 1614 | 709 |
| Essendon | 6 | 283 | 10 | 51 | 106 | 29 | 270 | 85 | 15 | 143 | 260 | 120 | 288 |
| Fitzroy | 7 | 447 | 96 | 1815 | 277 | 527 | 14 | 38 | 20 | 482 | 36 | 1359 | 268 |
| Footscray | 8 | 1960 | 2142 | 145 | 370 | 179 | 3110 | 304 | 688 | 1649 | 161 | 725 | 2251 |
| Hawthorn | 9 | 89 | 47 | 266 | 309 | 617 | 118 | 224 | 90 | 248 | 92 | 366 | 529 |
| Heidelberg | 10 | 209 | 9 | 29 | 308 | 376 | 100 | 50 | 100 | 692 | 526 | 484 | 139 |
| Malvern | 11 | 117 | 30 | 300 | 152 | 105 | 40 | 62 | 0 | 186 | 47 | 210 | 110 |
| Melbourne | 12 | 3665 | 866 | 2363 | 477 | 7913 | 1962 | 454 | 912 | 1530 | 2225 | 1761 | 1430 |
| Moorabin | 13 | 1567 | 296 | 157 | 976 | 1124 | 479 | 189 | 117 | 1572 | 3201 | 2021 | 1768 |
| Northcote | 14 | 353 | 151 | 482 | 385 | 1523 | 169 | 326 | 35 | 1349 | 76 | 911 | 392 |
| Oakleigh | 15 | 720 | 132 | 157 | 705 | 957 | 707 | 600 | 528 | 2204 | 2036 | 3901 | 388 |
| Port Melbourne | 16 | 1869 | 91 | 21 | 511 | 214 | 245 | 204 | 184 | 713 | 13020 | 1379 | 121 |
| Prahran | 17 | 846 | 238 | 767 | 275 | 340 | 14 | 52 | 16 | 249 | 41 | 610 | 177 |
| Preston | 18 | 1079 | 733 | 572 | 598 | 774 | 334 | 228 | 46 | 942 | 1129 | 2313 | 744 |
| Richmond | 19 | 1020 | 238 | 913 | 383 | 641 | 481 | 219 | 119 | 1237 | 400 | 2189 | 1031 |
| St Kilda | 20 | 133 | 21 | 120 | 62 | 97 | 59 | 15 | 9 | 99 | 20 | 60 | 283 |
| S Melbourne | 21 | 673 | 152 | 224 | 505 | 946 | 169 | 142 | 102 | 1119 | 497 | 2912 | 1700 |
| Sunshine | 22 | 2917 | 743 | 80 | 495 | 756 | 2641 | 424 | 1168 | 2505 | 488 | 4763 | 1027 |
| b) *Female employment* | | | | | | | | | | | | | |
| Brunswick | 1 | 259 | 590 | 5947 | 89 | 264 | 49 | 33 | 24 | 285 | 54 | 255 | 203 |
| Camberwell | 2 | 31 | 29 | 424 | 12 | 57 | 18 | 3 | 20 | 34 | 30 | 78 | 34 |
| Caulfield | 3 | 103 | 50 | 411 | 24 | 64 | 31 | 1 | 1 | 73 | 10 | 82 | 53 |
| Coburg | 4 | 55 | 327 | 1112 | 20 | 224 | 105 | 61 | 35 | 361 | 25 | 970 | 69 |
| Collingwood | 5 | 691 | 1669 | 4493 | 72 | 355 | 62 | 38 | 11 | 196 | 80 | 610 | 450 |
| Essendon | 6 | 173 | 61 | 549 | 22 | 16 | 61 | 84 | 15 | 28 | 130 | 32 | 214 |
| Fitzroy | 7 | 154 | 132 | 3884 | 55 | 88 | 9 | 2 | 6 | 136 | 4 | 106 | 195 |
| Footscray | 8 | 418 | 1552 | 466 | 48 | 95 | 1144 | 87 | 73 | 394 | 70 | 252 | 525 |
| Hawthorn | 9 | 42 | 70 | 903 | 57 | 162 | 8 | 2 | 7 | 76 | 57 | 89 | 378 |
| Heidelberg | 10 | 51 | 6 | 51 | 63 | 176 | 34 | 6 | 10 | 133 | 258 | 254 | 122 |
| Malvern | 11 | 65 | 20 | 524 | 28 | 48 | 16 | 18 | 0 | 54 | 9 | 94 | 67 |
| Melbourne | 12 | 817 | 543 | 8426 | 105 | 2201 | 963 | 86 | 474 | 295 | 249 | 397 | 589 |
| Moorabin | 13 | 1290 | 353 | 693 | 200 | 856 | 257 | 27 | 11 | 453 | 1383 | 642 | 819 |
| Northcote | 14 | 305 | 96 | 1331 | 39 | 203 | 198 | 85 | 10 | 406 | 38 | 214 | 299 |
| Oakleigh | 15 | 105 | 302 | 750 | 204 | 475 | 361 | 117 | 47 | 537 | 742 | 1856 | 469 |
| Port Melbourne | 16 | 1256 | 20 | 61 | 47 | 143 | 57 | 51 | 4 | 83 | 1124 | 94 | 115 |
| Prahran | 17 | 696 | 169 | 3266 | 45 | 203 | 6 | 7 | 15 | 134 | 6 | 353 | 275 |
| Preston | 18 | 391 | 549 | 1251 | 58 | 338 | 150 | 22 | 1 | 88 | 347 | 525 | 391 |
| Richmond | 19 | 464 | 302 | 2886 | 124 | 246 | 323 | 70 | 10 | 364 | 158 | 691 | 820 |
| St Kilda | 20 | 64 | 13 | 736 | 12 | 37 | 41 | 10 | 10 | 35 | 5 | 18 | 157 |
| S Melbourne | 21 | 475 | 117 | 550 | 132 | 329 | 75 | 54 | 19 | 155 | 121 | 596 | 342 |
| Sunshine | 22 | 768 | 494 | 164 | 46 | 189 | 522 | 20 | 88 | 377 | 3889 | 750 | 311 |

diversification indices *significantly* different for male and female employees? Use these same data in your crop combinations program to produce industry combinations. Map the indices of diversification and the industrial combinations and compare the results.

6 The production of life tables and the expectation of life at birth is a fundamental calculation in demographic analyses. Given the following tables for the probabilities of dying ($n^q x$) by age group x of the 1938-9 and 1958 populations of the Soviet Union calculate life tables for the two populations. The life table contains
  (i) the number of persons living at the beginning of the age interval ($l_x$)
  (ii) the number dying during the age interval ($d_x$)
Assume an initial population of 10,000. If the expectation of life at birth is given by

$$e_0 = \tfrac{1}{2} + \tfrac{1}{10} [2\tfrac{1}{2}.l_1 + 4\tfrac{1}{2}.l_5 + 5(l_{10} + l_{15} \ldots l_{65}) + (e_{70} + 2\tfrac{1}{2}) l_{70}]$$

where $e_{70}$ for 1938-9 population is 9 years
      $e_{70}$ for 1958 population is 11 years
and $l_i$ is the **number of people** living of age i,
calculate the life expectation in the two populations.
This exercise is based on problem 3 (page 47) of Chapter 4 in Pressat (1974). There are numerous other demographic problems in this book which lend themselves to solution by computer methods.

| Age | $n^q x$ 1938-9 | $n^q x$ 1958 |
|---|---|---|
| 0 | .1850 | .0406 |
| 1 | .1074 | .0143 |
| 5 | .0272 | .0055 |
| 10 | .0129 | .0040 |
| 15 | .0169 | .0065 |
| 20 | .0218 | .0090 |
| 25 | .0232 | .0109 |
| 30 | .0267 | .0129 |
| 35 | .0335 | .0154 |
| 40 | .0398 | .0203 |
| 45 | .0498 | .0267 |
| 50 | .0668 | .0393 |
| 55 | .0818 | .0532 |
| 60 | .1154 | .0813 |
| 65 | .1616 | .1113 |

## 9.6 Further Reading

Anderson, T.R. 'Potential models and the spatial distribution of population', *Proceedings, Regional Science Association*, 2, 175-82 (1956)

Cox, K. 'The application of linear programming to geographic problems', *Tijdschrift voor Economische in Sociale Geografie*, 56, 228-36 (1965)

Dawson, J.A. and Thomas, D. *Man and his world* (1974)

Howe, G.M. 'The moisture balance of England and Wales based on the concept of potential evapotranspiration', *Weather*, 11, 74-82 (1956)

Pincus, H.J. 'The analysis of aggregates of orientation data in the earth sciences', *Journal of Geology*, 61, 482-509 (1953)

Pressat, R. *A workbook in demography* (1974)

Seddon, B. 'Late-glacial cwm glaciers in Wales', *Journal of Glaciology*, 3, 94-9 (1957)

Thornthwaite, C.W. 'An approach toward a rational classification of climate', *Geographical Review*, 38, 55-94 (1948)

Williams, K. *Linear programming* (1969)

Yeates, M. *An introduction to quantitative methods in economic geography* (1968)

# Chapter 10
# Further Input/Output Commands

10.1 Additional FORMAT Specifications
In Chapter 4 we introduced the simple READ and WRITE statements necessary to input and output information. Each has an accompanying FORMAT statement that rigorously defines either the layout of the data on the cards to be read or the layout of the results on the printer paper. Discussion in the preceding chapters has been restricted to input and output involving the format types I, F and H. In this chapter we introduce three new format types, X, A and E together with a facility that enables data to be compiled directly into programs using the DATA statements.

The FORMAT statement consists of a statement number, cross-referenced from the input-output command and the word FORMAT followed by a specification enclosed by brackets and conventionally starting with a space. An I specification deals with integer variables, an F specification deals with real variables and an H specification allows strings of specified characters to be output. F and I are frequently used for input and output but H is more usually used for output where it is often used for printing table headings. In this way extremely long H specifications can occur with many blank characters (*see the worked examples in Chapter 8*). These layouts can be made simpler using the X specification to indicate that blank spaces are desired. Its form is (dX) where d is an integer constant, as in FORMAT (5X) which specifies 5 spaces. Obviously the X character is only rarely used on its own; no discernible output occurs on the lineprinter from a statement such as FORMAT (5X) but used in association with other formats, X becomes very useful. For example, if two table headings are required, one dealing with data for 1961 and the second 1971 then the following statement sequence might be used:

215

```
      WRITE (NOUT, 5)
      DO 100 I = 1,15
  100 WRITE (NOUT, 10) J(I), K(I)
       ⋮
    5 FORMAT (3X, 5H 1961, 10X, 5H 1971)
   10 FORMAT (3X, I5, 10X, I5)
```
The 1961 data are held in vector array J and the 1971 data in array K. The first column of results is indented three spaces and separated from the second column by 10 spaces. If X specification had not been used the following formats would have been needed to produce the same result
```
    5 FORMAT (23Hbbbb1961bbbbbbbbbb1971)
   10 FORMAT (I8, I15)
```
(where b indicates a blank character). The chance of miscounting the large number of blanks in this second version is large and makes it a very error-prone statement.

The use of X is not limited to outputs; it may also be used in association with READ instructions. At input it instructs the machine to skip the specified number of card columns before reading a value. A statement
```
              FORMAT (10X, I4)
```
associated with a READ command would result in transfer of a 4-digit integer number punched in card columns 11-14 with anything punched in other columns being ignored.

The use of H specification as an output format was discussed in Chapter 4 but it is also possible to use it as for input where it can be used to read characters into an existing H field within the FORMAT statement. For example the program sequence
```
         READ (NIN, 10)
      10 FORMAT (13Hbbbbbbbbbbbbb)
         WRITE (NOUT, 10)
```
associated with data cards
           bDATAbBLOCKb1
would have the effect of transferring the data ' DATA BLOCK 1' into FORMAT 10 (containing 13 spaces) and then writing out FORMAT 10 producing output of the characters DATA BLOCK 1. In this example the data are input into a format space containing blanks; if anything had been present in FORMAT 10 it would have been overwritten with the new data. This H FORMAT is sometimes useful for inputting table headings as data and has the advantage over A specification (*see below*) that the characters are not stored in a storage location but are input directly to the FORMAT statement.

A more usual form of manipulating text data is by the

use of A FORMAT. The general form is Aw where w is an
unsigned integer defining the total number of characters
of data to be placed in each storage location. However,
w has a maximum value for each computer installation and
values commonly vary between 4 and 10 depending upon the
word length (*see Chapter 2*) of the computer used.
Characters may be input into either real or integer
locations of scalar or array types. For example, the
program sequence:
     READ (NIN, 2∅) A,B,C
   2∅ FORMAT (A4,A4,A2)
with associated data
     bREGIONb∅1
would transfer the data characters REGION 01 to the three
variables A,B,C with bREG being stored in A; IONb in B
and ∅1 in C. The command
     WRITE (NOUT, 2∅) A,B,C
would result in
     REGION ∅1
being output. However, WRITE (NOUT, 2∅) B,A,C would
result in IONbbREG∅1. In this way titles may be input as
data, as for example with a series of town names associated
with data for each town. Each data card could consist of
a placename followed by the data. For example, if three
pieces of data for each of 25 towns were to be input then
a program sequence
    DO 1∅∅ I = 1,25
   1∅∅ READ (NIN, 1∅)(A(I), I = 1,6)
   1∅ FORMAT (A4,A4,A4,3F 1∅.3)
would transfer data off the first 42 columns of the card
with the first 12 columns specifying the town name.
Answers could be output in similar fashion. To simplify
FORMAT 1∅ the repetition of the A4 specification could be
written as 3A4 with the number preceding the A indicating
the repetition of the basic field description.

 Before using A format it would be as well to check how
many characters will fit into a storage location on the
machine being used. If, for example, the computer will
take only 4 characters yet A8 is specified with the word
bEXAMPLE being input, then effectively only MPLE will be
stored. On output the first four characters will have
been lost. On the other hand if the computer will
accept 8 characters per variable location and the word
bEND is input, bENDbbbb will be stored (*see Chapter 12*).

 In general terms if Aw is the specification and g is the
number of characters which the computer is capable of

               217

storing in a single location then if w is less than g the
input quantity is stored left justified.  If w is greater
than g the right most g characters are stored and the
remaining characters are ignored.

A further format specification utilises the letter E.
The format specification E is used on *real numeric* data
which is in exponent form.  The real number 123.45 could
be written as .12345 x $10^3$.  The use of E format assumes
the value 10 in the multiplying value and would output the
number as .12345E+03 (*see also Chapter 4*).  The precise
format to produce this number would be E10.5.  In
a general form the format is Ew.d where w is an integer
defining the total number of characters in the field and
d is the number of digits to the right of the decimal
point.  The total of w includes the signs, zeros, letter
E, decimal point and significant digits.  If w is greater
than the total of characters then blanks are printed at
the left end of the number.  If w is less than the total
number of characters then usually a * is printed to denote
the specification is not the correct size.  Generally
w is at least d + 6 for negative numbers and d + 5 for
positive values.  The same rules govern the input of
numbers in E format.  An integer preceding the letter E
in the FORMAT statements defines the number of times the
field is to be repeated.  The letter E in the data to be
input is optional so that, in the form of data, the two
numbers
$$-1.23E+02$$
and $$-1.23+2$$
are identical (and refer to the number -123.) but the
first would require an input FORMAT (E9.2) while the
second requires FORMAT (E7.2).

The presence of a letter E in the input data or output
values must not be confused with the letter E of the
format specification.  The two are very different.  The
following short statement sequence (*Fig 10.1*) shown with
associated data indicates the type of situation when the
E format specification is useful.  It shows a possible
start of a program to analyse certain characteristics of
large American cities.  Nine variables are input for each
of the 148 SMSAs with a population of over 200,000.  The
data are prefaced by the name of the SMSA.  The first
three data cards are shown and the nine variables are:

   (i) Land area in sq miles
   (ii) Population, 1970

```
C PROGRAM FOR CALCULATION OF URBAN INDICES                              USA   5
      DIMENSION NAME(148,3),IREA(148),POP(148,4),DWE(148),IFAC(148),    USA  10
     *FAC(148,2)                                                        USA  15
C SET INPUT OUTPUT CHANNELS                                             USA  20
      NIN = 5                                                           USA  25
      NOUT = 6                                                          USA  30
C INPUT DATA                                                            USA  35
      DO 1 I = 1,148                                                    USA  40
    1 READ(NIN,100)(NAME(I,J),J=1,3), IREA(I), (POP(I,J),J=1,4),DWE(I), USA  45
     *NFAC(I), (FAC(I,J),J=1,2)                                         USA  50
  100 FORMAT ( 3A4,I4,E7.3,F4.1,2E6.3,F5.1,I5,E8.1,E8.3)                USA  50


AKRON       09050.679+612.2.174+2.084+2015.6008770101.0+301.401+9
ALBANY      22160.721+609.6.169+2.108+2011.6006330065.0+300.893+9
ALBUQUERQUE 11690.316+620.4.189+2.062+2029.1002710008.0+300.094+9
```

Fig 10.1 Sample program to show use of E format

(iii) Percentage population change, 1960-1970
(iv) Birth rate per thousand resident population, 1969
(v) Death rate per thousand resident population, 1969
(vi) Percentage change in the number of dwellings, 1960-70
(vii) Number of manufacturing plants, 1969
(viii) Number of employees in manufacturing, 1969
(ix) Value added by manufacturing, 1969

The E mode format is used to input very large values such as value added in manufacturing while F and I modes are used for the smaller values and the SMSA name is transferred in A mode. FORMAT 1ØØ specifies the data layout. Note that the DO loop label is attached to the READ statement not the continuation card. Although the continuation card is physically the last card in the loop, logically it is merely an extension of the previous card which is therefore the last executable statement in the DO loop.

Six field descriptors for inclusion in format statements have now been dealt with. These, in general form are:
rFw.d
rEw.d
rIw
rAw
nHh$_1$, h$_2$ ... h$_n$
nX

where w and n are the width of the field (number of characters) as specified by a positive non-zero integer constant

219

      d is the number of digits after the decimal point in the defined format
      r is the number of times the field is to be repeated
      h is a permissible character and $h_1$ is usually a blank

The first three format specifications in the list, F, E and I, deal only with numeric quantities. The format X deals only with the character *blank* while A and H may be used to input or output any permissible character, alphabetic, numeric or special (*see Chapter 4*). There are other letters used for other generally available field descriptors. Some of these are mentioned briefly in Chapter 13, but these are confined to elaborate programs beyond the scope of this book. In addition, specific computer installations have introduced other individual format types. A common one is the *free format* used to input numbers of any field width using a space to terminate each number. The precise specification of free format differs widely among computer centres but it is a useful format for inputting large amounts of raw data (*see also Chapter 13*). The publications of individual computer centres deal with such local variations in format form and function but for reasons to be outlined in Chapter 13 these local 'dialects' of FORTRAN should be treated with caution.

## 10.2 Repetition of FORMAT Specifications

Additional to the format specifications are a number of commands to help in the design and layout of output and to a lesser extent input. These additional features fall naturally into two types; first those concerned with repeating specifications and, secondly those controlling the spaces between lines of output on a lineprinter or the use of additional cards on input.

    Specifications in a format statement may be *repeated* by inserting a repeat count integer before the letter of the specification. Thus 14A4 indicates fourteen alphanumeric fields each of length 4. This repeat feature has been mentioned above and in earlier chapters. In this case the repetition is limited to only one field whereas frequently it would be useful to repeat a whole sequence of fields. Such repetition of compound format specifications is possible using brackets preceded by an integer defining the number of times the sequence is to be repeated. For example,

             FORMAT (A4,A4,I2,F8.2,I2,I2,F8.2,I2)
could also be written as
             FORMAT (2A4,I2,F8.2,2I2,F8.2,I2)
or           FORMAT (2A4,2(I2,F8.2,I2))
In addition to the brackets enclosing the whole specification
a maximum of two levels of nested brackets is permissible.
   In writing FORMAT statements it is possible to specify
more fields for input or output than these values to be
transmitted, as in the sequence:
             WRITE (NOUT, 15) I,A,B,J,C
          15 FORMAT (I3,2(2F6.3,I2))
Five values I,A,B,J and C are to be output. The contents
of I are specified as I3, A and B each as F6.3, J as I2,
(the compound specification is then repeated) and C is in
F6.3. The FORMAT 15 is therefore larger than the output
list - an F6.3 and I2 remain to be satisfied. FORTRAN will
ignore these unsatisfied fields. This is a useful facility;
it enables programs to be written that allow variable
numbers of values to be transmitted in different runs.
   Alternatively, if the number of items in the input/output
list exceeds the specification of the FORMAT statement then
the numbers are output until the FORMAT is exhausted. To
input/output the remaining items in the list, the FORMAT
statement is automatically scanned from right to left and
a closing bracket is sought. If none is found the whole
specification is repeated, but if a closing bracket is
found the scan continues to the left to the next opening
bracket and to the field separator before the opening
bracket. The format specification is then repeated from
this field separator. An example shows this situation:
             WRITE (NOUT, 20)(A(I),I = 1,8)
          20 FORMAT (F6.2,2(F8.1,F9.2))
Eight values are to be output. The results will be:
            A(1)   F6.2
            A(2)   F8.1
            A(3)   F9.2
            A(4)   F8.1
            A(5)   F9.2
            A(6)   F8.1
            A(7)   F9.2
            A(8)   F8.1
If the FORMAT statement had been:
          20 FORMAT (F6.2,F8.1,F9.2,F8.1,F9.2)
the result would have been
            A(1)   F6.2
            A(2)   F8.1

```
            A(3)    F9.2
            A(4)    F8.1
            A(5)    F9.2
            A(6)    F6.2
            A(7)    F8.1
            A(8)    F9.2
```
because in the second case the scan would have returned to the beginning of whole format specifications.

## 10.3 Line Control Specifications

A second group of character combinations allows new output lines and pages to be started and allows items to be input off new cards. A slash, /, anywhere in a FORMAT specification indicates that a new line of output is to be started or input is from a new card. Slashes may be used in combination either by repeating the character several times

```
                WRITE (NOUT, 21)
             21 FORMAT (  /////)
```
or by bracketing
```
                WRITE (NOUT, 22)
             22 FORMAT ( 5(/))
```
In the above two examples five new lines are introduced into the output; in this case five blank lines are placed between any previous and all subsequent lineprinter output. If used on input they would signify that at this point five blank cards are present data in the data. Usually this new line character is used within mixed FORMAT specifications. In outputting results it could be used to separate table headings from numerical results as the following statements to output 12 values of population change from arrays A and B show:
```
          25 FORMAT(18H POPULATION CHANGE/8H 1951-61,
             *2X, 8H 1961-71/ 12(2F8.2/))
                WRITE (NOUT, 25)(A(I), B(I),I = 1,12)
```
A table of two columns would result from this statement sequence. The same result would be obtained by a revised FORMAT statement in which the last section was (2F8.2). The final / and the repeat integer 12 are optional. If omitted, the first passage will cause A(1) and B(1) to be output and the final closing bracket will cause a new line to begin. A reverse scan (*see above*) of the FORMAT specification will then begin and halt at the opening bracket before 2F8.2, two values in F8.2. This reverse scan will be repeated until all 12 pairs of values have been printed.

As seen in this example, when a slash is used to separate fields, a comma is not necessary. It must be remembered that each time the end of a FORMAT specification is reached a new record, card or line, is dealt with so that the final closing bracket of any FORMAT specification is also effectively a slash. The statement sequence:

      READ (NIN, 3∅)(A(I),I = 1,1∅)
      3∅ FORMAT (6F8.2)

will mean that the first 6 of the data elements to be read into A are contained on the first card and the second 4 elements on a second card.

  A second way of controlling the occurrence of blank lines is through a direct command to the carriage of the lineprinter rather than through using / characters. This direct command is carried on every output format statement. When a formatted record is prepared for printing the first character of the record is not printed but is used by the printer to determine the vertical spacing as follows:

    blank  one line
    ∅  two lines
    1  to first line of next page
    +  no advance

It will be recalled that a convention has been adopted in the use of H FORMAT statements in which a space is left between the letter H and the characters to be output. This convention has practical application in that succeeding WRITE statements appear on successive lines. Supposing for example a space was not left as in the following H FORMAT statement:

      25 FORMAT (19H1951-61 POP CHANGE )
      WRITE (NOUT, 25)

The result of omitting the space is that the first character is a 1 which signifies to the printer that the output is to appear on a new page. The characters 951-61 POP CHANGE would then appear heading a new page of output. The output has been wrongly positioned and the first character lost. This rule governing the vertical positioning of lines of output is also worth remembering when devising output formats in F or I mode. Outputting the number 100 in I3, when it is the first number in an output sequence, will cause a new page to be initiated. Similar problems can arise with A format mode. To overcome these problems, FORMAT statements may begin with a (1Hb sequence if normal printing is desired or (1H+ if overprinting is required. A corresponding sequence is taken for the other vertical spacing indicators. These commands can be used to good effect and in a positive fashion as shown in the program at the end of this chapter, to produce choropleth map plots. Correct vertical spacing is important in many programs producing graphical output but the control symbols sometimes cause unexpected problems in more conventional lineprinter output. Extreme care should be taken in the use of the vertical spacing symbols as mistakes can be very wasteful of paper. In particular the sequence 1H1 can occur in FORMAT statements enclosed within a DO loop. With a DO loop of say 50 items inclusion of 1H1 would produce for the programmer a vast quantity of waste paper, and probably a rebuke from the computer manager.

10.4 Alternative Output Channels

For the most part this chapter has dealt with the output of information to a lineprinter but much can relate equally well to output to a card punch. Instead of obtaining the results typed on paper it is possible at some installations to get results punched onto 80-column cards with the aim of using them for further processing. The output channel number held in NOUT would be different if card output is required but probably it is easier to set up a further location NPUN to contain the channel number of the card punch. NPUN can then be substituted for NOUT when results are required on the card punch. A card contains only 80 columns of punch positions but the usual lineprinter has at least 120 possible column positions. Because of this difference, a program sequence both to output a table to a lineprinter and to produce card output of the same numbers will probably need two format specifications. For example the sequence:

```
      DO 1 I = 1,1∅
      WRITE (NOUT, 5)(A(I,J),J = 1,15)
    1 WRITE (NPUN, 5)(A(I,J),J = 1,15)
    5 FORMAT (15F7.2)
```

to output the contents of array A to both lineprinter and card punch would result in card punched output in which only part (the first 3 characters) of the twelfth number appeared on the card. To produce usable card output a different FORMAT statement - say

```
   1∅ FORMAT (11F7.2/4F7.2)
```

would be necessary together with associated modification of the WRITE statement to

```
      WRITE (NPUN, 1∅)(A(I,J),J = 1,15)
```

For output onto other types of machinery, for example

a paper tape punch or teletype, other modifications to
FORMAT statements may become necessary to allow for the
different number of printing positions.

## 10.5 Input and Output of Arrays

Having considered the layout of input and output in some
detail we will now consider briefly what happens when
particular forms of input and output lists are used in READ
and WRITE statements.  A two-dimensional array can be
input or output in several ways which cause differences in
the numbers placed in a particular location or in the output sequence of values from memory locations. To appreciate fully the possibilities for array input and output
it is necessary first to consider how arrays are stored in
the computer memory.  Arrays are stored in ascending order
of their locations.  The value of the first subscript increases most rapidly and the value of the last increases
least rapidly.  For example if the array A(3,2) has two
dimensions, the first of three elements and the second of
two elements, it would be stored as follows
    A(1,1), A(2,1), A(3,1), A(1,2), A(2,2), A(3,2)
This array storage is shown further in Fig 10.2 which shows
the three-dimensional array A(3,3,3) with the rows, columns
and slices located as indicated.

Some methods of manipulating arrays have been mentioned
in Chapter 8.  If the transfer of a whole array is intended the simple command
           WRITE (NOUT, 10) A
where 10 is a suitable format would result in the complete
contents of array A being output.  The output sequence
would be that shown above the first output of all the rows
for column 1 and then for each succeeding column.  The
same result could be obtained by using two implied DO loops
       WRITE (NOUT, 10)((A(I,J),I = 1,N),J = 1,NN)
where N and NN contain the row and column dimensions of the
array.  The short hand form assumes that the whole array
is filled with significant numbers and that the array
declared in the DIMENSION statement is not larger than that
actually used.  The implied DO loop version could also
have been written with normal DO loops as
           DO 1 J = 1, NN
           DO 1 I = 1, N
         1 WRITE (NOUT, 10)A(I,J)
In this latter case each item of the array would be written
on a separate line, because each element of the array is
effectively dealt with by a different WRITE statement.

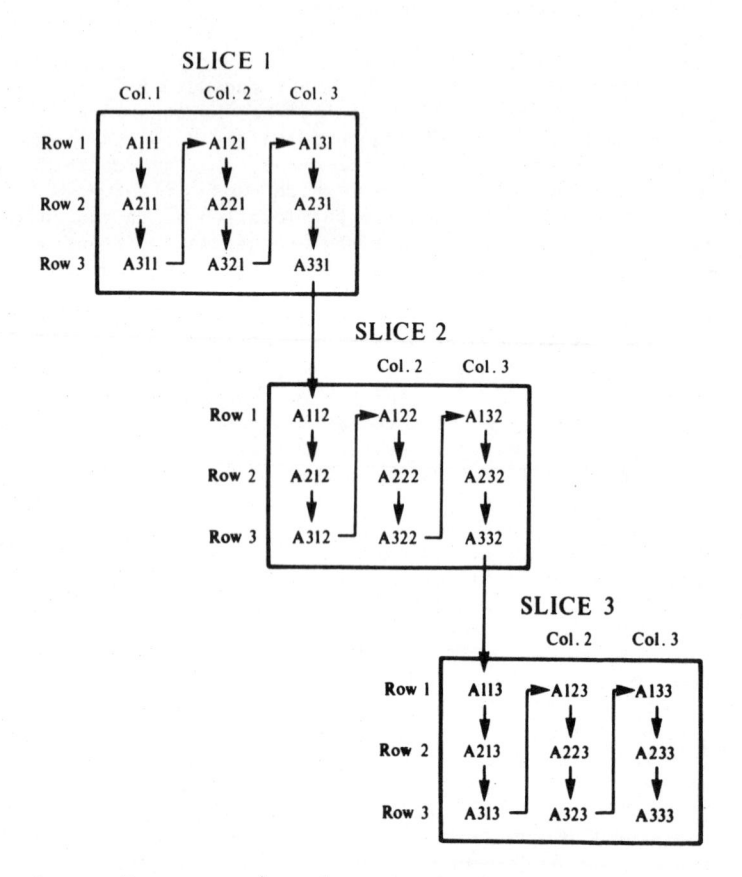

*Fig 10.2* Representation of array structure

With the implied loops case the whole array is dealt with by one WRITE statement and layout is completely controlled by the FORMAT statement.

A slight variant of the implied DO loops is shown in the following output statement
WRITE (NIN, 15)(((A(I,J),B(J,K),I = 1,4),K = 1,5),J = 1,4)
In this case data are transmitted as follows. First the

I loop is completed with K and J fixed at 1, ie
A(1,1); B(1,1); A(2,1); B(1,1); A(3,1); B(1,1); A(4,1); B(1,1)
K is then incremented and I varied again
          A(1,1); B(1,2); A(2,1); B(1,2) etc.
When the K incrementation is complete J is varied. It is
difficult to imagine a situation when this sort of layout
would be required! Quite complex input and output sequences
can be dealt with in implied DO loops involving more than one
array and controlled by several subscripts. A useful form
is a simpler version as in
          READ (NIN, 20)(A(I), B(I),I = 1,N)
in which pairs of values are dealt with. Programs involving the use of X and Y co-ordinates may utilise such an
input statement with each co-ordinate pair punched on individual cards. Each different type of array transfer has
its own particularly useful traits but it must be remembered
that the way the input is punched onto the card or the way
the print out appears on the page can be controlled to
a considerable extent but not entirely by the FORMAT statement. The FORMAT statement is an integral part of all
array transfer statements.

## 10.6 The DATA Statement

READ and WRITE statements are not the only methods of transferring data into a computer program. One of the simplest
statements in FORTRAN is the assignment of a constant value
to a memory location. A further way to introduce characters into a program is by the DATA statement. The general
form of the DATA statement is
          DATA $K_1/d_1/$, $K_2/d_2/$, ... $K_n/d_n/$
where K is a list of variable names and array elements and
d is a list of constants (either numeric or alphabetic or
consisting of special characters). The DATA statement provides an alternative to a whole sequence of simple arithmetic
statements at the start of a program and is slightly more
efficient when it is executed. It also allows particular
symbols, perhaps letters or special characters, to be entered
into locations, for example the plotting symbols are input
by the DATA statement in the worked examples which follow.
A series of examples will serve to show the various possible
forms of the DATA statements:

|      | Statement | Effect |
| --- | --- | --- |
| (i)  | DATA A/17.2/ | transfers constant 17.2 to A |
| (ii) | DATA A,B,C/17.2, 17.2, 15.0/ | transfers 3 constants to A,B and C respectively |

|   | Statement | Effect |
|---|---|---|
| (iii) | DATA A,B,C/2*17.2, 15.0/ | same effect as (ii) (note use of * for multiple entries) |
| (iv) | DATA A,B/2*17.2/, C/15.0/ | same effect as (ii) |
| (v) | DATA A,I,B/17.2, 5,17.2/ | includes allocation of integer number |
| (vi) | DIMENSION X(2,10) DATA X(1,4), X(2,5), X(1,1)/3*0.0/, X(2,4), X(2,10)/ 2*1.0/ | allocates real values to specified array locations |
| (vii) | DIMENSION Y(5) DATA Y/1.0,2.0, 3.0,4.0,5.0/ | allocates real values to all locations in array Y |
| (viii) | DIMENSION V(5), C(2,30) DATA (V(I),I = 1,4)/ 1.0,2.0,3.0,4.0/, (C (I,J),J = 1,30), I = 1,2)/30*0,30*10/ | allocates values to groups of array elements and uses implied DO loops |
| (ix) | DATA NP/1H+/, NM/1H-/, NMU/1H*/, ND/1H// | allocates special symbols +, -, *, and / to integer storage locations |
| (x) | DIMENSION M(5) DATA M/1HA, 1HB, 1HC, 1HD, 1HE/ | allocates letters to locations |
| (xi) | DIMENSION J(5) DATA (J(I),I = 1,3)/ 3HYES, 2HNO, 3Hb/, /, X/20.4/, I/5̄7, NO/3HYES/ | allocates different symbol types to different locations all within one DATA statement |
| (xii) | DATAbFMTb̲/9Hb̲(7F8.2 )// READ (NIN,FMT) (A(I),I = 1,7) | allocates a format specification to a particular location. The location name may then be used in an input-output comment. Note the different meaning of the final two // in the data statement |

A number of rules are implicit in these examples. In particular five major rules govern the use of DATA statements.

(i) No variable may appear more than once in a definition.

(ii) The mode of each variable in the list must match that of the corresponding constant. Alphabetic and special character constants may be attached to either real or integer variables.
(iii) If all the elements in an array are to be defined then the array name may be used without subscripts. If this form is used the same limitations occur as with the use of this form in input-output lists. There must be as many constants as there are array elements and the order of assignment is the same as the order of array storage (leftmost subscripts varying most rapidly - *see Fig 10.2*).
(iv) A variable subscript for an item in the variable list may only be used within an implied DO loop.
(v) All subscripts must be in one of the forms
c, c*i, c*i±c, i, i±c
where c are unsigned integer constants and i is the index in an implied DO loop.

## 10.7 Alternative Reference to FORMAT Specifications

The DATA statement may be used to define a FORMAT specification as in the final example given above. The variable to which the allocation is made may then be substituted in the READ or WRITE statement in the position used for the FORMAT statement label. The statement sequence

DATA A/6H (2I3)/
READ (NIN, A) J,K

would first transfer to location A the format specification, complete with brackets and spaces. The format specification, held in variable A, is then transferred to the READ command and data values would be input to locations J and K.

A second method of holding format specifications in variable locations is by the use of A format. The format specification, complete with spaces and brackets, is input as data into an array. The array is then used in the READ/WRITE statement again in place of the format label. An example would be the following program sequence:

DIMENSION A(5), I(5)
READ (NIN, 1∅) A
1∅ FORMAT (5A4)
READ (NIN, A) I

with the data as:

(5I3)
001002003004100

The format specification effectively only takes up the
first locations of A with the remaining 3 being filled with
blanks. Inputting format specifications in this way is
useful for programs that are going to be used several times
with data of different format or in the production of
general purpose programs likely to be utilised by several
people. It is an advantage to have input formats as flexible
as possible to accept a wide variety of data layouts. One
way of achieving this flexibility is to input the format
specifications as data. There are two points to notice
in this connection. The first is the differing ability of
computers to hold different length A format. Usually it
is possible to use A4 but to be sure of complete generality
A1 should be used. Secondly, it is advisable to set to
zero the whole array which is to hold the format specification. This stops any characters accidentally left in
the array from becoming mixed with the desired effective
format specification. It is also good practice to leave
blank the first character in the A mode format.

## 10.8 Summary

In this review of more advanced input-output techniques
a number of new statements have been introduced. The
complete range of input-output statements has not been
covered but those not mentioned are statements relatively
infrequently encountered by geographers using the computer.
The following worked examples show some of the ways these
more advanced input-output forms are used in geographical
programs. Additional format modes discussed are:
      X  signifying spaces
      A  used for transfer of text
      E  used for numbers in exponent form
and these modes together with I, F and H mode specifications may be repeated by the use of brackets within the
format specifications. The layout of input or output may
be controlled by orders to the lineprinter and contained
in the first character of an H mode format as follows:
    1H$\underline{b}$  (where $\underline{b}$ signifies a blank) normal spacing
    1HØ  double spacing
    1H1  move to first line of next page
    1H+  overprinting
The chapter also includes details of the use of the DATA
statement which is used to assign constant values (either
numeric, alphabetic or special character) to storage
locations.

10.9 Worked Examples
(i) *Problem.* Whether as a precursor to choropleth mapping or to study their statistical form the geographer frequently needs to turn continuous data into frequency data. A useful graphical representation of frequency data is the histogram. The problem solved in this worked example is the preparation of a histogram.
*Algorithm and flowchart.* The algorithm classifies input data into specified classes and outputs a histogram with a * representing a specified number of data values. The flowchart is shown in Fig 10.3. Data frequency class intervals and various controls are input, data are classified and the histogram is output.
*Program.* The program listing is shown in Fig 10.4. The size of the arrays to be used is specified in the DIMENSION statement. A DATA statement is used to set array A to contain the symbol *. The title of the project in A format constitutes the first data card. The number of data values, and the format to be used to input the data are then read from the same card. The format specification is input into the array FMT using A mode. The array FMT is then used in the following READ statement to input the data. The array FMT is again used in a READ statement to input the frequency class dividers but not before its contents have been reset with the different format specification. After the class divider values the final input is two control variables; the first specifies the number of data values to be represented by one symbol and the second specifies the level of accuracy of the data. These two values are necessary as it may be important to scale the histogram if a large data block is input and additionally output can be clarified if the level of accuracy of the data is known.

With all the data residing in the computer, the scores are then sorted into classes in the nested DO loops 1∅∅ and 1∅5 with the number of data in each class being accumulated in array NC. Arbitrary limits of 5∅ classes and 999 data values have been imposed in the program but these can easily be changed by alteration to the type specification statements. The remainder of the program outputs the histogram line by line in DO loop 115. A check, however, is made that a datum occurs in a class before the corresponding line is printed. The FORMAT statements are collected together at the end of the program.

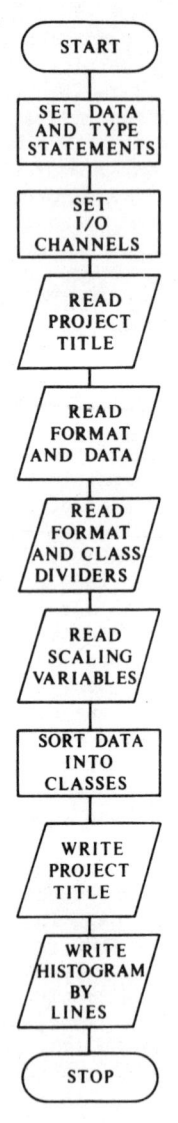

*Fig 10.3* Flowchart for a program to produce a histogram

```
C PROGRAM TO PRODUCE A HISTOGRAM OF A LIST OF DATA                  10
C                                                                   20
      DIMENSION CL(50),X(20),D(999),NA(80),FMT(10),NC(50)            30
      DATA NA/80*1H*/                                                40
      NIN = 5                                                        50
      NOUT = 6                                                       60
C CLEAR FMT PRIOR TO USE - NOT NEEDED ON ALL SYSTEMS                 70
      DO 75 I = 1, 10                                                80
   75 FMT(I) = 0.0                                                   90
C READ PROJECT TITLE INTO THE WHOLE OF ARRAY X                      100
      READ(NIN,30) X                                                 110
C READ NO OF DATA VALUES AND THEIR FORMAT IN I3,10A4                120
      READ(NIN,10) N,FMT                                             130
C READ DATA TO BE GROUPED FROM SUCCESSIVE CARDS                     140
      READ(NIN,FMT)(D(I),I=1, N)                                     150
C CLEAR FMT PRIOR TO USE - NOT NEEDED ON ALL SYSTEMS                160
      DO 70 I = 1, 10                                                170
   70 FMT(I) = 0.0                                                   180
C READ NO OF CLASS INTERVALS AND FORMAT IN WHICH THEY ARE PUNCHED ON 190
C THE FOLLOWING CARD                                                200
      READ(NIN,10) NN,FMT                                            210
C READ CLASS DIVIDE POINTS                                          220
      NNN= NN +1                                                     230
      READ(NIN,FMT)(CL(I),I=1,NNN)                                   240
C READ NO OF VALUES TO BE REPRESENTED BY ONE HISTOGRAM SYMBOL AND   250
C LEVEL OF ACCURCAY OF DATA IN I3,F8.5                              260
      READ(NIN,15) NSCALE, VALUE                                     270
C SET THE COUNTING LOCATIONS AT ZERO                                280
      DO 80 I = 1,NN                                                 290
      NC(I) = 0                                                      300
   80 CONTINUE                                                       310
      DO 100 I = 1,N                                                 320
      DO 105 J = 1, NN                                               330
      IF ( D(I) .GE. CL(J) .AND. D(I) .LT. CL(J+1)) GO TO 95         340
  105 CONTINUE                                                       350
   95 NC(J) = NC(J) + 1                                              360
  100 CONTINUE                                                       370
C OUTPUT TITLE FOLLOWED BY HISTOGRAM REQUIRED                       380
      WRITE(NOUT,20) X                                               390
      DO 115 J = 1, NN                                               400
      NM = NC(J) / NSCALE                                            410
      CLJ = CL(J+1) - VALUE                                          420
      IF ( NM .NE. 0) WRITE(NOUT,25) CL(J),CLJ,NC(J),(NA(I),I=1,NM)  430
      GO TO 115                                                      440
      WRITE(NOUT,25) CL(J),CLJ,NC(J)                                 450
  115 CONTINUE                                                       460
      WRITE(NOUT,40)    NSCALE                                       470
      STOP                                                           480
   10 FORMAT(I3,10A4)                                                490
   15 FORMAT(I3,F8.5)                                                500
   20 FORMAT(1H1 //15X,20A4)                                         510
   25 FORMAT(1H ,F7.2,3H I ,F7.2,I3,2H I,80A1)                       520
   30 FORMAT(20A4)                                                   530
   40 FORMAT(1H0,5X,23H EACH SYMBOL REPRESENTS,I6,12H DATA VALUES)   540
      END                                                            550
SUPERMARKET EMPLOYEE PRODUCTIVITY IN LARGE TOWNS IN DENMARK
051  ( 16F5.1)
343,2308,0294,7226,4241,0202,3164,1201,0191,0177,1413,5130,7271,7142,1258,8189,0
136,8200,4215,0293,6374,1169,4230,6247,3181,8178,7225,1237,2196,7259,7148,9267,9
135,7205,7142,1212,8341,4239,7220,1205,5307,9309,1289,0419,7297,1312,4228,1315,4
386,8250,6251,2
007  ( 8F4.0)
000,150,200,250,300,350,400,450,
00101,
```

Fig 10.4 (A) Program listing and (B) data for a program to produce a histogram

233

*Data and run.* The data are a series of scores, for large
towns in Denmark, relating to the average level of
employee productivity in supermarkets in each town. The
data are sales per employee measured in '000kr, and are to
be grouped into 7 classes. The full data series and
results for the program are shown in Fig 10.5. The output

```
              SUPERMARKET EMPLOYEE PRODUCTIVITY IN LARGE TOWNS IN DENMARK
    0.00 I  149.00   6  I******
  150.00 I  199.00   8  I********
  200.00 I  249.00  16  I****************
  250.00 I  299.00  10  I************
  300.00 I  349.00   7  I*******
  350.00 I  399.00   2  I**
  400.00 I  449.00   2  I**
         EACH SYMBOL REPRESENTS      1 DATA VALUES
```

*Fig 10.5* Histogram of supermarket employee productivity
of Danish towns

begins with the title of the project and then shows for
each class its range, the number of data in the class and
finally the histogram for the class. The reason for
having the accuracy of the data specified in the data is
seen in the printing of the class ranges. The results
show the modal class to be 200-249 thousand kroner but
that a considerable range of values exists. The next step
in an analysis could well be the plotting of this range of
values onto a map of Denmark.

(ii) *Problem.* In spite of the increasing application of
sophisticated statistical models in geography, a great
deal can still be learnt about phenomena simply by mapping
their occurrence over space. One of the most common
techniques of mapping involves the use of *choropleth*
symbols. Normally 6 or 7 shades or symbols will be used.
The production of these maps is a tedious and time-
consuming job that is well suited to computerisation, and
this case study presents a simplified program to do the
work. It will produce shaded maps on a standard lineprinter with 7 shades applied at standard deviation
intervals for up to 10 variables measured over a maximum
of 40 data zones, and illustrates a number of advanced
input-output features such as character input and output
in A format and the use of carriage control to produce
overprinting.

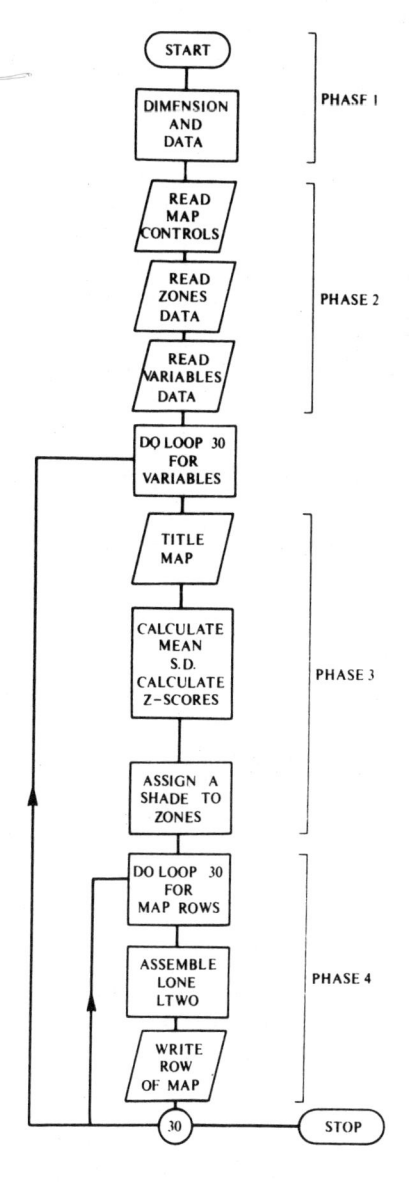

*Fig 10.6* Flowchart for a program for choropleth mapping

The use of the program is illustrated by a simple exercise in which six variables associated with the 'Welshness' of Wales (Bowen, 1959; Carter and Thomas, 1969), measured over the lattice of the 13 counties that existed prior to 1974, are mapped but the program is likely to be of more general application than this.
*Algorithm and flowchart.* A summary flowchart to perform the necessary operations is presented in Fig 10.6. The flowchart gives rather less detail than those given previously but has been designed to show the major strcutural units of the algorithm. These are the DIMENSION and DATA statements, the input of the data to be mapped, the assignment of shading categories to each print position and the final output of the complete map.
*Program.* The program listing is shown in Fig 10.7. The first block of program specifies the storage required and has a DATA statement which gives the characters that are to be used to shade the maps. The functions of the various arrays that appear in these statements are critical to an understanding of the program:

| | |
|---|---|
| NAMES(10,20) | - holds alphanumeric information that gives the names of each of up to 10 variables to be mapped. Each name is held in the 20 storage locations that make up a row of this array in the FORMAT(20A4). There can thus be up to 80 characters in each name. |
| LONE(100) and LTWO(100) | - are designed to hold a line of output of the finished map in the FORMAT(100A1). |
| D(40,10) | - holds the original data to be mapped. Each zone occupies a *row* and each variable a column of this array. |
| MAP(100,100) | - MAP is a very important feature of the algorithm. In it are held a series of numbers giving the reference numbers of the zones to be mapped *as they will occur* on the completed lineprinter map. There may be up to 100 print positions across the map and up to 100 lines down it. |
| NP(7),MP(7) | - These arrays hold the characters used in plotting the various shade levels on the completed map. Their contents are assigned, using a DATA statement, as: |

```
C A PROGRAM TO CHOROPLETH MAP AREA DATA
C DATA INPUT IS DESCRIBED IN THE PROGRAM AND IN THE TEXT
C
      DIMENSION NAMES(10,20),NCAT(40),NP(7),LONE(100),LTHO(100),
     1    MAP(100,100),D(40,10),NCHAR(16),NZ(16),HP(7)
C THESE ARRAYS HOLD THE RUN-TIME FORMAT INFORMATION
      DIMENSION VFT1(20),VFT2(20)
C DATA STATEMENTS THAT SPECIFY THE PLOTTING SYMBOLS TO BE USED
      DATA NP/1H ,,1H=,1H+,1HM,1HO,1H=,1H*/
      DATA MP/1H ,1H ,1H ,1H ,1H=,1HO,1HO/,NONE/1H /
C SET INPUT AND OUTPUT CHANNELS
      NIN = 5
      NOUT = 6
C DATA INPUT STARTS HERE
C FIRST DATA CARD CONTAINING THE FOLLOWING INFORMATION
C      READ NO OF ZONES (NZONE), NO OF VARS (NVAR),NO OF COLUMNS
C      IN THE MAP (NCOL) AND NO OF ROWS (NROW) IN FORMAT 4I3
      READ(NIN,90) NZONE,NVAR,NCOL,NROW
      Z = NZONE
C SECOND DATA CARD GIVES THE FORMAT FOR THE STRIPLINE CARDS THAT
C SPECIFY THE LAYOUT OF THE AREAS TO BE MAPPED
      READ(NIN,92) (VFT1(I),I = 1,20)
      DO 10 I = 1,NROW
      NLIN = 0
C THIRD AND SUCCESSIVE CARDS GIVE THE AREA LAYOUT
      READ(NIN,VFT1) (NCHAR(J),NZ(J), J = 1,16)
      MCOL = 1
C ASSIGN ZONE NUMBERS TO THE MAP ARRAY
      DO 14 J =1,16
      NLIN = NLIN + NCHAR(J)
      NPOSIT = NCHAR(J) + MCOL -1
      DO 16 K = MCOL,NPOSIT
   16 MAP(I,K) = NZ(J)
      MCOL = NPOSIT + 1
   14 CONTINUE
C CHECK THAT MAP INPUT LINE HAS REQUISITE NUMBER OF CHARACTERS
      IF (NLIN .NE. NCOL ) WRITE(NOUT,98) I
   10 CONTINUE
C READ IN VARIABLE NAMES ONE PER CARD IN 18A4
      DO 20 I = 1,NVAR
   20 READ(NIN,92) (NAMES(I,J),J = 1,20)
C READ IN FORMAT CARD FOR THE DATA THEMSELVES
      READ(NIN,92) (VFT2(I),I = 1,20)
C READ IN DATA CARDS
      DO 25 I = 1,NZONE
   25 READ(NIN,VFT2) (D(I,J),J = 1,NVAR)
C END OF DATA INPUT
C
C ENTER PHASE TWO OF THE PROGRAM *************
C
      DO 30  I = 1, NVAR
C OUTPUT MAP TITLE
      WRITE(NOUT,94)(NAMES(I,J),J = 1,20 )
C ASSIGNMENT OF SHADING CATEGORIES TO THE AREAS
C DETERMINE MEAN AND STANDARD DEVIATION FOR THIS VARIABLE
      SUM = 0.0
      SUMSQ = 0.0
      DO 40 J = 1, NZONE
      X = D(J,I)
      SUM = SUM + X
   40 SUMSQ = SUMSQ + X*X
      AVERAG = SUM /Z
      STDEV = SQRT ((SUMSQ/Z) - AVERAG*AVERAG)
      WRITE(NOUT,95) AVERAG, STDEV
C ASSIGN SHADES TO ZONES
      DO 50 J = 1, NZONE
      X = (D(J,I) - AVERAG )/STDEV   + 3.5
      NX = INT(X) + 1
      IF ( NX .GT. 7 ) NX = 7
      IF ( NX .LT. 1 ) NX = 1
   50 NCAT(J) = NX
```

237

```
      C OUTPUT LINES OF MAP                                            720
            DO 30 K = 1, NROW                                          730
            DO 80 M = 1,NCOL                                           740
               MZONE = MAP(K,M)                                        750
               IF(MZONE .NE. 0 ) GO TO 70                              760
               LONE(M) = NONE                                          770
               LTWO(M) = NONE                                          780
               GO TO 80                                                790
         70 NSHADE = NCAT(MZONE)                                       800
               LONE (M) = NP(NSHADE)                                   810
               LTWO(M) = MP(NSHADE)                                    820
         80 CONTINUE                                                   830
      C LONE HOLDS THE FIRST PRINTER LINE, LTWO THE OVERPRINT CHARACTERS 840
            WRITE(NOUT,96) (LONE(J),J=1,NCOL)                          850
            WRITE(NOUT,97) (LTWO(J),J = 1,NCOL)                        860
         30 CONTINUE                                                   870
            STOP                                                       880
         90 FORMAT(4I3)                                                890
         92 FORMAT(20A4)                                               900
         94 FORMAT(1H1,20A4)                                           910
         95 FORMAT(1H0,8H MEAN = ,F7.3,21H STANDARD DEVIATION =,F12.3//) 920
         96 FORMAT(1H ,100A1)                                          930
         97 FORMAT(1H+,100A1)                                          940
         98 FORMAT(45H **WARNING**INCORRECT NO OF CHARACTERS ON ROW ,I5) 950
            END                                                        960
```

*Fig 10.7* Program listing for a program for choropleth mapping

| Array | Location | 1 | 2 | 3 | 4 | 5 | 6 | 7 |
|---|---|---|---|---|---|---|---|---|
| NP |  | . | − | + | M | O | = | * |
| MP |  |   |   |   | − | 0 | 0 |   |

At output these characters are superimposed on each other to give a graded series of seven levels:

. − + M θ θ ℮

The remaining arrays are less critical to an understanding of the program and will be dealt with as they are used.

The second major block in the program reads in the data to be mapped. The first READ, according to the FORMAT (4I3), takes in values for the variables NZONE = number of zones for which data exist, NVAR = number of variables to be mapped, NCOL = number of columns of print in final map and NROW = number of lines of output in each map. The next series of cards specify the layout of the data zones. In order to produce these cards the outlines of the area and the boundaries of the data zones are laid out at a suitable scale on a grid similar to that illustrated in Fig 10.8. The grid is drawn to correspond to the print positions on a standard lineprinter so that one cell represents a single print position and there are 10 such

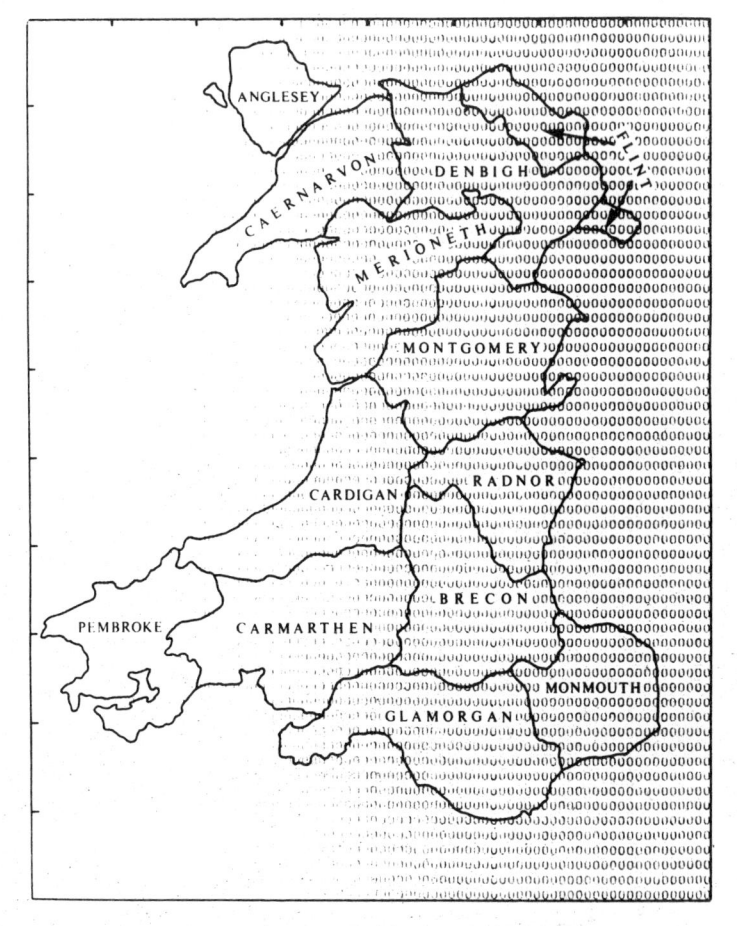

*Fig 10.8* Grid for data preparation for choropleth program

positions per inch across the page and 6 lines per inch down it. Each zone is now given a number, with 0 denoting sea or other blanks on the map, and this number is recorded in the appropriate cell of the grid as illustrated. Inevitably this digitisation process involves some generalisation of the zone boundaries but at the end of the process we have an array of numbers up to 100 x 100 which can

239

act as a key for the mapping process. In the program it
is the array called MAP and is read into the machine in
a shorthand form. Beginning at the upper map edge, each
row of MAP is read off a single card on which are
successive pairs of values for NCHAR and NZ. NCHAR is the
number of characters and NZ the code number assigned to the
zone. Thus the first row of the map in the illustration
has 80 blanks and would be specified
      80 0
The fifth row (card 5) has 21 blanks followed by 2 1's
which correspond to Holyhead Island, 1 blank, 9 more 1's
(Anglesey), 8 blanks, 3 6's corresponding to Denbighshire,
7 blanks, 6 7's (Flint) and ends with 23 blanks. In the
data this would be coded
  21 0 2 1 1 0 9 1 8 0 3 6 7 0 6 7 23 0.
All this sounds, and is, rather tedious to prepare but
once a set of zones has been coded in this way the cards
can be used time and time again to map any data sets that
have been collected on the same areal units. The read
statement to input these cards is
   READ (NIN,VFT1)(NCHAR(J),NZ(J),J = 1,16)
This type of *implied DO loop* we have met before, but the
format number looks odd. In its position we have written
VFT1 and three statements higher in the program will be
found the command:
   READ (NIN, 92)(VFT1(I),I = 1,20)
   92 FORMAT ( 20A4 )
What is happening here is that immediately before reading
some cards the format for these cards is input as data
(*see section 10.7*).

The instructions which follow these inputs reconstruct
the array MAP and at the same time count the number of
characters in each line as a check on the accuracy of the
data. The next series of cards to be read are again
composed of alphanumerics, in this case NVAR cards each of
which has punched on it the name of a variable to be mapped
in 20A4 format, and finally the data array D is input
according to the format information specified in VFT2 which
is again read in as data. There is at least one card per
zone and variables must be in the same numeric order as the
original codes used for MAP and the variables names cards
respectively.

These input operations having been completed the program
now takes each variable in turn and performs the following
operations:

(i) Write out the variable name using FORMAT statement
94 which reads:
94 FORMAT (1H1, 2∅A4)
It will be seen that the carriage control character
1 is used to start off a new page of printer paper.
(ii) Calculate the mean and standard deviation for the
variable and print these out.
(iii) For each zone a standard score on the variable is
found using the relation

$$z_i = \frac{(x_i - \bar{x})}{SD_x}$$

$z_i$ = standard score
$x_i$ = observed score
$\bar{x}$ = mean of variable x
$SD_x$ = standard deviation of x

Whatever the initial data values the standard scores
will have zero mean and unity standard deviation and
are in effect the number of standard deviation units
a particular datum is above or below its overall
mean but these values cannot be directly turned into
mapping categories. In order to do this we must
remove the negative values and convert the results
into positive integers that correspond to the 7
available shades. This is done by adding 3.5 to
all values and then using the function INT to
truncate them to the highest integer value less than
the numbers. Finally, to remove zeros a one must
be added to all the resulting integers. Thus
a standard score of -2.354 indicating a very low
value more than two standard deviations below the
mean would be converted as follows:

Adding 3.5 ... -2.354 + 3.5 = 1.15
Using INT ...                = 1
Adding 1   ...     1 + 1     = 2

indicating that it should be assigned to the second
shade category (-). Taken collectively these
statements will produce the following categories
and shadings:

| Standard deviation range | Category | Shade |
|---|---|---|
| greater than 2.5 | 7 | ● |
| 1.5 to 2.49 | 6 | ● |
| 0.5 to 1.49 | 5 | ⊖ |
| +0.5 to -0.5 | 4 | M |
| -0.51 to -1.5 | 3 | + |

241

| Standard deviation range | Category | Shade |
|---|---|---|
| -1.51 to -2.5 | 2 | - |
| less than -2.5 | 1 | . |

The shade numbers for each of the NZONE zones are stored in the array NCAT by the assignment statement labelled 50.

(iv) Finally the computer map is assembled and printed line by line using the information in the arrays MAP and NCAT. MAP gives the location of the zones according to each print position and NCAT gives the category of shading to be used for each zone. Each line of output is stored in the integer arrays LONE and LTWO into which are placed the appropriate characters from the arrays NP which is used to fill LONE and MP, used for the overprint characters held in LTWO. LONE is written out under the control of

96 FORMAT (1H , 1∅∅A1)

which specifies a line feed followed by 100 alphanumerics and LTWO is then superimposed on it using the statement

97 FORMAT (1H+, 1∅∅A1)

The carriage control character + suppresses the normal line feed that occurs between individual WRITE statements.

*Data and run.* The program was run using the data in Fig 10.9 relating to the 13 old counties of Wales. A typical output map, that for percentage Welsh speaking at the 1961 census, is shown in Fig 10.10. There is an obvious and very strong concentration of high values in the counties of Anglesey, Caernarvon, Merioneth, Cardigan and Carmarthen which are remote from the English border, surrounded by a zone of intermediate values at or near the overall mean and with a periphery of values less than this mean. Carter and Thomas (1969) suggest that these represent the 'core', 'domain' and 'sphere' of the Welsh culture region as would be expected from Meinig's general model of culture regions (Meinig, 1965). Comparison with the other maps (not shown) also reveals that the core counties are non-urban, non-manufacturing and have a high proportion of their land devoted to rough grazing. They are also the counties that in 1968 voted to retain Sunday closing of licensed premises.

```
   13   6  80  60
(16(12,13))
80   0
80   0
24   0  7  149   0
24   0  9  122   0  2  723   0
21   0  2   1  1   0  9   1  8   0  3   6  7   0  6  723   0
22   0  1   1  1  012   1  4   0  3   3  8   6  9  720   0
25   0  9   1  2   0  7  310   6  9   7  1   0  1  716   0
25   0  7   1  1  010  313   6  9  715   0
27   0  3   1  1  013  313   6  8  715   0
30  013  316   6  5   7  2  614   0
29  014  317   6  2   7  5  613   0
29  015  324  612   0
27  011   3  1   9  5   3  8   6  3  912   6  1  712   0
25  011   3  6   9  1   3  1   6  5   9  1   6  7   9  8   6  6   7  9   0
22  013  322   9  6   6  4   0  4   7  9   0
21   0  7   3  6  020   9  7   6  8   0  1  710   0
19   0  6   3  9  017   9  3  11  6  620   0
18   0  3   3  2   0  2   3  9  014   9  8  11  4  620   0
35  013  913  1119   0
35   0  2   9  1  018  916  1116   0
37  011  916  1116   0
35   0  7  921  1117   0
34   0  8  920  1118   0
35   0  6  920  1119   0
35   0  3   9  2  421  11  1   0  2  1116   0
36   0  6  421  1117   0
35   0  9  410  11  4  1322   0
35   0  9   4  7  11  7  1322   0
34  013  415  1318   0
33  012  420  1315   0
32  013  419  1316   0
30  014   4  3   2  1  13  3  212  1317   0
27  017   4  8  218  1318   0
24  020   4  9   2  8  1319   0
23  021  410   2  7  1319   0
18  026  410   2  7  1319   0
16   0  2  1215   4  2   5  1   4  8  510   2  6  1320   0
14   0  6  1211  414  511   2  3  13  2  219   0
 7   0  3  12  1  012  1223  516  218   0
 7  016  1222  517  218   0
 3  019  1223  517   2  1  1017   0
 1  016  1227  518   2  8  1010   0
 5  012  1227  519  210  10  7   0
 6  013  1224  519  211  10  7   0
 6   0  6  12  2   0  6  12  5   5  3  011   5  5   0  1   2  3   8  6   2  3  817  10  6   0
 3  010  12  1   0  5  12  8   0  9  522  816  10  6   0
 4   0  2  12  6   0  6  1210   0  8  523  815  10  6   0
 8   0  9  1217  025  815  10  6   0
 9   0  4  1216  010   8  3  017  614  10  7   0
29   0  8   8  7  018   8  9  10  9   0
31   0  2  011  010   8  2  1016   0
45  017  818   0
48  013  819   0
49  010  821   0
54   0  1  825   0
80   0
80   0
80   0
80   0
80   0
PERCENTAGE WELSH SPEAKING AT 1961 CENSUS
PERCENTAGE VOTING TO RETAIN SUNDAY CLOSURE OF PUBS 1968
PERCENTAGE EMPLOYED IN AGRICULTURE 1961 CENSUS
PERCENTAGE EMPLOYED IN MANUFACTURE 1961 CENSUS
PERCENTAGE LAND IN ROUGH GRAZING
PERCENTAGE LIVING IN TOWNS OF POP GREATER THAN 10000
(6F6.2)
75,50  66,10  16,30  10,50  12,60  20,10
28,10  31,30  13,20  22,30  60,00   0,0
68,30  61,90   8,50  12,20  59,40  35,40
74,80  64,40  25,80   2,10  36,80  19,20
75,10  53,80  13,00  19,70  22,30  25,70
34,80  40,10   8,60  18,10  33,30  33,70
19,00  20,40   5,90  42,80   9,20  30,80
17,20  28,70   1,10  27,80  40,60  81,00
75,90  66,20  14,20   6,90  71,70   0,0
 3,40  10,30   2,20  33,90  18,60  76,60
32,30  40,60  30,50   9,40  40,70   0,0
24,40  42,90  14,20   5,80  18,00  28,00
 4,50  22,90  34,60   3,30  42,30   0,0
```

*Fig 10.9* Data for the choropleth program

*Fig 10.10* Percentage of population which is Welsh speaking
as output from the choropleth program
Mean of distribution is 41.023
Standard deviation is 27.472

## 10.10 Worksheet

1 Which of the following FORMAT statements are invalid, and why?
- (i) FORMAT (6I2,3F4.1,2(I2,2X))
- (ii) FORMAT (6I3,6F3)
- (iii) FORMAT (3E16.2,E14.9)
- (iv) FORMAT (3F6.2,6(I3,A40)
- (v) FORMAT (2(I6,2X,I7)3(I2))
- (vi) FORMAT 8I2/6F8.3//2I
- (vii) FORMAT (//2I3//)
- (viii) FORMAT (2F3.2,2I6.1)
- (ix) FORMAT (E16.2,3HAB(/(F5.2,2I1)
- (x) FORMAT (X99,2F6.1,6I6)

2 Write a program to calculate and print out population pyramids given raw data in the following form. Make the program print out as in the partial pyramid shown in Fig 10.11.

Population of Greenland in 1965

| Age | Males | Females |
|---|---|---|
| 0- 4 | 3877 | 3760 |
| 5- 9 | 3222 | 3048 |
| 10-14 | 2227 | 2152 |
| 15-19 | 1551 | 1477 |
| 20-24 | 1637 | 1527 |
| 25-29 | 1769 | 1470 |
| 30-34 | 1537 | 1276 |
| 35-39 | 1260 | 1124 |
| 40-44 | 831 | 780 |
| 45-49 | 708 | 642 |
| 50-54 | 568 | 580 |
| 55-59 | 426 | 501 |
| 60-64 | 328 | 377 |
| 65-69 | 225 | 256 |
| 70-74 | 126 | 166 |
| 75-79 | 35 | 64 |
| 80-84 | 14 | 27 |
| 85-89 | 1 | 11 |
| 90-94 | 2 | 0 |
| unknown | 10 | 8 |

3 Write a program to produce a scatter diagram, with appropriately scaled axes, of the following x,y data. Design your program additionally to express the expenditure in per-head terms and replot the scatter diagram.

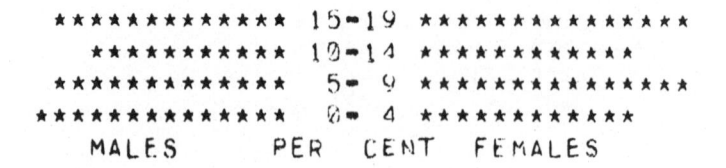

```
************ 15-19 ***************
************ 10-14 **************
************** 5- 9 ****************
*************** 0- 4 **************
     MALES      PER  CENT  FEMALES


        SCALE : ONE * EQUIVALENT TO .....
```

*Fig 10.11* Sample output for program to calculate population pyramids

Amount spent on health and housing by county councils (excluding urban areas) in the Republic of Ireland, 1962-3.

|                   | Amount spent (£) |          |            |
|-------------------|------------------|----------|------------|
| County authority  | Health           | Housing  | Population |
| Carlow            | 263357           | 84580    | 25634      |
| Cavan             | 503803           | 91285    | 53386      |
| Clare             | 654718           | 97156    | 65142      |
| Cork              | 937654           | 712469   | 222395     |
| Donegal           | 755555           | 181269   | 105227     |
| Dublin            | 317431           | 1003803  | 133092     |
| Galway            | 1486350          | 129808   | 122148     |
| Kerry             | 920554           | 151853   | 96051      |
| Kildare           | 498202           | 173070   | 56555      |
| Kilkenny          | 595181           | 126648   | 51509      |
| Laoghis           | 394175           | 88845    | 45069      |
| Leitrim           | 271620           | 34487    | 33470      |
| Limerick          | 334891           | 202166   | 82553      |
| Longford          | 296934           | 56563    | 27085      |
| Louth             | 574149           | 76605    | 30503      |
| Mayo              | 916048           | 89169    | 108939     |
| Meath             | 549460           | 186272   | 57560      |
| Monaghan          | 450369           | 52104    | 36901      |
| Offaly            | 408539           | 89571    | 42069      |
| Roscommon         | 571716           | 102771   | 59217      |
| Sligo             | 459523           | 39627    | 40416      |
| Tipperary N.R.    | 439046           | 114412   | 41179      |
| Tipperary S.R.    | 626578           | 93706    | 47451      |
| Waterford         | 254122           | 92168    | 38035      |
| Westmeath         | 485117           | 113141   | 43237      |
| Wexford           | 658943           | 106122   | 61732      |
| Wicklow           | 385315           | 102761   | 38270      |

(*Hint*. Set up a two-dimensional array to represent the whole graph area and insert symbols at appropriate row by column locations. See Day, 1972)

4 Write a program to calculate and output a pareto line graph for the following data. The x axis represents size and the y axis the cumulative ranked percentages. Use the number of the region and a * symbol for plotting the exact calculated values and a + symbol for the interpolated line plot between the exact values. Assume a linear interpolation between each pair of calculated values. Write the program to allow for a logarithmic transformation of the size variable and produce a second pareto line graph utilising this transformation. The data provided refer to wholesale sales in the French regions in 1966.

| Region | Wholesale sales (F) |
|---|---|
| 1 Paris | 485,703,975,000 |
| 2 Champagne | 30,897,254,000 |
| 3 Picardie | 27,781,128,000 |
| 4 Normandie | 45,658,092,000 |
| 5 Centre | 50,496,345,000 |
| 6 Nord | 110,213,681,000 |
| 7 Lorraine | 42,836,062,000 |
| 8 Alsace | 48,758,660,000 |
| 9 France Comte | 15,353,339,000 |
| 10 Basse Normandie | 25,724,647,000 |
| 11 Pays de la Loire | 61,387,934,000 |
| 12 Bretagne | 65,674,749,000 |
| 13 Limousin | 18,140,514,000 |
| 14 Auvergne | 23,820,868,000 |
| 15 Poitou-Charrentes | 36,563,176,000 |
| 16 Aquitaine | 66,461,036,000 |
| 17 Midi-Pyrenees | 43,562,715,000 |
| 18 Bourgogne | 36,457,541,000 |
| 19 Rhone-Alpes | 110,887,492,000 |
| 20 Longuedoc | 48,773,884,000 |
| 21 Provence Cote d'Azur-Corse | 98,824,276,000 |

5 Write a program to read two pieces of information about one point specified by an x,y co-ordinate. Sort the data and output at the correct co-ordinate point a particular symbol depending on the appropriate class into which the data falls. The following data of a 20% sample of self-service shops in Derby in 1964, may be used. The co-ordinates relate to ½-mile squares and the origin of the origin of the grid is the north-west corner of the map.

| Grid location |  | Size of | Type of operation |
|---|---|---|---|
| east of origin | south of origin | shop (ft$^2$) |  |
| 7 | 11 | 3000 | Co-op (Derby) |
| 6 | 4 | 720 | Co-op (Derby) |
| 4 | 5 | 650 | private |
| 7 | 8 | 6000 | multiple (Fine Fare) |
| 12 | 12 | 790 | vol chain (Mace) |
| 5 | 7 | 300 | vol chain (Mace) |
| 10 | 12 | 700 | vol chain (Mace) |
| 13 | 7 | 80 | private |
| 4 | 9 | 1750 | Co-op (Derby) |
| 5 | 4 | 450 | Co-op (Derby) |
| 6 | 6 | 670 | Co-op (Derby) |
| 3 | 6 | 2400 | Co-op (Derby) |
| 6 | 9 | 680 | Co-op (Derby) |
| 7 | 9 | 780 | Co-op (Derby) |
| 7 | 7 | unknown | private |
| 5 | 11 | 1000 | Co-op (Derby) |
| 3 | 7 | 4180 | Co-op (Derby) |
| 1 | 10 | 700 | Co-op (Derby) |

Use the following size classes: unknown, 0-499, 500-1499, 1500 and over. (Remember usually 10 columns is equivalent to 6 lines but first check if this is correct for your own machine.)

## 10.11 Further Reading

Bowen, E.G. 'Le Pays de Galles', *Transactions, Institute of British Geographers*, 26, 1-24 (1959)

Carter, H. and Thomas, J.G. 'The referendum on Sunday opening of licensed premises in Wales as a criterion of a culture region', *Regional Studies*, 3, 61-71 (1969)

Day, A.C. *FORTRAN techniques with special reference to non-numerical applications* (1972)

Meinig, D. 'The Mormon culture region', *Annals, Association of American Geographers*, 55, 191-220 (1965)

# Chapter 11
# Program Segmentation

11.1 Subprograms

In earlier chapters we have seen that a major attraction of the computer is its utility to the geographer as a means of reducing the effort of processing information. The time saved can be spent on larger study areas, more variables, and larger sample sizes. A major advantage of the computer is its ability to perform repetitions of the same operations with different data at each cycle. One way of repeating calculations is to use the DO statement introduced in Chapter 7, but there are more advanced statements allowing the programmer to segment programs into distinct modules or *subprograms* for use several times within the same program or by different programs that require the same basic operation. Unlike the DO command, which is controlled by a strictly defined numerical sequence, subprograms can be used, or *called*, in any suitable order at irregular intervals. Their use has a number of benefits. They allow long and complex programs to be broken down into separate logical entities whose internal flow of control and statement numbering do not affect overall consideration of the program. They provide a means of avoiding duplicating statements within a program and so lead to economies in program size. In addition, a subprogram developed in one context by one programmer will often prove transferable without alteration to other problems and programmers. Most university installations will maintain valuable libraries of subprograms to perform common but often complex operations, and these can be incorporated directly into the user's programs (*see also Chapter 13*).

Each subprogram is identified by its initial statement and must terminate with an END, whereas the main program that calls it, whose first statement is the first to be obeyed, contains no such special first line. With some exceptions to be noted, the order of subprograms within a program is not significant and there is no limit to the number that may be incorporated. In this chapter consideration is given to three basic types of subprogram that a programmer may create, the *statement function*, the *function subprogram* and the *subroutine subprogram*, together with two associated statements called COMMON and EQUIVALENCE.

11.2 Statement Functions

In Chapter 7 and many of the subsequent examples, a series of standard functions supplied as a part of the FORTRAN language were introduced and used (*for a full list see Appendix A*). These allow the programmer to calculate quantities such as $\sin(x)$ and $\log_e(x)$ simply by writing a function name together with its arguments anywhere within an assignment statement:

$$X = SIN(X)$$
$$Y = ALOG(X) * 123.9$$

At execution, functions return in place of their name the actual values required and thus may be placed almost anywhere in a program where they are needed. Although FORTRAN provides a great many useful standard functions, some programs may need to evaluate a particular nonstandard expression many times and at different points with different arguments each time. One simple way to do this would be to write separate statements for each operation, but a much more efficient and logical way to do this is to define one's own functions using the statement function.

There are two distinct aspects to the use of such user-defined statement functions. The first concerns the definition of the function and the second its actual use with specific data. A statement function is defined *within* the program or subprogram in which it is used. Definition is by a *single* statement similar in form to an arithmetic assignment. There may be any number of such definitions but in any program or subprogram unit these definitive statements must precede the first executable statement and must follow any specification statements such as DIMENSION and DATA. The formal definition of a statement function is:

$$\underline{f}(a_1, a_2, \ldots, a_n) = \underline{e}$$

In this $\underline{f}$ is any function *name* invented by the programmer according to the same rules that apply to variable names. It must consist of up to six letters or digits, the first of which is a letter, and must be punched in columns 7 to

72 of the card. For obvious reasons the chosen name should not be the same as any of the standard functions already supplied. $\underline{e}$ is any arithmetic expression containing references only to simple scalar variables (not array elements), intrinsic functions and previously defined function subprograms. This expression defines the operation to be carried out and whose result will be the value assigned to the function. It should be noted that it must not contain any reference to its own function name. The $a_i$ are distinct simple variable names called the *arguments* of the function and are enclosed within parenthesis in the statement. Because these do not have any actual numerical values when the function is defined, their names serve only to indicate the number, type (INTEGER or REAL) and order of the arguments and may be the same as any variable names used elsewhere in the program. For this reason they are called dummy arguments. In the defining statement, but not when the function is actually called, these arguments may not be subscripted variables.

To illustrate function definition, consider a climatological program in which it is frequently necessary to convert temperatures from degrees Fahrenheit to their Celsius ($^oC$) and Absolute ($^oK$) equivalents. The Absolute temperature scale has the same interval as the Celsius but has its zero the lowest possible temperature, absolute zero, which is $-273.16^oC$. Functions to perform these conversions might be inserted into the program immediately after any DIMENSION statements as:

FTC(F) = (F - 32.0) * 0.555555           (1)
CTK(C) = C + 273.16                       (2)
FTK(F) = (F - 32.0) * 0.555555 + 273.16   (3)

Function (1) will convert from $^oF$ to $^oC$, (2) from $^oC$ to $^oK$, and (3) from $^oF$ to $^oK$. In these examples the function names are FTC, CTK and FTK and the initial letter indicates that they are all of type REAL; the dummy arguments are the variable names F and C, and the expressions involving these arguments are written on the right-hand sides of the assignments.

On reaching a point in the program where a conversion is needed, the functions are called simply by writing their names and arguments as part of an expression in precisely the same way that standard functions are called. These new arguments will be the names of variables already defined or constants and are referred to as the *actual* arguments. They must agree in number, type, and order with the corresponding dummies in the function definition.

251

What happens when the statement containing the function call
is executed is that these actual arguments are associated
with the dummies in the statement definition and the expression $\underline{e}$ is evaluated using their numerical values.
The result of the evaluation is then made available to the
expression that contained the function call. It can be
seen that this mechanism of association between actual and
dummy arguments enables the function to be called as and
when required with different variables or constants as the
actual arguments.

In our example of a climatological program it might be
wished to calculate the intensity of radiation and wavelength of maximum emission from a black body radiator at
a temperature of say TA($^o$F) using the Stefan-Boltzmann
relation and Wein's displacement law. Mathematically,
these relate the radiation intensity and maximum emission
wavelength to the absolute temperature as

    intensity                  = $5.6697 \times 10^{-8}$ $T^4$ (W/m$^2$)
and
    maximum emission wavelength = 2898/T (micrometers)

To evaluate these expressions we could write FORTRAN
assignments as follows. For the radiation intensity
               TA   = 212.∅
               TENS = 5.6697 E-8 * FTK(TA)**4

This makes use of the function FTK defined previously but
has as its actual argument the current value of the
location TA, in this case 212.0. Later in the same program the same function might be used with a different
argument to calculate the maximum emission wavelength from
an object at a temperature of say TB set at 100°F:
               TB   = 1∅∅.∅
               WAVE = 2898.∅/FTK(TB)

In both cases the effect is the same as if we had written
an assignment to convert from °F to °K instead of the
function name.

A very simple consolidated program (*Fig 11.1*) to find
either the radiation intensity or the maximum emission
wavelength given the temperature in °F is shown below.
Each data card contains a variable NCALC that specifies
the operation to be performed and the input temperature.
The statement function FTK is defined immediately after
the introductory comments and, depending upon the value of
NCALC, is used to calculate either the radiation intensity
(statement 1∅) or the maximum emission wavelength
(statement 2∅).

    Statement functions can be a very powerful aid in

```
C  CONSOLIDATED PROGRAM TO FIND EITHER THE RADIATION INTENSITY (W/SQ.M.)
C  OR THE MAXIMUM EMISSION WAVELENGTH (MICROMETRES) GIVEN THE TEMPERATURE
C  IN DEGREES FAHRENHEIT
C  STATEMENT FUNCTION
      FTK(F) = (F -32.0 ) * 0.55555 + 273.16
      NIN = 5
      NOUT = 6
   5  READ(NIN,100) NCALC, TEMP
C  IF NCALC = 1 - FIND INTENSITY
C  IF NCALC = 2 - FIND MAXIMUM EMISSION WAVELENGTH
C  IF NCALC = 3 - STOP
      GO TO ( 10,20,40) , NCALC
  10  ANS = 5.6697E-8 * FTK(TEMP) ** 4
      WRITE(NOUT,110) ANS
      GO TO 30
  20  ANS = 2898.0 / FTK(TEMP)
      WRITE(NOUT,120) ANS
  30  GO TO 5
  40  STOP
 100  FORMAT (I1,F6.2)
 110  FORMAT( 13H INTENSITY = ,F12.3)
 120  FORMAT( 13H L=MAXIMUM = ,F12.3)
      END
```

*Fig 11.1* A program to calculate radiation intensity or maximum emission wavelength

shortening FORTRAN programs but they are limited in that only *one* statement can be repeated and that definition is *internal* to the particular main or subprogram that uses them.

## 11.3 Function Subprograms

A function subprogram allows a number of expressions to be evaluated to yield a single value and is defined externally to any program that uses it. Physically, this means that unlike the statement function which is defined within any one program module a function subprogram definition forms a distinct and separate module or *series* of cards *after* the main program in the card deck. The advantages are again that effort is not duplicated and storage space is conserved. In addition, a function subprogram may be compiled and tested independent of any main program.

Function subprogram *definitions* must have at least 4 major elements, a *title* specifier, a series of statements forming the subprogram *body*, at least one and possibly more RETURN commands, and an END to signify the end of the subprogram body. The first statement must be a title specifier of the general form:

<u>t</u> FUNCTION <u>f</u> ($a_1, a_2, \ldots, a_n$)

In this <u>t</u> is an optional type declarator (INTEGER, REAL and so on), and <u>f</u> is the symbolic name of the function to be

253

defined and invented by the programmer in the same way as variables and statement functions are named with between 1 and 6 letters or digits of which the first must be a letter. If the optional type declarator is omitted then the first letter of the symbolic name must be chosen according to the usual FORTRAN convention. As usual the whole statement is punched in columns 7 and 72 of the punched card. The chosen name must not appear in any non-executable statement within the program body but elsewhere in the defining subprogram it *must* appear on the left hand side of an assignment. As before, the $a_i$ are the subprogram's dummy arguments and must be distinct non-subscripted variable or array names. Examples of valid function subprogram names would be

```
          FUNCTION F2(M,P)
          REAL FUNCTION NF2(M,P)
          FUNCTION TOTAL (A,N)
          INTEGER FUNCTION SUM (A,N)
```

The optional type declarators REAL and INTEGER appearing in the titles of functions NF2 and SUM indicate that they are to have values that are of types REAL and INTEGER rather than the types that would be expected were the normal naming convention to be used. The use of these type declarators is discussed in more detail in Chapter 12.

Following this definition there must be a *program body* which serves to calculate the value to be associated with its symbolic name and which must contain at least one RETURN command. A RETURN directs the flow of control back from the subprogram to the main calling program and consists simply of the word RETURN punched within columns 7 to 72 of the card. The program body must terminate with an END statement just as is required in the main program.

As an example, suppose that it is frequently necessary in a statistical program to sum the contents of a real, singly-dimensioned array or arrays. A function subprogram to do this could be inserted into the card deck after the main program as follows:

```
          REAL FUNCTION TOTAL (A,N)
          DIMENSION A(100)
          TOTAL = 0.0
          DO 10 J = 1,N
       10 TOTAL = TOTAL + A(J)
          RETURN
          END
```

To call a function subprogram such as this, all that need be done is to write the function name with actual arguments

in place of the dummies as part of an expression. The
actual arguments may be constants, variable names, array
names or even the names of other external functions but in
all cases they must agree in number, order and type with
the dummies used in the definition. The effect of a call
is much the same as that for statement functions. The
actual values of the arguments are associated with the
dummies and the program segment is executed until
a RETURN is met. At this point control is returned to
the calling main program but with the actual value of the
function in place of the function name. Our function
called TOTAL might be called in a simple program to find
the average of a column of 20 numbers:
```
      DIMENSION D(2∅)
      M = 2∅
      NIN = 5
      NOUT = 6
      READ (NIN, 9∅) D
      AVER = TOTAL (D,M)/FLOAT(M)
      WRITE (NOUT, 95) AVER
      STOP
   9∅ FORMAT(F6.∅)
   95 FORMAT( 7H AVER =,F1∅.3)
      END
```
These statements would be followed by those defining REAL
FUNCTION TOTAL (A,N). In this example there is little to
be gained from using a function, but the same function
could be used many more times in the same or in different
programs.

Although function subprograms are usually designed to
return a single value in place of their name, it is also
possible to include in the program body statements that
also alter the value of one or more of the arguments.
This is perfectly permissible and will result in the
return of these altered values to the main program. In
effect, this device enables function subprograms to return
more than one function value to the main program but it is
a programming technique that must be used with some care.

11.4 Subroutine Subprograms
We now have means for instructing the computer to repeat
single statements and groups of statements, but in both
cases are restricted to subprograms that are fundamentally
only able to return one value to the main program. More
frequently than might be imagined, we may want to create
self-contained modules that handle a great many values or

repeatedly execute long and complex operations for which function subprograms are not really suited.  Instead, FORTRAN allows the use of any number of SUBROUTINE subprograms which are a powerful and often-used feature of the language.  Subroutines are similar to function subprograms in that they are defined externally to the main program, and physically follow it in the program deck.  Again a strict logical distinction must be maintained between the definition of the subroutine and its actual use or call.

At definition, a SUBROUTINE is identified by its first statement which can take one of two possible forms:
                SUBROUTINE $\underline{s}$
or          SUBROUTINE $\underline{s}(a_1,a_2,...,a_n)$
Both are punched between columns 7 and 72 of the punched card.  In both cases $\underline{s}$ is the symbolic name of the subroutine to be defined.  It is chosen using the rules for variable naming except that the first letter does not carry its usual significance as to the type of quantity involved, whether integer or real.  It is perfectly possible to have a subroutine with a name such as MATRIX which deals entirely with real numbers or one called AMULT dealing with integers.  Once selected, this name must *not* appear in any statement in the subroutine itself.  The $a_1$-$a_n$ enclosed within parenthesis are the familiar dummy arguments whose names are selected according to the normal rules of type and length and which are used within the body of the subroutine just like ordinary variables.  For reasons to be explained, it is not necessary for every subroutine to have arguments and in such cases it is the first form of the initial statement that is used.

Following this first card is the main body of the subroutine, written just like any normal FORTRAN program and using the dummy argument names (if present).  This may consist of any number of DIMENSION and executable statements, including input and output operations.  It can also introduce names for variables that are used purely locally within itself and have no significance outside the routine.  Indeed it may even introduce variable names that are the same as those used in the main program but whose storage and manipulation will be totally independent of them.  Each subroutine must terminate with an END and must contain at least one RETURN command.  SUBROUTINE subprograms are similar in many respects to FUNCTION subprograms but there is one vital difference which is crucial to understanding their manner of operation.  It will be remembered that the FUNCTION subprogram name must always appear in the defining body on the left hand side of an assignment and that at execution the returned value is associated with this function *name* written as part of an expression in the main program. A SUBROUTINE's name has *no* such value associated with it and must *not* appear in any statement within the defining subprogram. This has a number of important consequences. First, all the values it returns to the main program must be by way of its arguments instead of by way of its name. Secondly, a SUBROUTINE cannot be brought into action simply by writing its name but must be referenced using a special statement written at the appropriate place in the main program. This is the CALL command whose general form is:

$$\text{CALL } \underline{s} \ (a_1, a_2, \ldots, a_n)$$

It is punched between columns 7 and 72 of the punched card and s̲ is the symbolic name of the required subroutine. In the CA̲L̲L command the dummy arguments of the subroutine definition are replaced by actual arguments which at the first level of simplicity must agree in number, type and order with those used as dummies. A third consequence of the lack of a value being associated with a SUBROUTINE's name is that already noted; the name itself carries no significance in terms of the type of value used.

It is possible for one SUBROUTINE to call another and so on to create a chain of dependencies but FORTRAN does not allow a subroutine to call itself either directly or indirectly by way of some other subprogram.

In detail a great deal of programming lore is associated with using the SUBROUTINE effectively and some extremely subtle tricks can be accomplished but the *essential* features of their use can be illustrated by the simple example shown in Fig 11.2. This is a very short program to read in an array of real numbers, sort them into ascending numerical order, and then print out the sorted list. For convenience, these tasks have been arranged into two program modules. The first is the main program which handles the input and output of the list while the second is a general purpose SUBROUTINE called SORT designed to sort *any* such list and which is called by the main program. It will be noticed that the cards that define SORT follow the main program in the card deck and that there is a RETURN and an END within the body of the SUBROUTINE. In the definition the dummy arguments are A which represents a real array as dimensioned and N which represents a simple integer number giving the number of

```
C  MAIN PROGRAM TO SORT UP TO 100 NUMBERS IN ASCENDING ORDER
C  THE MAIN PROGRAM READS THE DATA, CALLS SUBROUTINE SORT, AND THEN
C  OUTPUTS THE ANSWER
      DIMENSION B(100)
      NIN = 5
      NOUT = 6
      READ(NIN,90) NO
      READ(NIN,95)(B(I),I = 1,NO)
      CALL SORT(B,NO)
      WRITE(NOUT,100)(B(I),I=1,NO)
      STOP
   90 FORMAT(I3)
   95 FORMAT(F4.0)
  100 FORMAT(1H ,F7.0)
      END


      SUBROUTINE SORT(A,N)
C  ARGUMENTS ARE A = REAL ARRAY TO BE ARRANGED INTO ASCENDING ORDER
C              N = NUMBER OF ELEMENTS IN A (MAX IS 100)
      DIMENSION A(100)
      M = N-1
      DO 10 I=1,N
      DO 20 J = 1,M
      IF ( A(J) .LE. A(J+1) ) GO TO 20
      HOLD = A(J+1)
      A(J+1) = A(J)
      A(J) = HOLD
   20 CONTINUE
   10 CONTINUE
      RETURN
      END
```

*Fig 11.2* A program to sort numerical values in an array

items in the list. The actual sorting routine is
extremely simple and the reader should have little trouble
in understanding it.

This SUBROUTINE is brought into action, or called, by the
statement

CALL SORT(B,NO)

in the main program. In this B and NO are the actual
arguments. B is the name of the array in the main program
that is to be sorted and NO is its number of elements.
As the SUBROUTINE is executed, every reference within it to
the dummy arguments A and N is treated as a reference to
these corresponding actual arguments and at this level of
programming this reference leads to the results that one
would intuitively expect - the array B in the main program
is arranged into ascending numerical order. It should be
noted at this point that occasions do arise when the
results are not quite so obvious and that a skilled programmer can use this arguments mechanism to produce quite
cunning results! Although in this simple example SORT

is used for only one operation a more complex program might
have occasion frequently to call this type of general
purpose subroutine to sort many different arrays and array
lengths.

Most university and college installations maintain
a library of subprograms, usually as subroutines, many of
which will be already loaded into the machine. To use
these all that need be done in any one program is to write
appropriate CALL commands and provide appropriate arguments.
This is an attractive way to build up a complex program
which ensures that programming labour is not duplicated
unnecessarily but we would suggest that a great deal of
care is taken to ensure that such library routines are
carefully checked for suitability and accuracy before they
are accepted into a program. In addition, routines
available at one installation may not be available at
another so that there is a considerable loss in program
portability.

A major aid in writing such general subroutines is the
programming facility called *adjustable dimensions*. In
Fig 11.2, the arrays B and A were both dimensioned to have
100 elements, a number which then forms the maximum
allowed on any one call, yet within SORT only N elements
are ever used. As in our example, N will usually be
considerably less than 100, resulting in a waste of the
extra storage locations. An alternative would be to
write the DIMENSION statement *in the subprogram* (never in
a main program) that has an integer variable name or names
in place of the usual integer constant(s) that give the
dimensioning information:

    SUBROUTINE SORT (A,N)
    DIMENSION A(N)

The necessary dimensioning information can then be passed
into the subroutine through the arguments mechanism and
our array A is now an *adjustable array*. Such an array
can only appear in a subprogram, and the values or variable
names used to convey the array size must appear in the
argument list and must have values assigned to them before
the routine is called. In addition these arguments cannot
be redefined during execution of the subprogram and the
maximum size of the array may not be exceeded.

11.5 The COMMON Statement

As already intimated, a FORTRAN subprogram is a distinct
entity whose only association with other subprograms and
the main program is by the symbolic names used in the CALL

command and by the transfer of arguments. The names of any variables used within such a subprogram have no meaning outside it. Physically, this means if any variable is given the same name in two or more subprograms, the computer will normally reserve different storage locations with independently varying values for each, and the use of the name in one subprogram will in no way affect its use in another. If simple scalar variables are being used there will be little waste of storage, but if large arrays are declared independently in the main program and in all its subprograms there may be a considerable waste of store. FORTRAN provides a way of overcoming this by allowing each subprogram to define a number of data areas, called *common blocks*, and to share these storage locations. Each program module is given an extra specification statement called the COMMON statement which serves to define the data storage locations to be used and to give the variable names that are to be used *in that module* for the quantities stored in these locations. In this way the *same* storage locations can be used independently by different subprograms, and variables in different subprograms with different names can be associated together if this is required. In both cases the net effect is to conserve storage use and in a large program this may prove to be essential if it is to be fitted into the available computer store.

All the specified common blocks except one are given names or labels and are called *labelled common*; the other block does not carry a name and is called *unlabelled common*. In order to set up this shared store, a specification statement similar to the DIMENSION statement already met is used and is inserted into each and every subprogram that requires to use the shared locations. The specification is by one of two possible forms according to whether or not the block is to be unlabelled or labelled common. The simplest is for the first, or unlabelled, common block:
 COMMON $a_1, a_2, a_3, \ldots, a_n$
As usual this must be punched somewhere between columns 7 and 72 of the punched card and is inserted into the subprogram prior to any executable statements. In unlabelled common each $a_i$ is a variable name (BIG,LITTLE), array name (DATA,MATRIX) or even an array declarator (ARRAY(1∅,1∅), MIT(124∅)). In the latter case the effect is to reserve the specified number of locations for each array in precisely the same way that a DIMENSION statement reserves storage, but if an array name is written in a COMMON

statement without any subscripting information then this same name must also appear in a DIMENSION statement. An unlabelled common specification like the above will reserve one single *area* of data storage but we may often need to reserve more than one such area using a labelled common specification statement which is of the general form:

$$\text{COMMON } /x_1/a_1,a_2,\ldots,a_n/x_2/b_1,b_2,\ldots,b_n$$
$$\underbrace{\qquad\qquad\qquad}_{\text{optional}}$$

The only difference in this form is the inclusion of the /x/ in the statement. Each x is called the block name and serves simply as a symbolic identification label for that block of common storage. Block names bear no relationship to any variables given the same names in the program and must consist of between 1 and 6 letters or digits, the first of which must be a letter. Certain sequences of letters are not allowed as COMMON block names, notably those sequences that are also used in FORTRAN to identify other types of statement such as FORMAT, GO TO or READ, but in most cases the programmer is free to chose any suitable name. Having given each labelled block its symbolic name, the $a_1 \ldots a_n$ list the variables to be associated with that block. A given name may appear more than once in a COMMON statement; the processor will simply string together in each block all the entities assigned to it in order of their appearance in the statement.

All this may sound rather complicated but can best be understood by looking at what actually happens when common storage is set up. The effect of a COMMON statement in any two subprograms is simply to associate variables with *different* names in *different* subprograms with the *same* locations in the computer's store. It cannot be emphasised too strongly that it is the storage locations that are being shared and not any of the variable names associated with it. An unlabelled COMMON statement in the main program might be written

COMMON A,B,M

This would cause the first (unlabelled) common block to be shared by the variables A,B and M. In any subsequent subprogram, the statement

COMMON T,V,MY

would assign the variables called T,V and MY in that subprogram to the same storage locations as were used by A,B and M in the main program. Each now has its own variable names, and in principle these are distinct, but they now have been made to occupy the same storage locations and so will have the same values.

A more complicated sequence of COMMON specifications that makes use of both labelled and unlabelled common might be:

```
       SUBROUTINE SUBA
       COMMON ZEE, Q(1ØØ)/ABC/X,Y
       :
       :
       SUBROUTINE SUBB
       COMMON /ABC/S,T
       :
       :
       SUBROUTINE SUBC
       COMMON ZEE
```

In this, SUBROUTINE SUBA has an unlabelled first common block in which the variables ZEE and Q are located, together with a second labelled block to which the programmer has given the symbolic name ABC and in which X and Y are located. The subsequent COMMON statement in SUBROUTINE SUBB will associate S in SUBB with X in SUBA and T in SUBB with Y. In SUBROUTINE SUBC, ZEE is associated with a logically distinct variable also called ZEE in SUBA.

Because these associations are made by storage blocks and not by variable names, there is no restriction that the apparent type of the variables, whether integer or real, in an unlabelled block be the same. It is perfectly possible to associate integer data in one subprogram with real data in another and with a clear understanding of how the machine allocates its store this type change can be used to some advantage. In general, however, it is not recommended because different data types may occupy different numbers of store locations in different computer systems and can give peculiar, non-standard, results. If labelled common is used, the definitions must be such that all corresponding positions are identical in type and allocate the same number of variables to each block.

11.6 The EQUIVALENCE Statement

A final specification statement to be considered is the EQUIVALENCE statement. Like the COMMON statement this does not lead directly to any processing but has the effect of associating variables that have different names with the same storage locations. Unlike the COMMON statement, which is used to associate variables in different subprograms, an EQUIVALENCE specifies an association of two or more variables with the same location within the *same* subprogram. The EQUIVALENCE statement required to do this is placed immediately before any executable statements in the subprogram and is punched in columns 7 to 72 of the punched card. Its general form can be written:

$$\text{EQUIVALENCE } (k_1), \underbrace{(k_2), \ldots, (k_n)}_{\text{optional}}$$

Each (k) is a list of variable or array names separated by commas and enclosed within parenthesis that are to be assigned to the same storage location or locations. The statement

EQUIVALENCE (M,MOD,KULE), (RADS,CIS)

placed within a program would instruct the processor to use the same storage location for the different program variables M,MOD and KULE and the same location for RADS and CIS. Each element in each of the lists that are enclosed within the parentheses is thus assigned to the same storage location or part of a storage location. The above example deals with simple scalar variables, but arrays can be equivalenced in the same way. When an array name is used on its own the effect is for that name to stand for the first element of the array. The specification statements

DIMENSION W(50),CRUMB(5,10)
EQUIVALENCE (W(1),CRUMB(1,1))

would cause the elements of both W and CRUMB to occupy the same 50 storage locations beginning with the first element in each. An equivalence can also be established with elements other than the first, for example

EQUIVALENCE (A(1),B(1)), (A(51),C(1))

On meeting EQUIVALENCE for the first time the immediate reaction is to ask why it should be necessary to associate variables with different names in the same subprogram with the same storage locations. After all, if the programmer wants them to mean the same thing then he can simply give them the same name and leave it at that. At first sight EQUIVALENCE is unnecessary. There are two primary reasons for using EQUIVALENCE in a subprogram. First, it may be necessary to conserve storage space by equivalencing variables that are actually distinct but which are never used at the same time within any one subprogram. The programmer will retain the advantage of having the variables with different names and avoid any confusion when they are referenced without using any further storage space. As an example, a program might use a 10 x 10 array called ITAB at one point and a totally separate

array called JTAB of the same size at another. Now it would be perfectly possible to write all the assignments involving the second array using the name ITAB to reference the original ITAB storage locations but many people might find this confusing, perhaps because these second set of locations are being used in a manner which is totally distinct from their original use as ITAB. Instead the statement
                EQUIVALENCE (ITAB(1,1), JTAB(1,1))
would allow distinct names to be retained for the same locations.

A second use of EQUIVALENCE is in the correction of inadvertent programming errors. After writing a long program the author may suddenly realise that he has accidentally changed the name of a variable half way through his coding, from say GEOG to GOEG. Rather than go back and change all the references to GOEG he can use the equivalence mechanism to make the changes for him by inserting the card
                EQUIVALENCE (GEOG,GOEG)
This will associate both variables with the same location in store so that any changes to GEOG made within the program will also be automatically made to GOEG as well. Notice that this equivalence mechanism is similar to common storage in that the association is made by way of the storage location involved, not the variable names. In addition to these primary uses, some programmers will use the equivalence mechanism to create identities between real and integer quantities and between two-dimensional arrays and one-dimensional vectors.

EQUIVALENCE will only rarely be used by the novice programmer and there are certain side effects that can occur which suggest that it be used with caution. Consider for example a subprogram that has three specification statements:
        DIMENSION C(5), MATRID(10,10)
        COMMON D(5)
        EQUIVALENCE (MATRID(2,3),Z), (C(1),D(2))
In writing these statements, the programmer might have intended that element (2,3) of the integer array MATRID have the simple variable name Z and that the locations C(1) and D(2) mean the same thing. Whatever the intention, these statements have effects that could give rise to trouble. First, the EQUIVALENCE mechanism cannot be used to associate single elements of an array. In writing (C(1),D(2)) the effect will be for the array called D in common storage to be such that D(I + 1) be the same as

264

each C(I). As the diagram below shows, the unlabelled
common block will be extended from 5 to 6 locations:
<u>unlabelled common</u>
D(1)
D(2) = C(1)
D(3) = C(2)
D(4) = C(3)
D(5) = C(4)
extra location = C(5)
To accommodate the extra location required by the
EQUIVALENCE it is necessary to extend the common block by
one location. This same mechanism could be used to
extend common by two, three or even more locations and in
the example given with unlabelled common this is perfectly
feasible. FORTRAN does not allow blocks to be extended
*below* the first location. Notice too that a member of
a common block cannot be equivalenced to a variable which
is itself a member of the same or other common block.
A second possible source of difficulty is that the 10 by
10 array MATRID is implicitly declared by the first letter
of its name to be of type integer, whereas Z, the
associated simple variable is of type real. The names
are thus not quite as synonymous as the programmer expects
and again this could lead to trouble.

## 11.7 Worked Examples
*Problem.* Among the most remarkable of the landforms left by
the Pleistocene Ice Sheets are the low, streamlined hills
known as drumlins. A number of mechanisms have been proposed to explain drumlin formation, but as yet no single
theory seems adequate to explain all the facts of drumlin
shape, composition, orientation and spatial distribution
and nobody has found a drumlin being formed by a con-
temporary ice sheet. To see what light it can throw upon
drumlin origin a number of recent studies have tended to
concentrate upon the analysis of drumlin distribution using
methods which compare the observed distribution with that
which might be expected were the forms simply the result
of spatially random emplacement (Vernon, 1966; Smalley
and Unwin, 1968; Hill, 1973). The present example will
consider each drumlin as a point in space and analyse the
distance from each point to its nearest neighbouring point
as an index of randomness of the spatial distribution.
*Algorithm and flowchart.* The fit of an observed point
pattern to that given by a random distribution can be
tested using the simple nearest neighbour statistic

originally developed by the ecologists Clarke and Evans (1954). This is defined as

$$R = 2 \bar{D}_{obs} \sqrt{(N/A)}$$

$\bar{D}_{obs}$ = average of the observed distances from each point to its respective nearest neighbour
N = number of points
A = area of study region

In calculating this statistic, the units in which the distances are measured are irrelevant as long as they remain consistent throughout. If the distribution is random, R has a value close to 1.0 while values less than this indicate a clustering and values more than this a regular distribution.

Attempting this sort of calculation by hand is both tedious and error-prone, involving both the detection of each nearest neighbour and the measurement of its distance, but it is a simple matter to write a program to do this work for us. Figures 11.3 and 11.4 show a flowchart and a suitable program to calculate R values for distributions of up to 100 points. The program makes use of a statement function and a function subprogram and can be explained as follows. The first executable statements set the input and output channels, NIN and NOUT, to the required values and cause an initial data card to be read in containing values for NS, the number of points, and AREA, the square area over which these points are spread. Notice that AREA must be specified in the same units as are used to give the spatial co-ordinates of the data points; if the points are located using a km grid, then AREA must be in $km^2$, if miles are used in $miles^2$ and so on. The next statement simply reads in the (X,Y) co-ordinates of the points as the arrays X and Y. From this information it is possible to calculate the distance of every single point to every other point using the Pythagoras theorem. If the distance between two points located at say $(x_1,y_1)$ and $(x_2,y_2)$ is thought of as the hypotenuse of a right-angled triangle which has these points as two of its apexes, then this distance is given by

$$D = \sqrt{((x_1-x_2)^2 + (y_1-y_2)^2)}$$

Each D value derived in this way can next be thought of as forming one element of an array which holds all such possible inter-point distances:

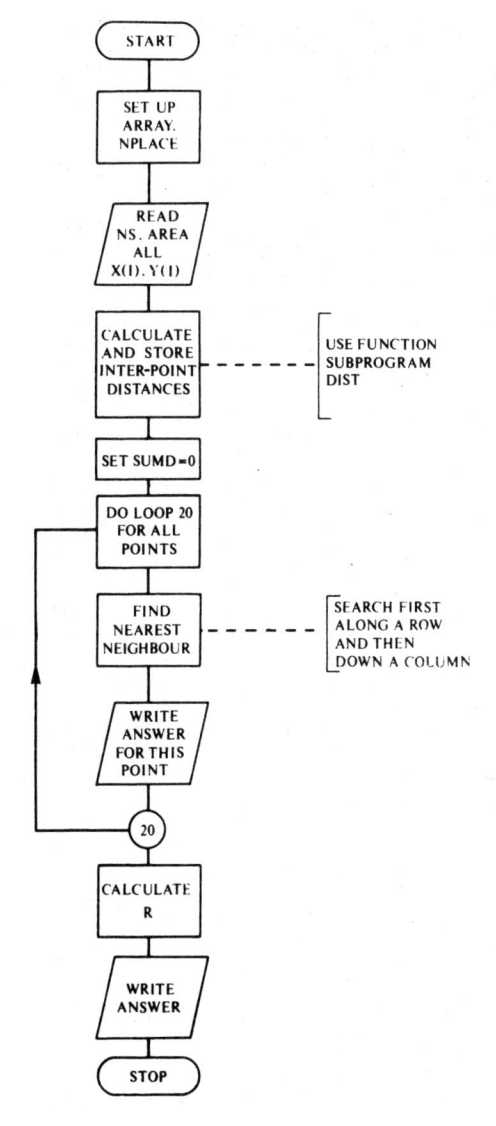

*Fig 11.3* Flowchart for a nearest-neighbour analysis program.

```
C  NEAREST NEIGHBOUR ANALYSIS OF A POINT PATTERN                    10
C  REFERENCE = CLARKE AND EVANS (1954) ECOLOGY, 35(4), 445-453       20
C  ILLUSTRATING THE USE OF FUNCTION STATEMENTS                       30
C                                                                    40
      DIMENSION X(100),Y(100),D(5050)                                 50
C  SUBSCRIPTING FUNCTION IS CALLED NPLACE                             60
      NPLACE(I,J) = I*(I-1)/2 + J                                     70
      NIN =5                                                          80
      NOUT = 6                                                        90
C  READ IN DATA AS DESCRIBED                                         100
C     NS = NUMBER OF POINTS,  AREA = STUDY REGION AREA               110
      READ(NIN,88) NS, AREA                                          120
      READ(NIN,90) (X(I),Y(I), I= 1,NS)                              130
C     CALCULATE ALL INTER-POINT DISTANCES STORING IN D               140
      DO 10 I = 1,NS                                                 150
      II = I -1                                                      160
      DO 10 J = 1, II                                                170
   10 D(NPLACE(I,J)) = DIST(X(I),Y(I),X(J),Y(J))                     180
      WRITE(NOUT,92)                                                 190
C  SEARCH FOR AND RETAIN EACH POINTS NEAREST NEIGHBOUR               200
      SUMD = 0.0                                                     210
      DO 20 I = 1,NS                                                 220
      AMIN = 1.0E6                                                   230
      III = I + 1                                                    240
      II = I -1                                                      250
C  SEARCH EQUIVALENT OF A ROW OF THE FULL DISTANCE ARRAY             260
C     ON FIRST CASE THIS IS OMITTED                                  270
      IF ( I .EQ. 1 ) GO TO 35                                       280
      DO 30 J = 1, II                                                290
      FAR = D(NPLACE(I,J))                                           300
      IF ( FAR .GT. AMIN ) GO TO 30                                  310
      NHOLD = J                                                      320
      AMIN = FAR                                                     330
   30 CONTINUE                                                       340
C  SEARCH EQUIVALENT OF A COLUMN OF THE FULL DISTANCE ARRAY          350
C     ON FINAL CASE THIS IS OMITTED                                  360
   35 IF (I .EQ. NS) GO TO 45                                        370
      DO 40 J = III,NS                                               380
      FAR = D (NPLACE(J,I))                                          390
      IF ( FAR .GT. AMIN ) GO TO 40                                  400
      NHOLD = J                                                      410
      AMIN = FAR                                                     420
   40 CONTINUE                                                       430
   45 CONTINUE                                                       440
C  WRITE OUT RESULTS FOR EACH CASE AND SUM IN DOBS                   450
      WRITE(NOUT,96) I,X(I), Y(I), NHOLD, AMIN                       460
      SUMD = SUMD + AMIN                                             470
   20 CONTINUE                                                       480
C  SUMMARY CALCULATIONS                                              490
      FNS = NS                                                       500
      DMEAN = SUMD / FNS                                             510
      R = 2.0 * DMEAN * SQRT(FNS/AREA)                               520
      WRITE(NOUT,94) DMEAN, R                                        530
      STOP                                                           540
   88 FORMAT (I3,F7.1)                                               550
   90 FORMAT(2F6.1)                                                  560
   92 FORMAT(41H POINT X-COORD Y-COORD NEIGHBOUR DISTANCE/)          570
   94 FORMAT (17H MEAN OBSERVED = ,F11.3/ 11H R=INDEX = ,F11.6)      580
   96 FORMAT(I4,2F7.1,4X,I4,2X,F11.3)                                590
      END                                                            600
      REAL FUNCTION DIST(X,Y,A,B)                                    610
C  DIST FINDS THE LINEAR DISTANCE BETWEEN POINTS (X,Y) AND (A,B)     620
      XX = X - A                                                     630
      YY = Y - B                                                     640
      DIST = SQRT( XX*XX + YY*YY)                                    650
      RETURN                                                         660
      END                                                            670
```

*Fig 11.4* Computer program for nearest-neighbour analysis

```
                    POINT
            1    2    3    4    5 ................NS
        1  0.0
        2  31   0.0
        3  20   19   0.0
POINT   4  17   28   5    0.0
        5   6    8   12   1    0.0
           ⋮                           ⋮
          NS                                       0.0
```

In this, the distance between point 2 and point 1 is 31
units, that from 5 to 2 is 8 units and so on. Because
the distance from point I to point J is the same as that
from point J to point I the array is symmetric, and each
point is distant 0.0 units from itself. We can use this
symmetry to save calculation and computer storage by only
dealing with the lower triangle of the array as shown.
Were these distances to be stored as the elements of a two-
dimensional array A, dimensioned from NS to NS, in all we
would reserve $NS^2$ storage locations but only ever use
about a half of these to store any numbers. If, for
example, we were dealing with 5 points this would result
in the reservation of 25 locations of which only 15 could
ever be used. With a 100 point problem this waste of
store is even more serious; we reserve 10,000 locations
yet only use 5,050 and so on. Instead of using a two-
dimensional array our program uses the single-dimensioned
array called D in which are placed only the *unique* inter-
point distances, row by row, and introduces a subscripting
function to access the correct location. The element (I,J)
in the symmetric two-dimensional array given above will
be stored as the $K^{th}$ element of the one-dimensional array
and K can easily be calculated from
$$K = I \ (I - 1)/2 + J$$
This calculation is used repeatedly at several points in
the program and is thus calculated in an integer statement
function called NPLACE. This is defined immediately
after the specification statements but before any processing
is done with dummy arguments I and J as
$$NPLACE \ (I,J) = I * (I - 1)/2 + J$$
It is referenced within the pair of DO loops that have as
their shared terminal statement the arithmetic assignment:
$$10 \ D(NPLACE \ (I,J)) = DIST(X(I),Y(I),X(J),Y(J))$$
When this statement is executed the actual arguments are
the current values of I and J which as explained are not

the same as the I and J used in used in the definition, and the effect is to evaluate the subscripting function to give the correct location in D. Hence with actual arguments of I = 4 and J = 2 we will be calculating the distance from the fourth to the second point and NPLACE will be evaluated as $4 * (4 - 1)/2 + 2 = 8$. The effect will thus be to select element 8 of the array D so that the left hand side of statement 10 in the program will be exactly equivalent to writing D(8).

The right hand side of the same assignment also contains a reference to a function, but this time it is to the function subprogram called DIST which is defined at the end of the main program as being of type real with the dummy arguments X,Y,A and B. DIST simply finds the straight-line distance between the points (X,Y) and (A,B). When it is referenced the actual arguments are the spatial co-ordinates of the $i^{th}$ point (X(I),Y(I)) and the $j^{th}$ point (X(J),Y(J)) and it returns in place of its name the required inter-point distance. On exit from these DO loops, the array D will have all these values in store. The remainder of the program simply takes every point in turn under the control of the DO loop that terminates at the statement labelled 2∅ and finds the number and distance of its nearest neighbour. At each cycle through this loop AMIN is first set to a very high value ($10^6$) and the distance to all other points tested against it. If any distance is less than AMIN, then AMIN has its value replaced by this closer distance and NHOLD is given the number of the point concerned. The location of the distances from each current point to all others must be found in two distinct loops. The first has the statement 3∅ as its terminator and examines all points lower in number than it while the second has statement 4∅ as its terminator and examines all those points in higher numeric order. Notice that the function NPLACE is again used within these loops to access the correct location. On exit from the loops AMIN contains the distance of the nearest neighbour to the current point and NHOLD its reference number. These data are written out with the distance values being summed in SUMD so that when the loop DO 2∅ I = 1,NS terminates there will have been produced a table of all nearest neighbours and nearest neighbour distances. A short final section evaluates the mean nearest neighbour distance and the R statistic.

*Data and run.* Data for the x,y co-ordinates of the centres of 50 drumlins in a relatively small ($40km^2$) area in the

| | |
|---|---|
| 5, 325, | |
| 11, | 7, |
| 2?, | 9, |
| 26, | 7, |
| 33, | 6, |
| 38, | 3, |
| 42, | 2, |
| 39, | 8, |
| 46, | 12, |
| 7, | 16, |
| 8, | 18, |
| 14, | 21, |
| 33, | 17, |
| 35, | 18, |
| 44, | 32, |
| 42, | 28, |
| 37, | 23, |
| 35, | 23, |
| 33, | 27, |
| 28, | 27, |
| 20, | 26, |
| 17, | 26, |
| 14, | 36, |
| 12, | 29, |
| 7, | 30, |
| 3, | 36, |
| 6, | 36, |
| 8, | 38, |
| 15, | 37, |
| 24, | 32, |
| 27, | 35, |
| 34, | 33, |
| 38, | 36, |
| 35, | 38, |
| 29, | 41, |
| 28, | 42, |
| 34, | 44, |
| 35, | 47, |
| 36, | 52, |
| 23, | 46, |
| 2, | 46, |
| 5, | 49, |
| 14, | 54, |
| 22, | 59, |
| 19, | 53, |
| 16, | 52, |
| 17, | 54, |
| 26, | 56, |
| 32, | 62, |
| 38, | 63, |
| 41, | 49, |

*Fig 11.5* Data for a point distribution — drumlins in the Vale of Eden

271

| POINT | X-COORD | Y-COORD | NEIGHBOUR | DISTANCE |
|---|---|---|---|---|
| 1 | 11.0 | 7.0 | 2 | 9.220 |
| 2 | 20.0 | 9.0 | 3 | 6.325 |
| 3 | 26.0 | 7.0 | 2 | 6.325 |
| 4 | 33.0 | 8.0 | 7 | 6.000 |
| 5 | 38.0 | 3.0 | 6 | 2.236 |
| 6 | 40.0 | 2.0 | 5 | 2.236 |
| 7 | 39.0 | 8.0 | 5 | 5.099 |
| 8 | 46.0 | 10.0 | 7 | 7.280 |
| 9 | 7.0 | 16.0 | 10 | 2.236 |
| 10 | 8.0 | 18.0 | 9 | 2.236 |
| 11 | 14.0 | 21.0 | 21 | 5.831 |
| 12 | 33.0 | 17.0 | 13 | 2.236 |
| 13 | 35.0 | 18.0 | 12 | 2.236 |
| 14 | 49.0 | 32.0 | 15 | 8.062 |
| 15 | 42.0 | 28.0 | 16 | 7.071 |
| 16 | 37.0 | 23.0 | 17 | 2.000 |
| 17 | 35.0 | 23.0 | 16 | 2.000 |
| 18 | 33.0 | 27.0 | 17 | 4.472 |
| 19 | 28.0 | 27.0 | 18 | 5.000 |
| 20 | 24.0 | 26.0 | 21 | 3.000 |
| 21 | 17.0 | 26.0 | 20 | 3.000 |
| 22 | 14.0 | 30.0 | 23 | 2.236 |
| 23 | 12.0 | 29.0 | 22 | 2.236 |
| 24 | 7.0 | 30.0 | 23 | 5.099 |
| 25 | 3.0 | 36.0 | 26 | 3.000 |
| 26 | 6.0 | 36.0 | 27 | 2.828 |
| 27 | 8.0 | 38.0 | 26 | 2.828 |
| 28 | 15.0 | 37.0 | 27 | 7.071 |
| 29 | 24.0 | 32.0 | 30 | 4.243 |
| 30 | 27.0 | 35.0 | 29 | 4.243 |
| 31 | 34.0 | 33.0 | 32 | 5.000 |
| 32 | 38.0 | 36.0 | 33 | 3.606 |
| 33 | 35.0 | 38.0 | 32 | 3.606 |
| 34 | 29.0 | 41.0 | 35 | 1.414 |
| 35 | 28.0 | 42.0 | 34 | 1.414 |
| 36 | 34.0 | 44.0 | 37 | 3.162 |
| 37 | 35.0 | 47.0 | 36 | 3.162 |
| 38 | 36.0 | 52.0 | 37 | 5.099 |
| 39 | 23.0 | 46.0 | 35 | 6.403 |
| 40 | 2.0 | 46.0 | 41 | 4.243 |
| 41 | 5.0 | 49.0 | 40 | 4.243 |
| 42 | 10.0 | 54.0 | 45 | 6.325 |
| 43 | 22.0 | 59.0 | 47 | 6.083 |
| 44 | 19.0 | 53.0 | 45 | 3.162 |
| 45 | 16.0 | 52.0 | 46 | 2.236 |
| 46 | 17.0 | 50.0 | 45 | 2.236 |
| 47 | 28.0 | 58.0 | 48 | 4.472 |
| 48 | 32.0 | 60.0 | 47 | 4.472 |
| 49 | 36.0 | 63.0 | 48 | 6.708 |
| 50 | 41.0 | 49.0 | 38 | 5.831 |

MEAN OBSERVED = 4.215
R-INDEX = 1.0450/1

Fig 11.6 Results of nearest-neighbour analysis

Vale of Eden are provided in Fig 11.5. The first card
specifies the number of points (50) and the study region
area measured in the same arbitrary units as were used to
code the drumlin locations (3250 sq units). The x,y
values relate to a rectangular grid with 65 divisions along
the y (south-north) axis and 50 along the x (west-east)
axis. The results of the analysis are shown as Fig 11.6.
It can be seen that the R statistic at 1.045671 is very
close to unity, indicating that *at this scale of observation*
the distribution is close to the random one predicted by
Smalley's dilatancy theory of drumlin formation (Smalley,
1966). Recently Hill (1973) has shown that drumlins exhibit significant departure from randomness at several
scales of observation across a drumlin field and it is
clear that there is a need for further observation and
identification of the various scales of randomness and
regularity in drumlin distributions before the full process
implications can be worked out.

(ii) *Problem.* The geographic interpretation of topographic
maps is usually carried out by visual inspection alone,
with the consequent danger that preconceived deterministic
relationships between the physical and human landscapes
will be suggested without any objective factual basis.
An alternative approach using correlation analysis has been
suggested by C. A. M. King (1969) who suggests that quantitatively measured variables that characterise the landscape can be related together using the product-moment
method to give indices varying from +1 to -1 according to
the strength and direction of any relationships between
them.  In this case study, correlations are examined
between six variables chosen to characterise a diverse
25km$^2$ area around Calver (SK 240745) in the English Peak
District.  Like many of the programs presented in the
case studies the methods used will be found to be applicable
in a large number of geographic analyses.
*Algorithm and flowchart.* The product-moment correlation
coefficient between a pair of variables $X_0$ and $X_1$ is defined
as:
$$r_{0,1} = [\Sigma X_0 . X_1 - (\Sigma X_0).(\Sigma X_1)/N] / [\Sigma X_0^2 - (\Sigma X_0)^2/N]^{\frac{1}{2}}.$$
$$[\Sigma X_1^2 - (\Sigma X_1)^2/N]^{\frac{1}{2}} \qquad (1)$$
The denominator in this expression is the product of the
respective standard deviations and the numerator is the
covariance.  This formula, generalised to deal with many
combinations of variables, may be represented by the
algorithm for a computer program shown in Figs 11.7 and 11.8.

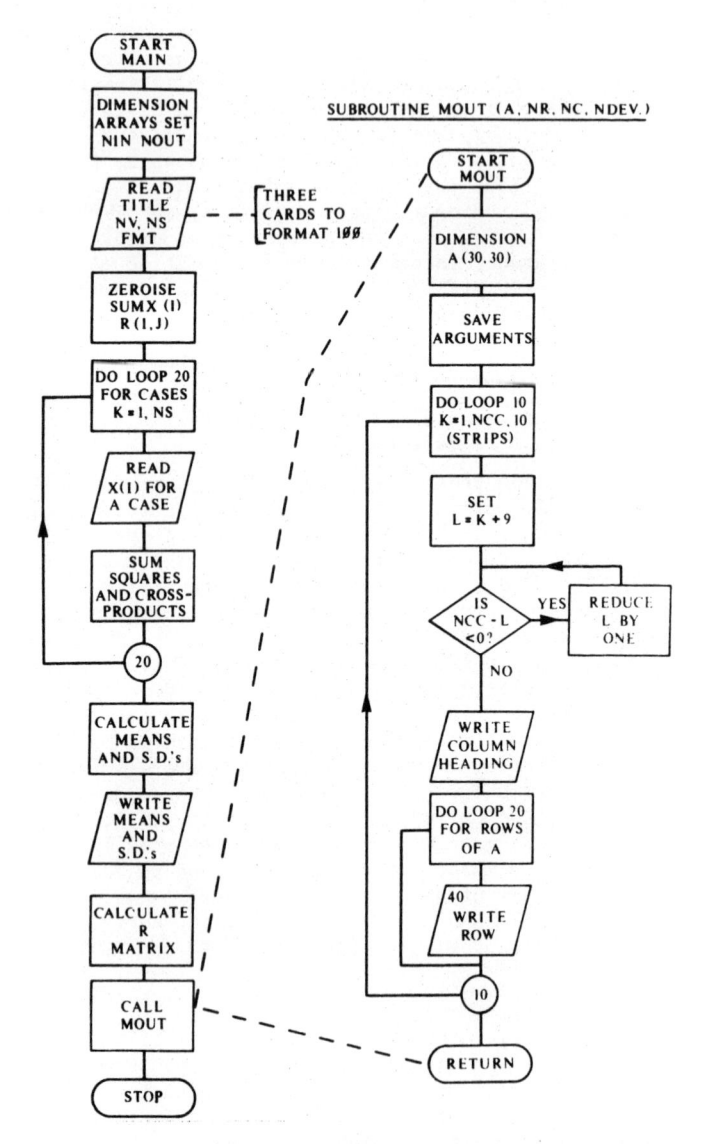

*Fig 11.7* Flowchart for correlation analysis program

```
C   CORRELATION ANALYSIS ILLUSTRATING SUBROUTINES                   10
        DIMENSION SUMX(30),X(30),R(30,30),TITLE(20),FMT(20)         20
        NIN = 5                                                     30
        NOUT = 6                                                    40
        READ(NIN,100) TITLE,NV,NS,FMT                               50
C   TITLE IS UP TO 80 CHARACTERS                                    60
C   NV = NUMBER OF VARIABLES                                        70
C   NS = NUMBER OF CASES                                            80
C   FMT = FORMAT FOR INPUT DATA CARDS                               90
        FNS = NS                                                    100
        WRITE(NOUT,110) TITLE                                       110
C   CLEAR COUNTERS                                                  120
        DO 10 I = 1,NV                                              130
        SUMX(I) = 0.0                                               140
        DO 10 J = 1, NV                                             150
        R(I,J) = 0.0                                                160
   10   CONTINUE                                                    170
C   READ DATA CARDS, ONE OR MORE PER CASE AND ASSEMBLE SUMS         180
        DO 20 K = 1, NS                                             190
        READ(NIN,FMT) (X(I),I = 1, NV)                              200
        WRITE(NOUT,120) (X(I),I = 1,NV)                             210
        DO 20 I =1, NV                                              220
        SUMX(I) = SUMX(I) + X(I)                                    230
        DO 20 J = 1, NV.                                            240
        R(I,J) = R(I,J) + X(I) * X(J)                               250
   20   CONTINUE                                                    260
C   COMPUTE MEANS AND STANDARD DEVIATIONS                           270
        WRITE(NOUT,125)                                             280
        DO 30 I = 1, NV                                             290
        SUMX(I) = SUMX(I) / FNS                                     300
        X(I) = SQRT((R(I,I)/FNS)-(SUMX(I)**2))                      310
   30   WRITE(NOUT,130) I, SUMX(I), X(I)                            320
C   COMPUTE CORRELATION MATRIX                                      330
        DO 40 I =1,NV                                               340
        DO 40 J = 1, NV                                             350
        R(I,J) = ((R(I,J)/FNS) -(SUMX(I)*SUMX(J)))/(X(I)*X(J))      360
   40   CONTINUE                                                    370
        WRITE(NOUT,140)                                             380
        CALL MOUT(R,NV,NV,NOUT)                                     390
        STOP                                                        400
  100   FORMAT(20A4/2I5/20A4)                                       410
  110   FORMAT(1H1,20A4)                                            420
  120   FORMAT(1H , 10F11.3)                                        430
  125   FORMAT(1H1,30H  VARIABLE   MEAN    STAN.DEVN)               440
  130   FORMAT(1H ,I5,5X,2F10.4)                                    450
  140   FORMAT(1H1,35H  MATRIX OF CORRELATION COEFFICIENTS/)        460
        END                                                         470


        SUBROUTINE MOUT (A,NR,NC,NDEV)                              480
C   MOUT WILL OUTPUT AN ARRAY A OF ORDER NR ROWS BY NC COLS ON OUTPUT  490
C   DEVICE NDEV                                                     500
        DIMENSION A(30,30)                                          510
C   SAVE ARGUMENTS FOR QUICK OPERATION                              520
        NRR = NR                                                    530
        NCC = NC                                                    540
        IOUT = NDEV                                                 550
C   PRINT OUT IN COLUMN STRIPS                                      560
        DO 10 K = 1, NCC, 10                                        570
        L = K + 9                                                   580
   20   IF ( NCC - L ) 25,30,30                                     590
   25   L = L-1                                                     600
        GO TO 20                                                    610
   30   WRITE(IOUT,90)(J,J = K,L)                                   620
        DO 40 I = 1,NRR                                             630
   40   WRITE(IOUT,92) I,(A(I,J), J = K,L)                          640
   10   CONTINUE                                                    650
   90   FORMAT(1H ,25X,17H COLUMNS OF ARRAY/3X,1HI12/5H ROWS)       660
   92   FORMAT(1H ,I4,10F10.4)                                      670
        RETURN                                                      680
        END                                                         690
```

*Fig 11.8* Computer program for correlation analysis

275

After the usual introductory statements, three data cards are read under the control of the first read command to take in a run title (TITLE), the number of variables (NV) and cases (NS) and a card giving the format for the succeeding data cards (FMT). The single-dimensioned array SUMX is intended to hold the sums of each variable ($\Sigma X_i$) and the two-dimensional array R the cross products ($\Sigma X_i X_j$) for all possible combinations of variables so that both these arrays must at the outset be initialised with zero values for every element. The data cards themselves are then read in one at a time according to the formatting information held in the array FMT under the control of a single DO loop which uses K as its control variable. Each card has punched in it the value of the NS variables for a single case and these are used to accumulate the raw totals in SUMX and the cross-products in R using nested DO loops with I and J as their control variables. These loops ensure that the cross-products are summed for every possible combination of variables. Notice also that the diagonal elements of the array, $R_{i,i}$, will contain the sums of squares for each variable ($\Sigma X_i^2$) needed to calculate the standard deviations.

When these summations are complete, a second phase of the program calculates and prints out the mean and standard deviation of each variable using the statements:

SUMX(I) = SUMX(I)/FNS    (forms the means)    (2)
X(I) = SQRT((R(I,I)/FNS) - (SUMX(I)**2))
            (forms standard deviations)    (3)

A final computational step assembles the matrix of all possible intercorrelations between pairs of variables using the formula (1) above. It is at this point that our program makes use of a general-purpose subroutine called MOUT to output the array of correlation coefficients held in R. In the definition, which is external to the main program, subroutine MOUT has the dummy arguments A, NR, NC and NDEV where A is the array to be output, NR the number of rows it contains, NC the number of columns and NDEV the number of the peripheral device to which the output is to be sent. It is called using the command

CALL MOUT (R,NV,NV,NOUT)

which automatically associates the actual array R with the dummy A and the current values of NV and NOUT with NR, NC and NDEV. The body of the subroutine is straightforward but incorporates two fairly useful programming devices. The first involves the arguments NR, NC and NDEV which are constantly used by the subroutine. It will be remembered

that on execution these dummy arguments are associated
with the actual arguments by way of the normal mechanism,
so that every time the computer needs to use one of these
arguments the dummy has to be associated with the actual
argument and a value for this actual argument retrieved
from the main program.  By computer standards this constant reference back to the main program to obtain values
for variables that do not in fact change is a time-
consuming business.  To stop it we have inserted the
statements

$$NRR = NR$$
$$NCC = NC$$
$$IOUT = NDEV$$

which copy the values for the arguments into local storage
locations and thus limit the back-reference to the main
program to just three operations, one each for NR, NC and
NDEV.  All subsequent operations in the subroutine use
these new local names.

The second device is much simpler and involves a mechanism
whereby large matrices of order greater than 10 by 10 can be
output in strips, each occupying a page of printer paper.
The exact mechanism used involves the DO loop

$$DO\ 10\ K = 1, NCC, 10$$

and the variable L; its working is left as an exercise for
the reader.

*Data and run.* Data for six variables measured for each 1km$^2$
grid square over a 25km$^2$ area around Calver have been
collected from the published 1:25,000 maps.  The major
features of this area and the data collection units are
shown as Fig 11.9.  To the east are barren uplands of
Millstone Grit, separated from an undulating upland cut
across Carboniferous Limestone in the west by the valley of
the River Derwent which is developed along shale strata.
The variables were selected in order to characterise the
visual landscape as follows:

(i) The *mean grid square height*, estimated as the sum
of the four corner heights plus four times the
value at the centre of each square, all divided by
8 (ft).

(ii) The *relative relief*, measured simply as the vertical
height difference between the highest and the lowest
points in the square (ft).

(iii) The *length of water-courses* (miles).  Each grid
square is of the same size so that these numbers
are also the drainage density.

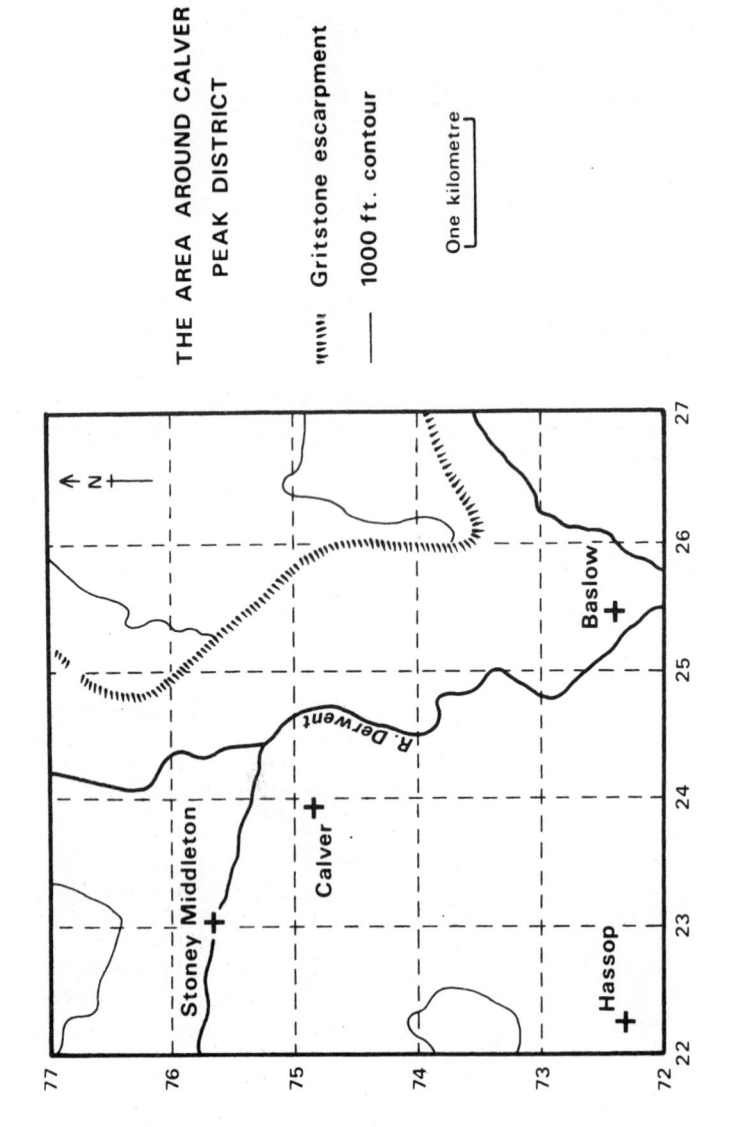

Fig 11.9 The area around Calver, Peak District

(iv) The *number of fields* per grid square.
(v) A *settlement index* giving an estimate of the number of permanent structures in each square.
(vi) The *amount of rough pasture*, estimated as the number of points on a regular 7 by 7 grid of points laid over each square that fall on land given over to rough pasture.

The data are shown as Fig 11.10 and the results as Fig 11.11. Figure 11.12 is an attempt to show the significant features of the resultant correlation matrix in a diagrammatic form. It should be noted at this point that the interpretation of these statistics in an *inferential* sense is complicated by the modifiable areal units used to collect the data as well as by the problem of spatial autocorrelation in each variable and the would-be correlation analyst should read the summary of these problems given in L. J. King (1969) before launching into an interpretation of what they mean. As *descriptive* indices the correlation coefficients give a very clear picture of the associations between the physical and the human landscapes in this area. Variable 1, the mean height, seems to be dominant and has moderate negative correlations with variables 3,4 and 5, all of which decrease as it increases. In the higher grid squares, the stream length, amount of settlement and number of fields are all less than in the lower squares along the Derwent Valley. The high positive correlation of mean height with the amount of rough pasture indicates that this increases markedly as one moves onto the high gritstone moorlands. Variable 2, the relative relief, does not correlate well with any of the other variables but as expected there are moderate positive relations between the amount of settlement, stream length and number of fields and a weaker negative relationship between settlement and rough pasture. Thus, unlike the area around Nottingham studied by C. A. M. King (1969), this area around Calver shows a reasonably close relationship between the physical and human landscapes. This is not to imply that any direct deterministics processes are at work in this area, but it does indicate and isolate associations upon which much of the geographical character of the Peak District depends.

11.8 Worksheet
1 Which of the following 16 statements are invalid in

```
CORRELATION ANALYSIS OF LANDSCAPE VARIABLES AROUND CALVER, PEAK DISTRICT
     6    25
  (5X,2F6.1,F5.2,F5.1,F5.2,F5.1)
  2272  600.0 375.0 0.00 31.0 4.20   0.0
  2273  919.0 550.0 0.00 27.0 0.00   0.0
  2274  822.0 375.0 0.00 30.0 0.00  19.0
  2275  759.0 350.0 1.50 63.0 8.20   0.0
  2276  969.0 525.0 0.00 67.0 7.00   4.0
  2372  578.0 300.0 0.00 34.0 0.40   3.0
  2373  628.0 475.0 0.00 39.0 0.40   8.0
  2374  697.0 475.0 0.00 54.010.00   5.0
  2375  547.0 325.0 1.50 46.011.20   0.0
  2376  759.0 675.0 0.00 30.0 2.60   1.0
  1472  444.0 175.0 0.75 62.0 9.00   0.0
  2473  466.0 325.0 2.00 42.0 2.40   3.0
  2474  463.0 175.0 2.50 64.016.20   3.0
  2475  513.0 200.0 2.75 55.012.80   6.0
  2476  597.0 525.0 2.25 41.011.40  15.0
  2572  447.0 300.0 1.50 46.037.00   0.0
  2573  572.0 550.0 0.25 64.0 2.60   8.0
  2574  716.0 600.0 0.00 63.015.00  24.0
  2575  916.0 575.0 0.00 11.0 0.00  38.0
  2576  975.0 250.0 0.25  3.0 0.00  49.0
  2672  538.0 375.0 3.00 34.0 1.80   5.0
  2673  753.0 550.0 1.50 34.0 0.60  34.0
  2674  966.0 200.0 1.00 28.0 0.20  27.0
  2675 1113.0 200.0 0.00 15.0 0.00  47.0
   676 1113.0 150.0 0.00  1.0 0.00  49.0
```

*Fig 11.10* Data for the Calver area

a *main* program and why?  What subprograms are required by each?

(i)     BA = SIN(NB)
(ii)    C = SQRT(ABS(D) + B)
(iii)   SUBROUTINE SUM(A,B)
(iv)    DAD = CALL SUM(PQ,X)
(v)     SUM(A,B) = A + B
(vi)    DIF(B,A) = A*A - B*ABS(NM)
(vii)   RETURN
(viii)  DIMENSION Y(2∅)
        CALL SUM(Y,Y(2))
(ix)    M = L(B) + INT(F)
(x)     POP = B**ABS(Y(Q))
(xi)    CALL MAIN (A(B(C,D)))
(xii)   FUNCTION BAT
(xiii)  ERS = SUB(A,B) - SUB(1,X)
(xiv)   DIMENSION B(2∅), A(2,5,2), C(1∅)
        X = SUM(SUM(B,A),C))
(xv)    Q = SUM(SUM(SUM(A,B),C))
(xvi)   P = SUM(SIN(X),COS(Y)) + ABS(X)

```
CORRELATION ANALYSIS OF LANDSCAPE VARIABLES AROUND CALVER, PEAK DISTRICT
  600.000    375.000    0.000    31.000     4.200     0.000
  919.000    550.000    0.000    27.000     0.000     0.000
  822.000    375.000    0.000    30.000     0.000    19.000
  759.000    350.000    1.500    63.000     8.200     0.000
  969.000    525.000    0.000    67.000     7.000     4.000
  578.000    300.000    0.000    34.000     0.400     3.000
  628.000    475.000    0.000    39.000     0.400     8.000
  697.000    475.000    0.000    54.000    10.000     5.000
  547.000    325.000    1.500    46.000    11.200     0.000
  759.000    675.000    0.000    30.000     2.600     1.000
  444.000    175.000    0.750    62.000     9.000     0.000
  466.000    325.000    2.000    42.000     2.400     3.000
  463.000    175.000    2.500    64.000    16.200     3.000
  513.000    200.000    2.750    55.000    12.800     6.000
  597.000    525.000    2.250    41.000    11.400    15.000
  447.000    300.000    1.500    46.000    37.000     0.000
  572.000    550.000    0.250    64.000     2.600     8.000
  716.000    600.000    0.000    63.000    15.000    24.000
  916.000    575.000    0.000    11.000     0.000    38.000
  975.000    250.000    0.250     3.000     0.000    49.000
  538.000    375.000    3.000    34.000     1.800     5.000
  753.000    550.000    1.500    34.000     0.600    34.000
  966.000    200.000    1.000    28.000     0.200    27.000
 1113.000    200.000    0.000    15.000     0.000    47.000
 1113.000    150.000    0.000     1.000     0.000    49.000

VARIABLE   MEAN       STAN.DEVN
   1      714.0000    205.5329
   2      383.0000    153.7400
   3        0.8300      1.0068
   4       39.3600     18.9418
   5        6.1200      8.1913
   6       13.9200     16.6323

MATRIX OF CORRELATION COEFFICIENTS
                      COLUMNS OF ARRAY
           1          2          3          4          5          6
   1    1.0000     0.0447    -0.5664    -0.6101    -0.5208     0.7270
   2    0.0447     1.0000    -0.3062     0.1367    -0.1180    -0.1581
   3   -0.5664    -0.3062     1.0000     0.3094     0.3767    -0.2815
   4   -0.6101     0.1367     0.3094     1.0000     0.5217    -0.6929
   5   -0.5208    -0.1180     0.3767     0.5217     1.0000    -0.3892
   6    0.7270    -0.1581    -0.2815    -0.6929    -0.3892     1.0000
```

Fig 11.11 Computer results for the Calver area

2 Which of the following statements or groups of statements are invalid and why?
    (i)    COMMON X,Y,Z,X(10)
    (ii)   COMMON COMMON
    (iii)  COMMON Z(9),Q(3)
           DIMENSION Z(6)

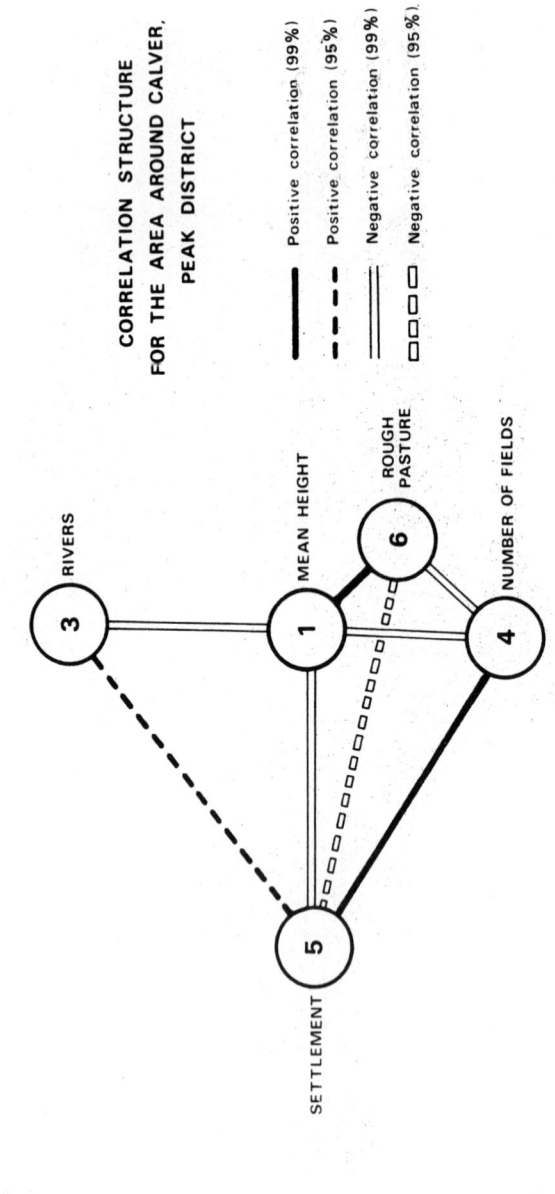

Fig 11.12 Correlation structure for the area around Calver

(iv) COMMON DIMENSION B(2∅),XY(1∅∅)
(v) DIMENSION Z(25)
    COMMON Z,B(3∅),C
    DIMENSION C(1∅∅)
(vi) COMMON X,Y,Z(26)
    DIMENSION X(1∅),Y(1∅),Z(26)
(vii) COMMON X,Y,Z,A
    COMMON A,B,C,D
(viii) COMMON B
    DIMENSION A(2∅)
    COMMON A

3 What are the resultant values of those variables changed as a result of executing the following statements, assuming the existence of the subroutine on the right?

```
COMMON X,K              SUBROUTINE DOIT(M,N,K)
CALL DOIT(2,I,P)        COMMON Q,MINE
                        Q = 3.∅
                        MINE = 29
                        NO = INR(2.∅*Q) - MINE + M*M
                        X = NO
                        RETURN
                        END
```

4 After studying the exercises of Haggett and Chorley (1969, pp35-43), write a program to input and then power any binary connectivity matrix until all the off-diagonal elements have been filled. Use a subroutine to perform the matrix multiplication, and try to make this as general as possible.

The following is the binary connectivity matrix for the UK telephone trunk network. There are 27 major exchanges as indicated by the 27 rows and columns of the matrix and the presence or absence of a direct link has been indicated by scoring a 1 where a link exists and a zero otherwise.
Run your program using the data and then answer the following:
  (i) at each step which exchanges are the best and worst connected?
  (ii) overall, which exchange is the best and which is the worst connected?
  (iii) what is the maximum number of steps necessary to transmit a call through the entire network?
  (iv) experiment with your program to discover the effects of adding additional links. Would, for example, a link between Cardiff and Shrewsbury increase the connectivity of the network more or less than one between Leeds and Birmingham?

| | | |
|---|---|---|
| 1 | Wick | 010000000000000000000000000 |
| 2 | Inverness | 101101000000000000000000000 |
| 3 | Aberdeen | 010011000000000000000000000 |
| 4 | Oban | 010001000000000000000000000 |
| 5 | Edinburgh | 001001001000000000000000000 |
| 6 | Glasgow | 011110100000000000000000000 |
| 7 | Carlisle | 000001011110000000000000000 |
| 8 | Stranraer | 000000100000000000000000000 |
| 9 | Newcastle | 000010100100000000000000000 |
| 10 | Leeds | 000000101010000000000100000 |
| 11 | Manchester | 000000100101000000000100000 |
| 12 | Liverpool | 000000000010110000000000000 |
| 13 | Shrewsbury | 000000000001011000000000000 |
| 14 | Birmingham | 000000000001100000000100000 |
| 15 | Gloucester | 000000000000100101000000000 |
| 16 | Oxford | 000000000000001010000100000 |
| 17 | Bristol | 000000000000000101100100000 |
| 18 | Cardiff | 000000000000001010000000000 |
| 19 | Exeter | 000000000000000010011000000 |
| 20 | Penzance via Plymouth | 000000000000000000100000000 |
| 21 | Southampton | 000000000000000000100100000 |
| 22 | London | 000000001100101100010111 00 |
| 23 | Cambridge | 000000000000000000000101000 |
| 24 | Norwich | 000000000000000000000110000 |
| 25 | Maidstone | 000000000000000000000100011 |
| 26 | Canterbury | 000000000000000000000000101 |
| 27 | Dover | 000000000000000000000000110 |

5(i) Write a program to simulate stream patterns using random walk techniques (Leopold and Langbein, 1962), printing out the evolving pattern using numbers to signify stream orders. In determining changes in river direction use a probability of 0.275 for turns to the right and to the left and 0.45 for the river to go straight on. Start by assuming 8 equally-spaced streams at the top of the page. To assist in this program you will need a random number generator. *Either* use any random number generating subroutine or function that is supplied by your computer unit *or* compose a subroutine to give two-digit numbers in the range 0-99 using the following sequence:

```
         RAND(1) = 94.0
         DO 10 J = 2,N
         RAND(J) = (10011.0 * RAND(J - 1)) * 0.001
         RAND(J) = (RAND(J) - INT(RAND(J))) * 100.0
      10 CONTINUE
```

In the above, N is the number of random numbers required

and RAND is an array in which they are stored. The
initial value of RAND is arbitrarily set at 94; any other
two-digit number would serve. Because these numbers have
been generated by a distinct sequence, they are not strictly
random; rather they are numbers that have most of the
properties of truly random numbers and for this reason they
are often called pseudo-random. To decide in which
direction to turn a stream, allocate the numbers 0-26 for
a left turn, 27-72 for straight on and 72-99 for a right
turn.

(ii) Analyse your simulated pattern using the Strahler
(1952) system of stream ordering. Does the law of stream
numbers hold for this simulated pattern?

(iii) Repeat the analysis using probabilities of 0.33,
0.34 and 0.33. How is the simulated pattern affected by
this change and what effect does it have on the fit to the
law of stream numbers?

(iv) Consider what modifications would have to be made
to the program to allow streams to turn in all directions
with a probability grid which is symmetrical about the
centre square.

6 Using the following data on small towns in North Island,
New Zealand write a program to:
  (i) Calculate reduced major axis coefficients for popu-
      lation plotted against sales and for population
      against sales per shop.
  (ii) Draw histograms of the residuals in each relation-
       ship and print out a ranked list of these values,
       labelling each with the town name.
For the method of calculation of reduced major axes see
Till (1973), and try to make your program as modular as
possible using functions and subroutines where appropriate.
The data are drawn from the Census of Distribution for 1968.

| Town | Population | No of shops | Total sales ($,000) |
| --- | --- | --- | --- |
| Eastbourne | 4620 | 28 | 1004 |
| Featherston | 1950 | 37 | 1923 |
| Greytown | 1730 | 23 | 1241 |
| Martinborough | 1450 | 26 | 1569 |
| Carterton | 3690 | 62 | 3540 |
| Otaki | 3690 | 69 | 3009 |
| Foxton | 2860 | 49 | 2048 |
| Pahiatua | 2590 | 67 | 3810 |
| Eketahuna | 720 | 14 | 934 |

| Town | Population | No of shops | Total sales ($,000) |
|---|---|---|---|
| Dannevirke | 5800 | 111 | 6919 |
| Woodville | 1540 | 33 | 1105 |
| Feilding | 9510 | 140 | 12051 |
| Havelock North | 6190 | 45 | 1713 |
| Taihape | 2890 | 59 | 4155 |
| Ohakune | 1380 | 25 | 1259 |
| Raetihi | 1360 | 32 | 1769 |
| Patea | 2010 | 36 | 1532 |
| Hawera | 8230 | 149 | 12047 |
| Eltham | 2310 | 43 | 2186 |
| Stratford | 5490 | 104 | 7906 |
| Inglewood | 2010 | 45 | 3010 |
| Waitara | 4900 | 66 | 3834 |
| Taumarunui | 6130 | 113 | 10178 |
| Taupo | 8730 | 131 | 7572 |
| Wairoa | 5240 | 100 | 7993 |
| Opotiki | 2570 | 64 | 3878 |
| Whakatane | 9230 | 150 | 10945 |
| Kawerau | 6080 | 51 | 3638 |
| Pataruru | 4530 | 89 | 6905 |
| Te Awamuta | 6820 | 155 | 11913 |
| Cambridge | 6130 | 104 | 8071 |
| Matamata | 3940 | 105 | 6325 |
| Te Puke | 3120 | 80 | 6578 |
| Mt Manganui | 7390 | 125 | 4553 |
| Ngaruawahia | 3800 | 43 | 1760 |
| Huntly | 5440 | 81 | 4735 |
| Tuakua | 1700 | 32 | 2196 |
| Thames | 5710 | 123 | 6582 |
| Paeroa | 3160 | 74 | 3589 |
| Waihi | 3170 | 64 | 2764 |
| Pukekohe | 6860 | 129 | 13069 |
| Waiuku | 2100 | 49 | 2961 |
| Morrinsville | 4555 | 110 | 6901 |
| Te Aroha | 3260 | 87 | 5935 |
| Helensville | 1320 | 30 | 1848 |
| Dargaville | 3970 | 95 | 8106 |
| Kaikohe | 3200 | 70 | 5294 |
| Kaitaia | 3170 | 85 | 6343 |
| Northcote | 8670 | 62 | 3650 |
| Newmarket | 1080 | 145 | 14549 |
| Henderson | 5860 | 122 | 6013 |
| Ellerslie | 4280 | 54 | 5838 |
| Glen Eden | 6270 | 51 | 2024 |

7 Markov chain analysis is proving to be a powerful tool in the analysis of change over time, as for example in the studies by Lever (1972) and by Clarke (1965). A simple introduction, with examples, has been given by Harvey (1967) to which reference should be made for the detail of the techniques required.
  (i) Write a computer program to read in and power a transition probability matirx. You should be able to make use of the general purpose multiplication routine developed in exercise 4 and perhaps also SUBROUTINE MOUT from the worked example earlier in this chapter. In addition, the initial state vector should be read in and the actual distribution at each time period printed out. Arrange your program to terminate when the unique fixed point vector has been found to an accuracy of 0.00001.
 (ii) Because the ratio of birth-rate to death-rate is roughly constant for all towns, the relative changes in urban population in the UK are dominantly the result of in- and out-migration. An estimate of the transition probabilities of towns in one size class at time $t_n$ being in the same or in a different size class at the end of a ten-year period $t_{n+10}$ is given by:

|  |  |  | Size Group at $t_{n+10}$ |  |  |  |
|---|---|---|---|---|---|---|
|  |  |  | A | B | C | D |
| Size Group at $t_n$ | A | .LT.10,000 | 0.90 | 0.09 | 0.01 | 0.00 |
|  | B | 10-20,000 | 0.02 | 0.81 | 0.17 | 0.00 |
|  | C | 20-60,000 | 0.00 | 0.02 | 0.93 | 0.05 |
|  | D | .GT.60,000 | 0.00 | 0.00 | 0.01 | 0.99 |

Given initial proportions in 1951 of 0.38 in A, 0.22 in B, 0.27 in C and 0.13 in D, what are the predicted distributions for 1961, 1971 and 1981?
(iii) Allow your program to run on to find the unique fixed point vector, and then repeat the analysis using initial proportions of 0.50, 0.25, 0.13 and 0.12. What does your result tell you about the nature of these Markov models?

11.9 Further Reading
Clarke, W.A.V. 'Markov chain analysis in geography: an application to the movement of rental housing areas', *Annals, Association of American Geographers*, 55, 351-59 (1965)

Clarke, P.T. and Evans, F.C. 'Distance to nearest neighbour as a measure of spatial relationships in populations', *Ecology*, 35, 445–53 (1954)

Day, A.C. *FORTRAN techniques with special reference to non-numerical applications* (1972)

Haggett, P. and Chorley, R.J. *Network analysis in geography* (1969)

Harvey, D.W. 'Models of the evolution of spatial patterns in human geography', in *Models in Geography* (ed. Chorley and Haggett, 1967)

Hill, A.R. 'The distribution of drumlins in County Down, Ireland', *Annals, Association of American Geographers*, 63, 226–46 (1973)

King, C.A.M. 'Map orientation of the Nottingham area by means of factor analysis', *East, Midland Geographer*, 4, 400–413 (1969)

King, L.J. *Statistical analysis in geography* (1969)

Leopold, L.B. and Langbein, W.B. 'The concept of entropy in landscape evolution', *US Geol. Survey, Professional Papers*, 500A (1962)

Lever, W.F. 'The intra-urban movement of manufacturing: a Markov approach', *Transactions, Institute of British Geographers*, 56, 21–38 (1972)

Smalley, I.J. 'Drumlin formation: a rheological model', *Science*, 151, 1379–80 (1966)

―― and Unwin, D.J. 'The formation and shape of drumlins and their distribution and orientation in drumlin fields', *Journal of Glaciology*, 7, 377–90 (1968)

Strahler, A.N. 'Hypsometric (area–altitude) analysis of erosional topography', *Bulletin, Geological Society of America*, 63, 1117–42 (1952)

Till, R. 'The use of linear regressions in geomorphology', *Area*, 5, 303–5 (1973)

Vernon, P. 'Drumlins and Pleistocene ice flow over the Ards Peninsula/Strangford Lough area, County Down, Ireland', *Journal of Glaciology*, 6, 401–9 (1966)

# Chapter 12
# Common Errors and Good Programming Style

12.1 Programming Style
The reader has now studied and used most of the commands making up standard FORTRAN and can write long, complex programs.  It will be apparent that some programs are 'good', others 'bad', but by what standards can computer programs be judged?  A good program probably will show four characteristics.  First, it will be free of FORTRAN coding errors and will provide correct solutions within the accuracy and precision specified by the programmer. This is termed *program integrity*.  Secondly, the program will be logically constructed and capable of being run without much modification on all machines on which FORTRAN is possible.  This feature is called *program design* and *program portability*.  Thirdly, efficient programming involves communication from programmer to programmer and with the outside world, so that a program should be easy to understand.  This involves clear *program presentation* and adequate *program documentation*.  Finally, a good program will execute without excessive use of processor time and storage.  This is *program efficiency*.  In their book on *Program Debugging*, Brown and Sampson (1973, p 35) liken a good program to a large amiable dog, 'not easily ruffled, slow to take offence and difficult to divert from its chosen course'.  They conclude that most programs are more like poodles, 'very finnicky about their food, demanding only the very best and tastiest tit-bits, very quick tempered, easily upset and generally more trouble than they are worth'.

This chapter aims to improve programming standards; the integrity, design, portability, presentation, documentation and efficiency of the programs we write.  We will also introduce three new data types to help in writing more effective programs and called DOUBLE PRECISION, COMPLEX and LOGICAL.

## 12.2 Program Integrity

More often than not, the geographer-programmer is more interested in the results of his analyses than in the niceties of how they are obtained. His first objective will be to produce a coding that gives correct answers and it is to this aspect that attention first turns. Errors in programs greatly increase the cost of computing by increasing the time spent by both the machine and the programmer; they lead to frustration of the programmer's creative talents as he spends most of his time debugging in an atmosphere of impatience instead of getting on with other, more useful, tasks. They also have an effect on their victims; an incorrect program is worse than none at all, and there is always someone ready to seize on a mistake and blame the computer.

There are three main types of error that can cause program failure. *Hardware failures* are now very rare and their existence usually obvious or the operating system is designed so that the user is unaware that a failure has occurred. Similarly, at the level of programming of this book, *systems software and compiler errors* are unlikely to be met very often. Most of our errors will be associated with the *programming* itself. We require programs to be correct in at least two senses, involving a correct coding together with an algorithm capable of giving sound answers when real numbers are used to solve real world problems.

Simple FORTRAN coding errors are usually detected by the compiler and can be removed during a 'dry run' and during program debugging as discussed in Chapter 3. Most computer centres produce a duplicated list of error messages produced by the compiler which expands on the messages that are written onto the computer output. There remain a number of very common errors which are not compiler-detected and which appear when the program is first run to give either nonsensical results, or worse, a catastrophic failure. This type of error causes most of the problems in debugging.

One of the most common errors is the accidental incorrect spelling of a variable name as in the sequence:

```
      C SET COUNTER AND TOTAL THE ARRAY A
        SUMM = 0.0
        DO 10 I = 1,100
        SUM = SUM + A(I)
     10 CONTINUE
```

The location accidentally spelt SUMM is intended to be the same as the SUM used to accumulate the required total. The result is that SUM is never initialised at zero and starts with random data. At exit, we have no guarantee that it will contain the true total, yet this error might only make itself felt by giving final, wildly inaccurate results. The characters Z, O (zero) and 1 (one) are often confused on coding forms with the similar 2, O and I and thus lead to variable's names being misspelt. As a matter of policy they should be written out according to some unambiguous convention as Z, ∅, 1, 2 Ō and I. It is also sound practice to avoid using them when naming variables.

A second common source of error arises when a program that works perfectly with test data fails dramatically when introduced to real data containing fatal conditions not anticipated in the program. A good program will always print out a mirror-image of its data and perform as many checks as feasible to ensure that they are reasonable. In commercial environments, computers often deal with a great deal of 'bad' data and are usually programmed to handle many error conditions, but in science we usually assume that all our data are 'good' and seldom test them adequately. Yet errors occur in coding, punching and specifying (by FORMAT statements) the input data. Whatever the cause, the result will often be to give totally implausible information to the program. Examples of this might be symmetric matrices (such as that used in exercise 11.9(4)) punched so as to be unsymmetric, percentage data that do not sum to as near to 100 as makes no difference, physically impossible negatives ($-10^O K$) and questionnaire codes that exceed the allowed range. Given these data, a good program will not press on regardless to produce nonsense or a catastrophic failure requiring the intervention of the operating system; it will detect errors, print out messages that identify them and then retire 'gracefully'. Alternatively, a short, separate program may be designed to test the data and run prior to the main program. The errors can then be corrected in the program re-run. One very common consequence of invalid data occurs if the computer attempts to divide by zero resulting in 'floating point overflow' as it reaches its own infinity – the highest number that it can store. Similar errors occur if standard supplied functions are called with invalid mathematical arguments as for example in the calls LOG1∅ (∅.∅) and SQRT(-1.345). Tests for zero and negative values are easily made with conditional statements prior to calling the supplied function.

Just as floating point overflow occurs when we try to use a number in excess of the highest that the machine can store, so subscript overflow occurs when a program tries to reference an array subscript exceeding the maximum size of the array given in the DIMENSION statement. Unfortunately, FORTRAN implementations are not supposed to check for this (many do) and it will pass undetected, resulting in random data being taken from storage locations that might have nothing to do with the program. Alternatively, an attempt to write into store outside array limits will overwrite locations belonging to other programs. At the extreme of nastiness these overwritten locations could be those used by the computer's own software, resulting in a serious system violation. Subscript overflow is normally detected by the occurrence of inexplicable values and a haphazard flow of control but is often difficult to locate precisely without running special checking programs. It seems to occur most frequently when a DIMENSION statement is set to the exact number of elements required then later in the program an attempt is made to increment a subscript one past this maximum:

```
          DIMENSION A(N)
             .
             .
             .
          DO 10 I = 1,N
       10 A(I + 1) = A(I + 1) + A(I)
```

Frequently, programs fail when their flow of control becomes disturbed, a fault often associated with the 'administrative variables' used to control it. A good program *never* relies on a test for equality involving real numbers such as

```
          IF (Z .EQ. 3.14159) GO TO 999
```

Z is very unlikely to be stored as exactly 3.14159 and there is no way of reaching the statement labelled 999. If this was

```
      999 STOP
```

we have created a program that may never cease to run and have wasted a great deal of processor time. Similarly consider

```
          GO TO (11, 13, 12), N  - what happens if N = 4?
and       DO 10 K = N1, N2      - is N1 ever less than
                                  1 or N2 less than N1?
```

and be watchful of 'one-out' errors in sorting programs resulting from the use of .GT. and .LT. when .GE. and .LE. were intended.

We could prolong this list to cover all sorts of unlikely errors, but as the programmer gains experience so he will learn by his mistakes, adopt standard ways of coding

operations that he knows are correct, and write programs
containing fewer and fewer simple coding errors. At the
same time it is likely that his programs will become
numerically more and more sophisticated and it is important
that he can detect the subtle errors that can occur as
a result of using an inadequate or even incorrect algorithm.

Our natural tendency is to assume that all arithmetic done
on a computer is reliable and more accurate than we could
achieve using pencil and paper. In many applications there
is abundant evidence that this is not the case and that
a high proportion of all computed results contain larger
errors than the author of the problem will know about or
admit. Often plausible, but incorrect, answers can be obtained and a program that works perfectly with one set of
data on one computer may not work with different data on
the same machine or with the same data on a different
machine. In the days of hand calculation, a human computer kept a record of the numbers used, avoided difficult
and error-prone operations, and carried as many significant
digits as seemed necessary to get a reasonable answer.
A digital computer with its fixed word length (*see Chapter
2*) provides a very different environment. Calculations
are carried out without intervention on quantities whose
representation is limited by a physical characteristic of
the machine. A digital machine cannot store quantities
such as $\frac{1}{3}$, $\pi$ and e without truncation and very large or
very small numbers without rounding off with a subsequent
loss of significant digits. Whether or not the uncertainties and errors introduced by round-off and truncation
are transmitted through to the solution in magnified or
reduced form depends very much upon the algorithm used.
Any serious scientific programmer working in any applications
field should be aware of these problems and should try to
implement only sound algorithms that will detect and warn of
these numerical difficulties. The study of numerical error
and the design of algorithms to avoid it form part of the
branch of mathematics called *numerical analysis* and are the
subject of a number of texts. Those by Fox (1964), Fox and
Meyers (1968) and Froberg (1970) are particularly recommended.
Some specific examples of the numerical problems in normal
least squares regression are dealt with in papers by Longley
(1967) and Wampler (1970) and in trend surface analysis by
Unwin (1975).

12.3 DOUBLE PRECISION, COMPLEX and LOGICAL Data
The designers of FORTRAN anticipated three particular

programming problems by providing three special data types to overcome them:

(i) DOUBLE PRECISION data to enable numbers to be stored with greater precision than is normal.
(ii) COMPLEX data to allow complex numbers to be used.
(iii) LOGICAL data to allow operations of the logical 'true' or 'false' type.

All the programs used in this book have involved three basic data types called REAL (denoted by the format specifiers F and E), INTEGER (denoted I) and alphanumeric (denoted A). The distinction between real and integer quantities has been implicit, using the first letter of any variable name to indicate its type. Thus the letters IJKLMN used as the first letter in a variable name signal that it is to be of type INTEGER. If, for any reason, we want to override this convention a *type declarator* can be inserted into the program. This is a non-executable statement, punched in columns 7-72 of the card, that must precede any executable statements and is of the general form:

$$\underline{t}\ v_1, v_2, \ldots, v_n$$

In this $\underline{t}$ indicates the data type and can be REAL, INTEGER, DOUBLE PRECISION, COMPLEX or LOGICAL. $v_1$ to $v_n$ are any variable names to be given the specified type. The pair of statements

INTEGER X,A,S
REAL ME, LONG, ITER

would override the normal typing convention and reserve integer locations for X, A and S, real locations for ME, LONG and ITER.

(i) DOUBLE PRECISION typing is used to represent ordinary real numbers, but instructs the computer to store them using more space than normal. As the name implies, this will usually be twice as many locations as used in single precision. These extra locations allow any datum declared to be of this type to be stored with more significant decimal digits than in single precision. Depending upon the computer used, a single precision location will hold anything from 7 to 20 decimal digits, a DOUBLE PRECISION location from 14 to 40 digits.

A constant can be written in DOUBLE PRECISION as

3.9370074D-2

in which the D replaces the normal E in exponent notation. Unless we actually need the extra digits there is little point in writing down constants in this form. A variable

294

can be declared to be of type DOUBLE PRECISION by writing
a type declarator
$$\text{DOUBLE PRECISION } v_1, v_2, \ldots, v_n$$
in which v signifies any variable name, including arrays.
Expressions involving double precision are easy to write
and when DOUBLE PRECISION constants and variables are mixed
with ordinary single precision real numbers will always
give a DOUBLE PRECISION result.  This is not true if the
single precision quantities are type integer.  In addition
certain supplied functions, notably DEXP, DLOG, DLOG1∅ and
DSIN, will return DOUBLE PRECISION solutions.

It is all very well to manipulate quantities with extra
accuracy within the machine, but we also need mechanisms
for input and output of these numbers.  FORTRAN allows use
of the D format in which fields are specified exactly as in
E format but using a D in place of the E.

DOUBLE PRECISION should always be used to guard against
significant round-off error in operations involving large
or small numbers, but there are two problems that deserve
mention.  First, contrary to commonly expressed opinion,
DOUBLE PRECISION is not a complete panacea.  It should
never be used to support a fundamentally unsound algorithm.
If a program genuinely needs this facility, the programmer
ought to think seriously of other methods of solution.
Secondly, the introduction of DOUBLE PRECISION can make
a program very machine dependent.  Regrettably, standard
FORTRAN lays down no minimum precision requirement, only
that 'the degree of approximation . . . must be greater
than that of type real'.  As a result DOUBLE PRECISION
means different things to different computers.  IBM
System 360 computers with a 32-bit word have, for example,
only a marginally more accurate representation if used in
DOUBLE PRECISION than a CDC 6000 Series machine with a 60-
bit word in single precision.

(ii) COMPLEX data are used to represent complex numbers
involving the imaginary quantity i = SQRT(-1).  A COMPLEX
datum is processor approximation to the value of a complex
number and consists of an *ordered pair* of variables.
These are two optionally signed real constants, separated
by a comma and enclosed within parenthesis as for example
   (2.∅, 3.∅)  equivalent to 2.0 + 3.0i
and  (4.E3, -1.3E2) equivalent to 4,000 - 130i
Like DOUBLE PRECISION, a COMPLEX variable must be declared
in a type declarator:
$$\text{COMPLEX } v_1, v_2, \ldots, v_n$$
Notice that the normal typing convention using IJKLMN does

not apply to COMPLEX data; once a variable is declared to be of this type, then COMPLEX it remains irrespective of the initial letter of the variable name. Input and output of COMPLEX data are easy. Remembering that the machine assigns *two* numbers for every COMPLEX variable or constant, all that need be done is to provide *two* appropriate fields in F or E format for each quantity. At output or input, the first number will be the real part, the second the imaginary. Although complex data are often used in engineering and electronics, we can think of very few geographic applications.

(iii) Finally, data may also be declared to be of type LOGICAL. A LOGICAL constant or variable is one that can take on one of two possible values, 'true' or 'false' according to some logical test. A FORTRAN logical constant is written simply as

.TRUE.

or .FALSE.

and a logical variable is declared using a type declarator of the usual form:

LOGICAL $v_1, v_2, \ldots, v_n$

LOGICAL constants and variables can be used with the relational and logical operators introduced in Chapter 6 to produce logical expressions similar in all respects to the more familiar arithmetic expressions:

arithmetic variable = arithmetic constant or arithmetic expression

logical variable = logical constant or logical expression

Hence we can write expressions such as

LOGICAL L5, ME
.
.
.
L5 = .TRUE.
ME = .FALSE.

which assigns the truth value 'true' to the location L5 and 'false' to ME. The right hand side of this expression might also be a logical expression as for example:

L5 = A .GT. 25.0

which sets L5 = .TRUE. if the value in the location A is greater than 25. Even longer relational expressions might be developed such as

L5 = AB .LT. EVAL .OR. ITER .GT. 40

which uses operators such as .LT.,.GT., and .OR. in a statement that can only be true or false according to the current values of EVAL and ITER. Input and output of logical

variables and constants uses the FORMAT specifier L together with the characters T and F.  The general form is
$$L\ w$$
in which w is the number of character positions to be used. The result is to set up a field of w-1 blanks followed by a T or an F in the remaining position according to whether the datum is .TRUE. (T) or .FALSE. (F).  Like the COMPLEX data type, LOGICAL data are seldom used in geographic applications.

## 12.4 Program Design and Portability

There is more to producing a good program than simply ensuring that it runs without error.  A good program will also have a robust, modular structure and be easily transferred from machine to machine.  To produce a good program structure, four basic rules should be followed:

(i) Never be in a hurry to code up your program into FORTRAN statements; work first on the logic using carefully drawn flowcharts for both the overall logic and for the fine structure of the coding.

(ii) Always carefully revise any coding after writing it down.

(iii) A good program is a *modular* program, made up of distinct and relatively self-contained units. Usually these will be FUNCTION and SUBROUTINE subprograms.

(iv) During its development phase a good program will also contain its own, programmed in, aids to debugging which can be removed for production runs without affecting the program's flow of control.

Correctly followed, these simple rules will eliminate trivial coding errors, reduce code duplication, reduce the core storage requirements and speed up program development and testing.  A point often forgotten is that general-purpose program modules developed in one context are often transferable without modification to a completely different problem and this reduces the time spent in developing new programs.  Eventually the programmer will have built up his own library of subprograms and will be able to write new programs simply by changing their order and supplying appropriate linking statements.

It is not easy to give definite rules for those operations which should be placed into FORTRAN subprograms, but as a guide you might produce modules for all segments whose functions are required at several places in a program (eg all matrix operations, print routines and error messages), all segments liable to frequent change as the program is developed, and all segments that will never change.  In a subroutine, 50 statements are a good average, but do not be afraid of much shorter segments.  One particularly helpful device is to anticipate future program developments and incorporate non-active 'dummy' subroutines into the original structure.  When you are ready to extend the program these routines can be filled with appropriate code without disturbance to the overall logic.

Fragmentation of a program into many distinct modules implies that we provide some way of transferring data from module to module.  As seen in Chapter 11, there are two ways of doing this, involving the COMMON list and the subprogram arguments.  It is important to use the correct mechanism.  A possible subdivision of subprograms is into *general purpose subprograms* used as a part of any program and *specific subprograms* unique to a particular application. Bearing this distinction in mind, we can suggest the following rule.  So that they can be incorporated without change into any program, general purpose subprograms should pass data through the arguments list, whereas specific subprograms should use COMMON.  Occasions will arise when we need to use both, but these should be the exceptions rather than the rule.  For simplicity, and to avoid undesirable side effects, try to place all, or nearly all, the program variables into COMMON and use only one version of the list in all the subprograms.  Avoid labelled COMMON, which is often implementation-dependent, and *never* vary COMMON across subprograms.

A well-designed program is also a portable program, and a great deal of programming effort would be saved if all programmers tried to write programs that will 'travel' from installation to installation without modification. FORTRAN is the most universal and the most portable of the programming languages.  Two of its most significant features are the simplicity of its essential commands and its consistency from machine to machine, so that moving a program is easiest when it is written in FORTRAN. Geographical programs tend to have long active lives and are frequently moved from installation to installation so that portability becomes a major program virtue.  Situations requiring portable programs are a change of institution, an institution changing its computer, a copy of a program being used at more than one institution, a program being published in the literature for general use, and even a situation in which the programmer has access to more than one machine. It is possible that the programmer will prefer to develop his programs using a small local machine and then actually run the programs with real data using a fast machine at a remote regional computer centre. Many of the case studies cited in this book have been run with only trivial modification on 4 or 5 different computer systems.

Again some fairly elementary rules can be suggested for producing portable programs:

(i) Portability implies generality. A general program will anticipate future changes and future applications. At a practical level, a good rule is always to use variables in preference to constants.

(ii) Confine yourself to what might be called essential FORTRAN statements - SUBROUTINE, FUNCTION, COMMON, DIMENSION, the various arithmetic and logical expressions, IF, GO TO, CALL, RETURN, DO, CONTINUE, STOP, READ, WRITE, FORMAT and END - and avoid using commands involving DATA, EQUIVALENCE, REAL, INTEGER, DOUBLE PRECISION, COMPLEX and LOGICAL. The list of essential commands should be sufficient to code most programs without losing too much efficiency.

(iii) Avoid writing programs that rely upon numbers above or below ±32767 in integer and $10^{38}$ with 7 significant digits in real. Many implementations place limits of this order on their numbers.

(iv) Do not use explicit device numbers in input/output operations. Do not write
```
       READ (7, 100) A
       WRITE (2, 11) A
```
use instead
```
       NIN = 7
       NOUT = 2
       ⋮
       READ (NIN, 100) A
       WRITE (NOUT, 11) A
```
NIN and NOUT can now be changed by simply altering two very simple assignments, rather than altering every single input or output statement. The effect is to make our communication with the outside world much more general; we could use different devices or could change machines very easily.

(v) Pay particular attention to input/output lists.

Some FORTRAN implementations allow free format input (usually signalled F$\emptyset.\emptyset$). Others will allow Hollerith strings in an output list to be written 'ABC' instead of 3HABC. Both facilities appear very useful and could save a lot of effort but both are implementation-dependent and should be avoided. A second source of trouble in input/output is the use of alphanumeric data in A format. Because it is possible to pack several alphanumeric characters into one word of computer store, as in A4 (4 characters per word) and even A1$\emptyset$ (10 characters per word), the temptation is always to save store by packing as many characters as possible into each word. This should be avoided. Although in most of our examples we have used A4, for complete generality use A1, and store all alphanumeric data in integer type locations.

(vi) Despite the FORTRAN standard of six letters or digits as an upper limit on each variable name, there is at least one common dialect (IBM 1130) that sets five as a limit so that one should try to restrict names to 5 characters.

(vii) Finally, many FORTRAN implementations will allow extensions to standard FORTRAN such as more than one statement per line, mixed mode arithmetic and so on. It goes without saying that these facilities are machine-dependent and should be ignored.

## 12.5 Presentation and Documentation

A good program is also easy to understand. Although all programs should be accompanied by an explanatory specification, it is good practice to make the code itself as self-explanatory as possible and there is a great deal that the programmer can do to improve the legibility of his efforts. The power of this method is that the annotation is done during programming and there is very little effort required on tedious documentation, usually a detested chore, after the testing is over. As a standard, we suggest that your programs should be self-documentary to the extent that they will be capable of being followed, in outline, by any non-programming geographer. It should not be necessary to have learnt the intricacies of FORTRAN to follow a computer program. Again there are some very simple rules to follow:

(i) Variable names should always begin with letters

that indicate their type (integers always i e IJKLMN and perhaps even D for double precisi ·· , C for complex and L for logical) and be long and meaningful. Variables named NIN, NOUT, SUMA, SUMA2, LOCQO and TABLE are fairly self-explanatory whereas names such as N, M, S, SB, LQ and T are not. If a document accompanies the program giving a mathematical development of the methods used, try to make variable names correspond to the notation used in the mathematical formulation.

(ii) Input and output should always be fully documented, and output clearly labelled. In these days of paper shortage output ought not to be excessive in its use of expensive printer paper.

(iii) Try to produce neat code. Spaces should be left around all operators, gather subroutines in alphabetic order, and gather all FORMATs together between STOP and END where they are easily found and will not clutter up the code.

(iv) Do not be afraid to indent statements to clarify a program structure as in the sequence:

```
C SUM THE ARRAYS A,B AND C
      SUMA = Ø.Ø
      SUMB = Ø.Ø
      SUMC = Ø.Ø
      DO 25 I = 1, NROW
           SUMA = SUMA + A(I)
           SUMB = SUMB + B(I)
           DO 3Ø J = 1, NCOL
                SUMC = SUMC + C(I,J)
 3Ø        CONTINUE
 25   CONTINUE
```

Notice that it is preferable to terminate every DO with a CONTINUE. This sequence is *much* clearer than the equivalent

```
      X = Ø.
      Y = Ø.
      Z = Ø.
      DO 2 I=1,N
      X = X+A(I)
      Y = B(I)+Y
      DO 2 J=1,M
 2    Z = Z+C(I,J)
```

(v) Statement numbers can confuse and should be used only where necessary. Try to put them in ascending order leaving gaps for future expansion.

(vi) Always provide an adequate commentary to identify the program, its programmer, the computing environment, all the input/output, and, as far as practical, all the variable names used. As a guide there should be about one comment line for every executable statement. The commentary and code can be laid out like a mathematics text book in which the commentary is the text and the code the equations. Use blank comment lines for clarity and always isolate particular self-contained program segments with appropriate annotation. Although comment cards are vital, there is little point in providing a self-evident commentary such as
```
      C INCREASE K BY 1.
        K = K + 1
```
which adds nothing to the code. A better comment should tell us something about K's function
```
      C INCREASE THE OUTPUT PAGE COUNTING LOCATION K
        K = K + 1
```

If due note is taken of the above guidelines, the program code ought to be more or less self-explanatory, but in addition there will be information that should be placed in a program specification. This must include a statement of what the program does, how and where it does it, all the relevant flowcharts, and details of a relevant set of test data with answers. If the program has been used for work subsequently published, or is based upon some published work, full references should be given. It is a great temptation not to write up a program, largely with the excuse that the author knows what it does and that it is not intended to be used by someone else. It is very easy to forget important program details; in effect next year you *will* be someone else! One useful exercise at this point would be to look back over the worked examples in this book, noting how and where the style could be improved. Many examples were produced to illustrate specific programming syntax rather than to act as final production programs, and we are all too aware that in many places they do not meet the exacting standards set in this chapter.

## 12.6 Efficiency and Use of Store

A good program is efficient in its use of computer time and store. In the early days machine time was costly and difficult to come by so that great stress was laid upon program efficiency. Programmers would spend hours trying

to shave fractions of seconds off run times. The
geographer-programmer whose primary interest is in the
results is likely to view speed of running as being of
somewhat diminished importance. Now computers are in
themselves very much faster and more efficient. The
costs of computation have fallen dramatically yet the costs
of the programmer's time have risen in like manner. It
should still be an aim to produce very efficient programs,
but not at the expense of clarity, simplicity and machine
independence. Above all else the temptation should be
resisted to be 'clever-clever'. Much of the coding of an
algorithm is essentially a mechanical task and although
short cuts are tempting they can lead to major errors.
There are in fact two aspects to machine efficiency, the
actual run time and the amount of store use. Often it is
said that these are in conflict, a fast program using a lot
of core and *vice versa*, but this is an oversimplification.
In practice machine-time consumption is often related to
the amount of core used by the code so that compact code
will also be fast code.

The basic rule is always to beware of duplication,
especially of computer operations which the programmer
does not see and within DO loops. Using this general
principle as a guide we can speed up program execution in
many ways, associated with basic operations, arithmetic
expressions, DO loops, and the DATA statement to enter
constants.

In *basic operations* a few simple rules can save a great
deal of machine time:

   (i) Addition and subtraction are faster than multiplication.
$$X = A + A$$
not $\qquad X = 2.0*A$

 (ii) Integer arithmetic is faster than real arithmetic
which is in turn faster than double precision.

(iii) Exponentiation is very slow, especially if non-
integer exponents are used
$$X = A * A * A$$
not $\qquad X = A ** 3$
and definitely not
$$X = A ** 3.0$$

 (iv) SQRT(A) is always faster than A ** 0.5

  (v) Multiplication is faster than division.
$$RECIP = 1.0 / CONST$$
$$A = B * RECIP$$

```
              X = Q * RECIP
              C = R * RECIP
    not
              A = B/CONST
              X = Q/CONST
              C = R/CONST
```
(vi) Any IF statement is slow and should be avoided if at all possible.

(vii) Looking up stored values in an array using their subscripts is slow.
```
              TEMP = A(I,J)
              X = B + TEMP
              Y = C + TEMP
              Z = D + TEMP
    not
              X = B + A(I,J)
              Y = C + A(I,J)
              Z = D + A(I,J)
```
which makes two unnecessary references to the storage location involved.

(viii) Similarly, references back from the body of a subroutine to its arguments are very slow. Instead, the device of using local variables to store the arguments introduced in case study 11.7(ii) should be used:
```
              SUBROUTINE XXX( AA,BB,CC,DD)
              A = AA
              B = BB
              C = CC
              D = DD
```
These local variables, A,B,C and D are now used throughout the body of the subroutine in place of the equivalent arguments.

In *arithmetic operations*, a moment or two of thought will often save the repetition of some calculations within a group of statements:

(i) If the same expression occurs more than once in a single statement, either bracket it or bring it out as a temporary variable:
```
              Z = A + B - C + 7.0 * (A + B)/D
```
can be written
```
              TEMP = A + B
              Z = TEMP - C + 7.0 * TEMP/D
```

(ii) Sometimes the repetition can occur over a group of statements as in the sequence
```
AA = X + 2.0 - A - 2.0 * T ** 3
BB = A + 2.0 * T ** 3 - B ** 2 + 7
CC = 2.0 * (A + 2.0 * T ** 3) - AA + 7.0
```
which is better re-written as
```
TEMP = A + 2.0 * T * T * T
AA   = X + 2.0 - TEMP
BB   = TEMP - B * B + 7
CC   = TEMP + TEMP - AA + 7.0
```
Two well-known examples of statement reformulation to avoid unnecessary arithmetic are as follows. In arithmetic involving polynomials, as in trend surface analysis, we might be tempted to code the arithmetic relation
$$y = ax^3 + bx^2 + cx + d$$
as  Y = A * X * X * X + B * X * X + C * X + D
This has six multiplications and three additions. A far better coding is
Y = ((A * X + B) * X) + C) * X + D
which involves only three multiplications and three additions. A second example is in the solution of quadratic equations using the formula
$$r = (-b \pm \sqrt{(b^2 - 4ac)})/2a$$
This can be inefficiently coded as
```
ROOT1 = (-B + SQRT(B**2 - 4.0 * A * C))/(2.0 * A)
ROOT2 = (-B - SQRT(B**2 - 4.0 * A * C))/(2.0 * A)
```
which involves 4 subprogram calls, 8 multiplications and divisions, 4 additions and subtractions, and 21 load and store instructions. A much better coding is
```
DENOM  = A + A
SDISCR = SQRT(B * B - 4.0 * A * C)
ROOT1  = (-B - SDISCR)/DENOM
ROOT2  = (-B + SDISCR)/DENOM
```
which reduces the work to 1 subprogram call, 5 multiplications and divisions, 4 additions and subtractions, and 16 load and store instructions.

A third area in which machine time can be saved is in the interior of DO loops. These contain operations that are executed repeatedly, so that a small saving in the centre of deeply nested loops will be multiplied many times over. Specifically, the programmer should *trim*, *jam* and *unroll* his DO loops. In DO loop *trimming*, we avoid repeating constant operations within a loop by placing the constant calculation outside it, storing its result as a temporary variable:

not
```
          DO 10 I = 1,100
          P(I) = Q(I) * ((A + B) + (C - D))
       10 CONTINUE
```
but
```
          TEMP = A * B + (C - D)
          DO 10 I = 1,100
          P(I) = Q(I) * TEMP
       10 CONTINUE
```
A similar example occurs in matrix multiplication. Consider the FORTRAN sequence:
```
          IF (MCOLA .NE. MROWB) STOP
          DO 10 I = 1,MROWA
          DO 10 J = 1,MCOLB
          C(I,J) = 0.0
             DO 20 K = 1,MCOLA
             C(I,J) = C(I,J) + A(I,K) * B(K,J)
       20 CONTINUE
       10 CONTINUE
```
At first sight this is a perfectly acceptable coding; it checks to ensure that the multiplication is defined and no actual arithmetic operation can be trimmed from it. If we remember that the processor has to use the subscripts of the array C to reference the particular storage location to be used, then it becomes apparent that this program makes a continual and unnecessary reference to subscripted locations. A better coding would be:
```
          IF (MCOLA .NE. MROWB) STOP
          DO 10 I = 1,MROWA
             DO 10 J = 1,MCOLB
             SUM = 0.0
                DO 20 K = 1,MCOLA
                SUM = SUM + A(I,K) * B(K,J)
       20       CONTINUE
             C(I,J) = SUM
       10 CONTINUE
```
DO loop *jamming* takes this principle even further by attempting to reduce the number of times a DO loop control variable is used. Instead of
```
          DO 20 I = 1,100
       20 A(I) = 0.0
          DO 30 I = 1,100
       30 B(I) = 0.0
```
it is obviously more efficient to write

306

```
          DO 20 I = 1,100
          A(I) = 0.0
          B(I) = 0.0
       20 CONTINUE
```
which jams the loops together under the same control variable sequence.

DO loop *unrolling* is more subtle, but again the objective is to reduce the total number of operations.  Its principle can be understood by examining two alternative codings
```
          DO 70 I = 1,4
          A(I) = 0.0
       70 CONTINUE
```
or better
```
          A(1) = 0.0
          A(2) = 0.0
          A(3) = 0.0
          A(4) = 0.0
```
The unrolled form gives explicit values to the subscripts and so reduces the total work to be done.  A more worthwhile unrolling might be the sequence:
```
          DO 70 I = 1,1000
          A(I) = 0.0
       70 CONTINUE
```
unrolled to give
```
          DO 70 I = 1,999,2
          A(I) = 0.0
          A(I + 1) = 0.0
       70 CONTINUE
```
or
```
          DO 70 I = 1,500
          A(I) = 0.0
          A(I + 500) = 0.0
       70 CONTINUE
```
Both alternatives are quicker than the original coding.

A final saving of machine to be discussed is the use of DATA statements to enter fixed constants into the program.  One possible coding would be using arithmetic assignments
```
          PI = 3.14159
          E  = 2.71828
```
but it is faster to use the DATA statement
```
          DATA PI,E /3.14159,2.71828/
```
This saving in run time is achieved only at the expense of using a DATA statement instead of a simple assignment and could give a loss in generality and portability as undesirable side effects.

The second aspect of programming efficiency is the use of core storage. By using modular techniques and by eliminating all duplication a compact *code* will already have been created, but *data* space can be conserved in a number of ways.

One technique that is easily learnt in FORTRAN, and is useful in programs having very large storage requirements which would otherwise be beyond the machine's capacity, is to pack small integer numbers into a single word in store. For example, if N-1 is the maximum possible value of N2, then N1 and N2 can be packed into the same location M using the assignment

$$M = N * N1 + N2$$

The same numbers can be unpacked using

$$N1 = M/N$$
$$N2 = M - N * N1$$

At some cost in generality, array storage space can be re-used, and thus conserved, by the EQUIVALENCE mechanism. Similarly, extensive use of COMMON will ensure that no locations are duplicated. Should these devices still prove inadequate to store the required data, it is sometimes possible to create more compact array structures. In case study 11.7(i), the nearest neighbour analysis program showed how a symmetrical, two-dimensional array can be stored as a single row or column vector containing only the unique locations and related to its two-dimensional equivalent by a subscripting function

$$NPLACE (I,J) = I * (I - 1)/2 + J$$

Similar functions can be developed for other regular arrays.

## 12.7 Summary

The above notes are intended to show how any program can be made error-free, of sound structure, portable, clearly presented and reasonably efficient in its use of machine time and store. These form our programming standards. In conclusion, it is proposed to illustrate how these standards can improve programs by giving a single case study involving two programs designed to do the same thing.

The problem is the calculation of the mean centre and standard distance of a weighted point pattern made up of a number of spatially distributed individuals using the formulae developed by Bachi (1963). Figure 12.1 shows a program written by a novice student programmer for a particular problem that involved 28 points. As coded, the student is to be congratulated on writing a program that

```
C         AN EXAMPLE OF BAD STYLE
C
          DIMENSION B(28),Q(28),R(28)
          G = 0.0
          H = 0.0
          F=0.0
          READ(7,119)(B(I),Q(I),R(I),I = 1,28)
    119   FORMAT(20X,F7.4,F8.4,F4.0)
          DO 123 I = 1,28
    123   G = G+B(I)*R(I)
          DO 124 I = 1,28
    124   F=R(I)+F
          DO 125 K = 1,28
    125   H=Q(K)*R(K)+H
          X0=G/F
          AY=H/F
          WRITE(2,79) X0,AY
     79   FORMAT( 2F10.3)
C         SECOND BIT
          ST = 0.
          SU = 0.
          DO 81 I =1,28
          ST = R(I)*(B(I)-X0)**2+ST
          SU = SU+(Q(I)-AY)**2 *R(I)
     81   CONTINUE
          SDL = SQRT(ST/F+SU/F)
          WRITE(2,23)SDL
     23   FORMAT(F11.4)
          STOP
          END
```

*Fig 12.1* An example of bad style: Mean centre and standard distance

will work but equally the coding contains a number of defects:

    (i) it is specific to 28 cases and to the input/output devices 7 and 2.
    (ii) the variable names used bear no relationship to the problem.
    (iii) the commentary is virtually non-existent.
    (iv) the input data are neither checked nor written out.
    (v) the output is not labelled in any way.

(vi) the spacing and layout are confusing.
(vii) the program uses single precision locations ST and SU to store numbers which are sums of squares. Depending upon the input data, these quantities could be very large indeed and serious round-off error might ensue.
(viii) Statement numbering is haphazard and does not leave room for expansion.

Figure 12.2 shows the same algorithm coded according to the principles ennunciated in this chapter. The coding is much clearer. It uses long and meaningful variable names, has been generalised to deal with any number of points up to 100, any data input format, and has numerous other detailed coding improvements. The commentary gives its reader a blow-by-blow account of what is happening and how to input data. True, there are more cards in the deck and the program takes very slightly longer to run, but the gains in clarity, generality and portability are well worth the extra effort.

In what order should the programmer tackle these various program qualities? Opinions differ, but we are inclined to agree with Kernighan and Plauger (1972) who recommend that you
- make it *right* before you make it faster
- make it *fail safe* before you make it faster
- make it *clear* before you make it faster.

Above all else, keep it *simple*. Frills, tricks, short cuts and non-standard FORTRAN are usually trouble.

## 12.8 Further Reading

Bachi, R. 'Standard distance measures and related methods for spatial analysis', *Regional Science Association, Papers and Proceedings*, 10, 83-132 (1963)

Brown, A.R. and Sampson, W.A. *Program debugging* (1973)

Fox, D.L. *Introduction to numerical linear algebra* (1964) __ and Meyers, D.F. *Computing methods for scientists and engineers* (1968)

Froberg, C. *Introduction to numerical analysis* (1970)

Kernighan, B.W. and Plauger, P.J. *The elements of programming style* (1972)

Longley, J.W. 'An appraisal of least-squares programs for the electronic computer from the point of view of the user', *Journal, American Statistical Association*, 62, 819-41 (1967)

```
C  AN EXAMPLE OF BETTER STYLE                                    STY010
C                                                                STY020
C  PROGRAM ABSTRACT                                              STY030
C  O.J.UNWIN, OCTOBER 1974 FOR CDC CYBER 72,ICL 4130 AND 1906    STY040
C  ENVIRONMENTS                                                  STY050
C                                                                STY060
C  PROGRAM TO FIND THE MEAN CENTRE AND STANDARD DISTANCE OF A PATTERN STY070
C  OF WEIGHTED POINT LOCATIONS                                   STY071
C     INPUT  CARD1   TITLE CARD OF ANY 80 CHARACTERS             STY080
C            CARD2   NP = NUMBER OF POINTS , FORMAT I3           STY090
C            CARD3   FMT = FORMAT OF ENSUING DATA                STY100
C            CARD 4 ET.8FQ. DATA CARDS EACH WITHX,Y AND W        STY110
C                    IN THAT ORDER FOR UO TO 100 POINTS          STY120
C                                                                STY130
C  REFERENCE = R.BACHI,(1963) STANDARD DISTANCE MEASURES AND RELATED STY140
C  METHODS FOR SPATIAL ANALYSIS                                  STY150
C  REG.SCIENCE,ASSOC,,PAPERS AND PROCEEDINGS,10,83=132,          STY160
C                                                                STY170
       DIMENSION TITLE(80),FMT(80),X(100),Y(100),W(100)          STY180
       DOUBLE PRECISION SUMX2, SUMY2                             STY190
C                                                                STY200
C  SET UP I/O AND SET COUNTING LOCATIONS AT ZERO                 STY210
C                                                                STY220
       NIN = 7                                                   STY230
       NOUT = 2                                                  STY240
       SUMX = 0.0                                                STY250
       SUMY= 0.0                                                 STY260
       SUMW = 0.0                                                STY270
       SUMX2 = 0.0                                               STY280
       SUMY2 = 0.0                                               STY290
C                                                                STY300
C  READ IN RUN TITLE FORMAT 80A1                                 STY310
C                                                                STY320
       READ(NIN,90)(TITLE(I),I=1,80)                             STY330
C                                                                STY340
C  READ IN NUMBER OF POINTS IN DISTRIBUTION(I3) AND DATA FORMAT  STY350
C                                                                STY360
       READ(NIN,95) NP                                           STY370
       IF ( NP .GT. 100 ) STOP                                   STY380
       READ(NIN,90) (FMT(I), I = 1, 80)                          STY390
C                                                                STY400
C  WRITE HEADING AT TOP OF PAGE                                  STY410
C                                                                STY420
       WRITE(NOUT,100) (TITLE(I),I = 1,80)                       STY430
C                                                                STY440
C  READ IN DATA CARDS AND WRITE OUT A MIRROR IMAGE CHECK         STY450
C  X AND Y HOLD THE SPATIAL CO-ORDINATES, W THE POINT WEIGHTS    STY460
C                                                                STY470
       READ(NIN,FMT) (X(I),Y(I),W(I),I=1,NP)                     STY480
       WRITE(NOUT,120)(X(I),Y(I),W(I),I=1,NP)                    STY490
C                                                                STY500
C  FIND MEAN CENTRE                                              STY510
C                                                                STY520
       DO 10 I =1, NP                                            STY530
       SUMX = SUMX + X(I) * W(I)                                 STY540
       SUMY = SUMY + Y(I) * W(I)                                 STY550
       SUMW = SUMW + W(I)                                        STY560
    10 CONTINUE                                                  STY570
       RECIP = 1.0/SUMW                                          STY580
       XMEAN = SUMX * RECIP                                      STY590
       YMEAN = SUMY * RECIP                                      STY600
C                                                                STY610
C  STANDARD DISTANCE CALCULATION                                 STY620
C                                                                STY630
       DO 20 I = 1, NP                                           STY640
       XD = X(I) - XMEAN                                         STY650
       YD = Y(I) - YMEAN                                         STY660
       SUMX2 = SUMX2 + W(I)*XD*XD                                STY670
       SUMY2 = SUMY2 + W(I)*YD*YD                                STY680
    20 CONTINUE                                                  STY690
       SDIST = SQRT(( SUMX2 + SUMY2) * RECIP)                    STY700
C                                                                STY710
C  WRITE OUT ANSWERS                                             STY720
C                                                                STY730
       WRITE(NOUT,130) XMEAN, YMEAN, SDIST                       STY740
       STOP                                                      STY750
    90 FORMAT(80A1)                                              STY760
    95 FORMAT(I3)                                                STY770
   100 FORMAT(1H1,80A1/30H       X=COORD    Y=COORD   WEIGHT/)   STY780
   120 FORMAT(1H ,3F10.3)                                        STY790
   130 FORMAT(33H CO-ORDINATES OF MEAN CENTRE ARE ,F10.3,9H (X) AND , STY800
      1       F10.3, 4H (Y)/30H WITH A STANDARD DISTANCE OF ,F11,4)   STY810
       END                                                       STY820
```

*Fig 12.2* An example of better style: Mean centre and standard distance

311

Unwin, D.J. 'Numerical errors in a familiar technique: a case study of polynomial trend surface analysis', *Geographical Analysis*, 7, 197-203 (1975)

Wampler, R. 'A report on the accuracy of some widely used least squares computer programs', *Journal, American Statistical Association*, 65, 549-65 (1970)

In preparing this chapter considerable use was made of *A manual of FORTRAN programming standards* (University of Leicester Computer Unit mimeo, 1974)

# Chapter 13
# Modern Computing

13.1 Introduction
The preceding twelve chapters have presented the basic techniques of FORTRAN programming as applied to common geographical problems. The methods and techniques considered and described have been available for over ten years to any geographer with access to a computer with a FORTRAN language compiler and a reasonable amount of core store. Many university computers, however, offer far more facilities than just standard FORTRAN.
Facilities differ considerably from university to university, often with special types of equipment concentrated into particular centres. The aim of this final chapter is to indicate some of the features which are available on this more specialised equipment or with the more advanced computers and operating systems now accessible to university geographers. Not everything considered in this chapter will be available in all computer centres but by this stage the reader should know his own computer installation quite well and be able to select from the following pages those sections relevant to his own needs as constrained by his local computational environment. Because we have gone beyond the point at which simple techniques solve our problems and have moved into the area where the operating system affects the method and approach to problem solution, this chapter will present no new general programs. More specifically, it is proposed to consider problems associated with:

(i) the processing of particularly large data blocks, for example using magnetic tapes holding census data,
(ii) the incorporation of other authors' programs into particular programs, for example program modules supplied by computer manufacturers or the use of

complete suites of programs such as SPSS or BMD,
(iii) the use of specialised input/output devices and
methods, for example graph plotting equipment,
(iv) using programming languages other than FORTRAN.

The chapter thus serves as an introduction to the more specialised manuals available in computer centres, which may look extremely fearsome when studied by a geographer who perhaps is only interested in drawing several hundred proportional cubes on a map!

## 13.2 The Use of Files

In many branches of the social and physical sciences the practitioner is often faced with the analysis of large amounts of data. Frequently, relatively simple analyses such as counts in questionnaires or percentage and ratio calculations in census-based studies are required. Elsewhere in this volume some programs for this type of analysis have been presented, but a major problem may arise with large data packs. Very simply, the computer store is not big enough to store all the data. A medium-sized shopping-centre interview study will probably yield at least 1000 questionnaires with perhaps 20 response questions and another 10 controls or interviewee observations. To store the raw data will require in excess of 30,000 storage locations. To manipulate it and perform calculations will probably need as much core again. Many university computers are physically not big enough to deal with such data blocks if they are held in the main core and the sheer bulk of cards begins to get unwieldy when 1000 data cards are added to 2-300 program cards. Yet more difficulties arise in passing these cards through a card reader. Four or five times is quite acceptable, but to pass 1300 cards through a card reader twenty times or more is inviting both the card reader and operator to complain. To overcome such difficulties, many computers have additional facilities to allow large programs or large data blocks to be kept, and accessed as needed, on an alternative medium to cards. Frequently such computers also use the idea of *files* which are areas of backing store either temporarily or permanently devoted to a particular job or an individual user.

In the input/output commands discussed in earlier chapters it has been envisaged that a READ command in some way communicates directly between the computer core and an input device, usually a card reader, which is specified by a number incorporated into the READ command. In reality this is usually not quite the situation. Communication is indirect, with program and data commonly being transferred to a temporary file in the backing store. The temporary file holds information on the job such as the name of the person running the job, as well as the job itself. In the backing store it is placed in a queue of jobs requiring running. After the job file has worked up the queue, at a speed which may depend on the estimate of running time (supplied by the programmer) and size of program and data, it then moves into the computer, is processed and then the temporary file is erased. A similar sort of arrangement occurs on exit from the computer. Output is temporarily stored until time is available on an output machine, for example a lineprinter, for the results to be produced in the required layout. Often the computer thus has a system of files, some providing input and some being filled by output. The control of these files is the job of the operating system (*see Chapter 2*).

While the program is operational within the computer it may also access files within the backing store. Some functions in the program may be kept in permanent files and available for access. Data may be accessed from permanent files in the backing store and results written back to a temporary file which is accessed again later in the same program. Files held on backing store allow some alleviation of the difficulties encountered in using large data files. First, the data may be placed onto a data file by a series of commands to the operating system and which usually define, among other things, a name by which the file is going to be known. Only one passage of the cards through the card reader is required and the data are then in a backing store medium (magnetic tape, disc, drum or card) accessible by a program usually without operator intervention and with a very rapid rate of access and transfer. Secondly the data are stored on a medium which unlike the main CPU store is easily capable of holding all the data items. It only remains to provide instructions in the program to access the data file.

The precise way of linking files to programs varies considerably with different computers and may be as simple as a command D = NAME placed in the job control commands at the start of the job to signify that the data are held on a file called NAME. Any READ commands related to these will automatically access the file. Alternatively, a far more complex series of job control commands may be necessary. In general we are concerned here with files of permanently held material, either a data block which is kept in a file rather than on cards or a program or part program which is used frequently and similarly stored on a file.

Sometimes, however, within a program we wish to write material temporarily to a file and later in the program to read all or selectively from this file. The life of the file is that of the program execution and it is thus termed a temporary or *scratch* file. In many ways scratch files have a role similar to a note pad used by someone doing calculations in their head. Occasionally, there will be a series of numbers which are not needed immediately in the mental calculations but which will be needed later. These are recorded on the pad and thrown away once the mental arithmetic is complete. This piece of paper is equivalent to the scratch file. In the computer program, the need for scratch file (or files) is specified in the job control commands and a number (numbers) is ascribed to it just as a card reader or lineprinter is associated with a device number. It is permissible to use several such files within a single program. The program can then be used to assign channel numbers to the scratch files just as input and output channel numbers are assigned. Thus a program may begin

$$\begin{aligned} \text{NIN} &= 5 \\ \text{NOUT} &= 6 \\ \text{NFIL1} &= 3 \\ \text{NFIL2} &= 4 \end{aligned}$$

Transfers to or from the scratch files may be made as normal WRITE and READ statements. Thus as

WRITE (NOUT, 10)(A(I),I = 1,500)

may transfer 500 values from array A to a lineprinter for output so

WRITE (NFIL1, 10)(A(I),I = 1,500)

transfers the same 500 values to a scratch file for use later in the program.

In addition to the familiar READ and WRITE statements there are two more statements which can be used in connection with scratch files. These two commands are used to reposition the file for future access. Imagine a pointer associated with each file and moving along the file as material is placed in it. At the start the pointer will be positioned at the start of the file. After a command such as

WRITE (NFIL1, 10)(A(I),I = 1,500)

it will have been advanced by 500 positions. If later

the command
$$\text{READ (NFIL1, 1\emptyset)(A(I),I = 1,5\emptyset\emptyset)}$$
is used it causes the values in the next 500 positions to be transferred into array A. There may well be nothing in these positions; what was really required was a transfer back of the original 500 values. To achieve this, the imaginary pointer must be repositioned at the beginning of the file prior to the READ transfer. This repositioning is achieved by the command
$$\text{REWIND X}$$
where X is the channel number (or a scalar variable containing this number) of the device containing the scratch file. For example:
$$\text{REWIND NFIL1}$$
Thus when a REWIND statement is executed the effect is that the next transfer command referencing the same channel will operate from the first record of the file. In file manipulation a second command is commonly available. The command BACKSPACE X, where X is again the channel number, moves our imaginary pointer back one space. This feature allows re-reading of a particular value. In some computer systems other commands are available to allow file manipulation but these are particular to individual installations and are best discovered by individual users by reference to their operating system manual.

The concept of files and the commands REWIND X and BACKSPACE X are common to all types of scratch files no matter whether they are physically contained on magnetic tape, disc, drum or other storage medium. From the user standpoint it seldom matters on exactly what type of equipment the file is stored. To the programmer it is just somewhere to store program and data. Usually the operating system allocates machinery and all the programmer need do is define his requirements. The more advanced operating systems allow the programmer to create files and selectively edit them by altering specified characters or, if the file contains program text, to add or delete program statements. This facility can be useful in debugging large programs; again it saves frequent passes through the card reader and saves on the number of cards being handled at any one time. Whether it is worthwhile to use files rather than cards for either data or program depends to a considerable extent on the size of the computer installation and the level of sophistication of its operating system.

## 13.3 Program Libraries and Data Banks

An understanding of file structure is important in the use of program libraries and data banks which may have been compiled by someone other than the geographer programmer and which are stored on the computer. Programs available to the geographer from program libraries tend to be one of three types. First there are libraries of procedures and subprograms which can be drawn together into a modular program incorporating some program statements provided by the geographer. Secondly, whole programs may be available either as published listings in books or periodicals or as computer-ready card decks which only require the addition of job control cards and data. Finally whole suites of programs are available in packages and parts of the package of programs are used as required.

Libraries of subprograms provide the most flexible form of the pre-packed program. Frequently, individual programmers gather together their own set of subprograms into a loosely structured library. Documentation of these is often far from adequate but it seems likely that even around British university geography departments many such libraries exist often duplicating each other. Perhaps of more use, and certainly more comprehensive and better documented, are published libraries of subprograms together with those supplied by computer manufacturers for use with particular types of computer. The published libraries tend to have a bias towards particular applications, thus for example the NAG (Numerical Algorithm Group) has a bias towards subprograms useful in numerical analysis while GINO and GHOST are libraries of subprograms producing graphical output. More general, however, are those supplied by manufacturers and documented in manufacturers' manuals. A comprehensive library of geography subprograms has yet to appear but individual subprograms in many of these more general libraries are of value to geographers. The use of these libraries depends first on the computer installation implementing the library and providing an advisory service. Secondly, the geographer programmer must write a program to call the required library routines which are usually kept in backing store in the form of subroutines or functions (*see Chapter 11*). Usually in programs developed in this way the analytical procedures are carried out by the supplied subprograms while input/output is provided by the programmer. Considerable flexibility is thus available in procedures such as data layout and even in the analysis. Often several subprograms exist for the same basic procedure but each

subprogram deals with a detailed difference in the procedure.
Such flexibility is not present in whole programs designed to perform a specific technique or solve a particular problem and made available on a general, published, basis. However, ease of operation may override considerations of flexibility and some well-known published programs such as the *'BBC' general survey analysis system* for editing, recoding and tabulating survey data are extremely easy to use so long as the data are in the correct format. Clearly there are advantages of economy in having pre-prepared programs to perform essentially standard techniques. At first sight the results of a survey conducted in 1968 among social scientists suggested a chaotic provision of *standard* programs for factor analysis. Twenty-seven different programs were available and in use among a sample of 60 users with no program having more than 5 users in the sample (National Computer Centre, 1972). Since 1968 it is likely that even more factor analysis programs have been written. But is this situation really as chaotic as it seems at first sight? Undoubtedly there is some duplication and wasted effort but in a technique as complex as factor analysis there are variations in the methods of calculation and different algorithms of numerical analysis may be used. No single program is the standard program and there is a clear need for more than one program to perform the analysis. If each program is adequately documented then a potential user can choose between the alternatives and take the program best suited to his needs. Even when different but well documented programs are used disagreement can still arise over interpretations of the results as in the debate between Mather (1971) and Davies (1971). Even more misunderstanding can arise when use is made of poorly documented standard programs or the only available standard program or if the user is not fully aware of what exactly the program does. Accusations of 'garbage in, garbage out' can quite justifiably be levelled against some uncritical use of complete programs.

This is not to argue that all complete packaged programs are without merit. There are several series of extremely useful geographic programs. In these, the algorithm is presented in detail and the program contains sufficient comment cards to follow how its logical structure relates to the algorithm. Publications of these whole programs take two forms. First, there are lists of programs available as in *Programs for Social Scientists*, (National Computer Centre, 1972), *Computers in the Environmental*

*Sciences* (Tarrant, 1971) and *Census Analysis* (LAMSAC, 1974). In these publications, broad details on program purpose, language, input/output and peripheral requirements, charge for use etc. are provided together with a source for the program from whom details of the algorithm and a listing or card deck are available. Secondly, there are publications containing all the details of the whole programs and often including test data (rather along the lines of the worked examples provided in earlier chapters). Into this category fall

(i) Serial publications, notably *Computer Contributions* (now *Geocom Programs*), *Computer Applications in the Natural and Social Sciences*, the various *Technical Reports* of the Computer Institute for Social Research, Michigan State University and the *Lund Studies in Geography* series C.

(ii) Books of programs, notably *Selected computer programs* (Tobler, 1970) and *A selection of geographical computer programs* (Baker, 1974).

(iii) Books dealing with a specific field but containing various listings of computer programs. Notable are Davis (1973), Balfour and Beveridge (1972), Veldman (1967).

(iv) Individual research papers and reports dealing with one or two solutions to a particular algorithm. Examples are Pitts (1965; 1969), Heap and Pink (1969) Ahnert (1971).

The whole programs reported in these various published forms at first sight seem to provide the novice programmer and non-programming geographer will all the geographical programs he needs. It can be argued quite strongly, however, that such whole programs should only be used by geographers who, if the program was not available, *could* have written it themselves. Such a level of expertise in the potential user will go some way to making sure such packaged programs are used correctly.

The final source of 'oven-ready' programs is the program suite. The most commonly used by geographers are *The Statistical Package for the Social Sciences* (Nie, Bent and Hull, 1970), *Biomedical Computer Programs* (Dixon, 1971), *Symap and Symvu* (Muxworthy, 1972) and *Grids* (Jaro, 1972). The first two are suites of statistical programs while the others are suites of mapping programs (see below). The two commonly used statistical suites differ slightly in that the building block of the *BMD* programs is the complete program while for *SPSS* it is the subprogram.

The Statistical Package for the Social Sciences (SPSS) is an integrated system of computer programs for the analysis of social science data. The system has been designed to provide the social scientist with a unified and comprehensive package enabling him to perform many different types of data analysis in a simple and convenient manner. SPSS allows a great deal of flexibility in the format of data. It provides the user with a comprehensive set of procedures for data transformation and file manipulation, and it offers the researcher a large number of statistical routines commonly used in the social sciences. (Nie, Bent and Hull, p1)

Each of these large program suites has an equally large explanatory manual to describe the methods used and give descriptions of the cards needed to perform particular analyses. The same sort of criticism as was levelled against the individual complete program can be levelled against these program suites. One particular danger is that their very ease of use promotes the misuse of techniques by workers who do not understand the various assumptions made.

With this limitation in mind, suites such as SPSS are extremely useful in geographical enquiry. They save considerable amounts of time and effort in the application of many statistical procedures. SPSS has, for example, groups of subprograms to produce

(i) Descriptive statistics and one way frequency distributions
(ii) Table displays of relationships between two or more variables
(iii) Bivariate correlation
(iv) Partial correlation
(v) Multiple regression
(vi) Scalogram analysis
(vii) Factor analysis

Furthermore, a suite is capable of having further subprogram groups added. The BMD suite has an even wider range of statistical procedures. Other suites are discussed in Blackman and Goldstein (1971). In all cases the programs and subprograms are statistical - not geographical. Geographers still await a suite of geographical subprograms along the lines of SPSS just as they still await a subprogram and function library. Only with the whole program package, the least useful of the three approaches to program libraries, is there material of an essentially geographical nature.

Somewhat similar to program libraries and equally dependent on file usage and backing store are data banks or *data archives*. These constitute a small part of the very large field of information storage and retrieval (Gurr and Painofsky, 1964; Dippel and House, 1969; Sharp, 1965) and in general are concerned with collecting, processing, updating and providing a user service of statistically based data (Bisco, 1967; UNESCO, 1972). The data mass is stored in computer readable form and accessed and sorted by data manipulation routines. During the 1940s and 1950s large amounts of statistical data were amassed but they lacked formal organisation. This organisation became possible with the advent of computerisation. Archives relevant to the human geographer usually contain social science type data (Rokkan, 1966) whereas organised computer-based data collections in physical geography have mostly developed at the individual user level with the geographer building up and structuring his own data collection. In social science data archives a number of mass data collections have developed which effectively provide a service for many geographers.

The organisation of such archives varies, but broadly they relate *either* to a specific topic such as consumer behaviour or the uses of leisure time and vacations or labour statistics (Mendelssohn, 1965) *or* to all types of material in a specific district (county, region, city, etc). An archive relating to consumer behaviour for example would contain 'analyses of survey data on trends and patterns of consumption by households and occupation groups, family budgets, financial decision-making, allocations of roles in the management of resources' (Rokkan, 1964), and the results of household and shop surveys on consumer behaviour. A geographer working on customer behaviour can then obtain access to the original survey sheets for a wide variety of surveys which would provide comparative material for his own survey work. The same aim is present in the larger social science data archives organised on a regional basis.

Typical of such archives are those held by local government planning authorities, some of which collect and store data by grid co-ordinate locations (Horwood, 1970). On a larger scale are the general purpose national archive systems. The first of these to be developed in Europe is that housed on the *zentralarchiv* of the University of Cologne. The archive began in 1960 and the acquisition policy is to include surveys conducted by German organisations and cross-national surveys including Germany as a surveyed country. Surveys have been collected from academic organisations, federal bodies and commercial institutions (Scheuch and Bruning, 1965). Such an archive is only possible with heavy dependence on computer facilities.

The administration of data archives is complex both from a computational and logical point of view (Berztiss, 1971) and in their general day-to-day operation. The computational theory of data structures, their storage needs and file ordering techniques, is largely beyond the interest of geographers and certainly beyond the scope of this volume. What is of concern however is how the computer techniques and operational procedures affect the utility of the archive for geographical use. Bisco (1970) indicates a number of problems that have to be solved before data archives become fully used as a social-science data source. The first problem is associated with a large number of archives that have developed during the late 1960s. Many potential users just do not know which archive holds the material they require. Each archive operates in a different way with different retrieval methods and storage structures. The data archive was conceived as unifying social-science data but much fragmentation remains. A second problem is that although a potential user knows a survey has been performed the results have not been placed into an archive. The survey data may have been destroyed or be sitting in a cupboard somewhere. A third problem arises out of the lack of generality of most surveys. Data are collected for specific purposes and by methods best suited to solve some specific problem. The results of such surveys are often of little value in comparison with surveys carried out to solve a related problem explored with a different approach and survey design. A final problem is associated with the coding procedures employed in the survey archive. Questionnaire material is often grouped or coded prior to analysis and the original survey responses are lost. Subsequent users may need a different grouping of responses but this is impossible if the archive holds only the coded responses. The emergence of data archives singe the mid-1960s has been valuable to computer-based geographical studies but major problems, somewhat similar to those occurring with whold program libraries, arise in their use as data sources.

The utility of program libraries of all types and of data archives is considerably enhanced by networks of computers. Computer installations may be joined into networks by two-way communication lines. A user at centre A can then effectively use all the other computers in the network and, more important, specialist libraries and data archives held at other computer centres comprising the network. - A number of such networks exist in North America and Europe some with line links of several thousand kilometres and connecting 10 or 15 computer installations. Such network systems allow individual computer centres to develop specialist library or archive facilities while still allowing users to obtain a 'normal' service. Undoubtedly there are major technological problems in developing networks but their savings, especially in the field of program library and data archive provision, are considerable.

13.4 Computer-graphics

Many computer operations, such as index number and areal density calculations, trend and factor analyses, yield results that are best presented to the user in map and diagram form (McCullagh and Sampson, 1972). The techniques, problems and potential associated with graphical output from computers are the subject of that branch of computer science called *computer-graphics* (Parslow and Green, 1971), and in this section we will briefly review the geographical applications that have been made of them. Like geography, many disciplines, notably engineering, geophysics and space research, have uses for graphics so that there is already a vast and expanding computer-graphics literature. Within geography, Tobler (1959, 1965) was one of the first to recognise potential applications but by 1976 a great many applications have been reported.

The advantages of graphical output are many. The speed of present-day computers poses a human problem in that the user has to assimilate his output before making deductions from it. The presentation of answers as graphs, diagrams and maps makes *visualisation* and hence assimilation much easier. A good example of this visualisation aspect is in numerical weather forecasting where the answers, usually a very large number of predicted atmospheric pressures at grid points over the region of interest, are displayed directly as a map of the predicted pressure field (Sawyer, 1960).  A second advantage lies in the *repeatability* of graphics.  Unlike the error-prone human draughtsman who would seldom if ever produce exactly the same map from the same information, our computer acts as a 'robot draughtsman' and will always produce the same output from the same data.  One particularly useful application that relies strongly upon this repeatability is the objective, automatic contouring by computer of a field of irregularly spaced data points.  Human cartographers are likely to produce very different contour maps from the same data according to their preconceived notions of how the map ought to look, but the suitably programmed computer will produce the same map at every trial with the same data. Graphical output is also *adaptable*.  Instead of making a fresh compilation whenever a new map or diagram is required, data stored on files can be drawn into maps and diagrams at different scales, on different projections and according to different mapping criteria.  The user can edit his diagrams, change their shape and size, 'zoom' in on features of special interest or select specific sub-areas or 'windows' for more detailed analysis.  Finally, there are a number of *computing* advantages.  As seen in Chapter 2 of this book, graphical output from electronic devices such as the cathode ray tube and microfilm plotter is extremely fast if compared to the slower electro-mechanical printers and punches.  Invariably, a plot of some information will be physically more compact than the numbers it represents.

There are two difficulties associated with using FORTRAN to produce graphical output.  In the early days when it was designed, computers did not have available to them the range of peripherals that exist now, so that the language was developed without any special provision for graphical output.  ANSI FORTRAN specifies no standard commands. Secondly, there is available a range of devices for graphics - lineprinters, graph plotters, CRT, microfilm plotters and so on - with widely different physical characteristics and manufacturers.  One result is that any graphics program, whether in FORTRAN or any other programming language, tends to be very system and hardware dependent.  Similarly, in an ideal world, our graphics programs might be expected to be device independent, capable of outputting on most output devices.  In practice, the physical characteristics of the

available output media exert a powerful influence on the types of programs that we write. Accordingly, we will confine ourselves to showing examples of the more commonly available systems and output media.
*Computer-graphics using the lineprinter.* Almost all computers have available to them a standard lineprinter, usually with 120 or more print positions on each line, which can be controlled using standard FORTRAN commands. There are thus strong arguments, chiefly those of portability of our programs, for attempting to produce graphical output on the lineprinter. In Chapter 10 two case studies are given to show how standard lineprinter characters may be used to build up histograms, scatter diagrams and maps by combining the normal range of symbols into blocks and by using the 1H+ overprinting command in output FORMAT statements (see for example Garfinkel, 1962; Monmonier, 1965; Kirk and Preston, 1971). Extensions of these methods to produce dendrograms, time slices from spatial simulation programs and simple trend surface maps (Pitts, 1965; Harbaugh and Merriam, 1968; Chorley and Haggett, 1965) are not difficult to program. However, the production of large, true-to-scale maps without excessive generalisation from a variety of input data is a far more difficult task. A number of systems have been described, such as the programs to produce simple choropleth maps reported by Sentance (1969), systems called CAMAP at the University of Edinburgh, and RGRID at the University of Michigan, LINMAP (Gaits, 1969) at the Department of the Environment and SYMAP. Of these, SYMAP, developed by H. T. Fisher and co-workers at the Harvard University Laboratory for Computer-graphics is the best known and most widely available (Muxworthy, 1972). It consists of a single, large FORTRAN program which will produce three basic types of map from almost any form of areally distributed data. Although most attention has been paid to the characteristics of the output from this program (Robertson, 1967; Rosing, 1969; Rosing and Wood, 1971), as geographers our main interest ought to lie in the spatial interpolation algorithm used to thread contours through the data points. This as described by Shepard (1968) and Rhind (1971) has given an interesting and comprehensive comparison between it and a number of alternative methods. A much simpler program, suitable for undergraduate implementation but based on similar mathematics, is presented in Davis (1973). Recently, much more powerful spatial interpolation techniques based upon the theory of regionalised variables

have become available (see for example Olea, 1974) but the basic SYMAP algorithm is likely to remain in use for many years.

The programmer need have little knowledge of FORTRAN to SYMAP his data. At many university centres the program is available as a library item and can be called using a single card. All that remains is to prepare suitable input data. Fig 13.1 shows a specimen of a simple data presentation to SYMAP. The map to be drawn is one of average annual accumulated temperature below $60°$ for the London area, 1951-1961, using data for fifteen meteorological stations given in Chandler (1965). The input is in a series of packages, each of which specifies particular features of the desired map. The cards under A-OUTLINE give the outline of the study area as a string of spatial co-ordinates, those under B-DATA POINTS give the locations of the sample points. C-OTOLEGENDS is used to locate text and other symbols and E-VALUES gives the accumulated temperatures at each of the stations. Finally, F-MAP specifies the type and size of map to be drawn together with various 'electives' that specify how the contours are to be represented. The resultant, objective isoline map is shown as Fig 13.2 which can profitably be compared with Chandler's hand drawn map of the same data. It shows relatively high values of accumulated temperature below the $60°F$ threshold over the country and outer suburbs with a distinct trough of lower values indicating warmer conditions over the city centre and elongated from west to east along the Thames axis. In the north, the heights of Hampstead stand out as an island of relatively cold air within this warm trough. It is apparent that the city centre is on average warmer than its surrounds, a reflection of the heat island effect.

Lineprinter graphics such as this are extremely useful for the fast production of thematic maps intended for use in research. For publication, they have many disadvantages. Aesthetically, the symbol blocks used leave much to be desired and the resolution is limited to that of the printer grid, normally 10 characters/inch across the page and 6 characters/inch down it. Normally, colour printing is impossible and the visual effect produced often depends strongly upon uncontrolled factors such as variations in printer pressure and inking. Lineprinter paper has to be fairly strong, but by normal cartographic standards it is unstable and cannot be relied upon not to warp with age. Except in very special circumstances, stable base detail

```
        A=OUTLINE
                    0.0         2.0
                    1.5         2.0
                    1.5         0.0
                    5.5         0.0
                    5.5         5.5
                    0.0         5.5
                    0.0         2.0
99999
B=DATA POINTS
                    2.7         2.8
                    2.2         2.8
                    2.7         2.5
                    2.5         2.7
                    3.2         1.7
                    3.8         1.2
                    3.0         3.7
                    3.0         0.7
                    3.8         3.9
                    3.3         5.3
                    4.3         3.1
                    1.3         2.9
                    4.9         0.6
                    4.3         3.4
                    2.1         2.5
99999
C=OTOLEGENDS
    12      P       1.3         2.9              3.0
   SOUTHGATE
     9      P       3.0         0.7              3.0
   AIRPORT
     9      P       4.9         0.6              3.0
   WISLEY
    10      P       3.8         3.9              3.0
   BROMLEY
    10      P       3.3         5.3             -10.0
   DARTFORD
    13      P       2.7         2.8              3.0
   WESTMINSTER
99999
E=VALUES
                  435.0
                  468.0
                  513.0
                  539.0
                  600.0
                  606.0
                  631.0
                  738.0
                  798.0
                  807.0
                  834.0
                  867.0
                  942.0
                 1023.0
                 1071.0
99999
F=MAP
C    ANNUAL ACCUMULATED TEMPERATURE BELOW 60 (F) , LONDON 1951 - 1961
C    ACTUAL VALUES LESS 3000 DEGREE DAYS HAVE BEEN MAPPED
C    SOURCE - CHANDLER,T.J. 1965 THE CLIMATE OF LONDON ,182-183
     1         12.5        12.5
     2          5.5         5.5
     8
    10
        DATA POINTS ARE METEOROLOGICAL OFFICE STATIONS
9999
99999
999999
```

*Fig 13.1* Data presentation to SYMAP

**AN EXAMPLE OF SYMAP OUTPUT**

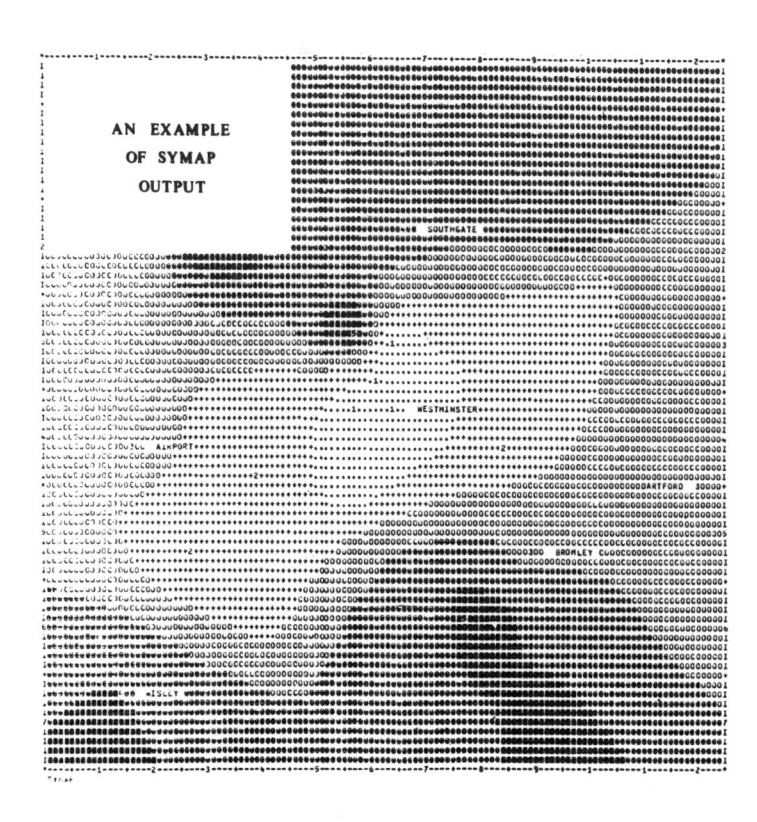

Labels visible on map: SOUTHGATE, WESTMINSTER, AIRPORT, DARTFORD 30000+, BROMLEY C_0000300, AISLEY

1.10 SECONDS FOR MAP
ELAPSED TIME (SPSS)=    1.25

C   ANNUAL ACCUMULATED TEMPERATURE BELOW 60 (F) , LONDON 1951 - 1961
U   ACTUAL VALUES LESS 3600 DEGREE DAYS HAVE BEEN MAPPED
C   SOURCE - CHANDLER, T. J. 1965 THE CLIMATE OF LONDON ,182-183
    DATA POINTS ARE METEOROLOGICAL OFFICE STATIONS

    DATA VALUE EXTREMES ARE    +35.00    1871.00

*Fig 13.2* Annual accumulated temperature below 60°F, London 1951-61

such as county outlines has to be added after the plot, usually as a transparent overlay of the type given in Rosing and Wood (1971). Finally, the printer lacks flexibility. With the exception of the overprint facility it cannot be sent back to areas already printed and the programmer must carefully sort his output into printing order before starting the output.

*Computer-graphics using the graph plotter.* The graph plotter overcomes many of these difficulties. In this an inked pen or other scribing implement is controlled by the program and directed to move across the paper in the X and Y directions. As explained in Chapter 2, *drum plotters* allow the pen to move only in the Y plane; changes in X are produced by moving the paper backwards and forwards beneath it. *Flat bed plotters* allow the paper to remain stationary and the pen to move in both planes. Modern plotters of both types are capable of high quality linework with a resolution better than 1/200in, and can be used to draw diagrams of any required complexity (Bickmore, 1967, 1968; ECU, 1971). Some plotters will allow the programmer to use lines of different colour on the same plot without any operator intervention.

There are no commands in standard FORTRAN to drive the plotter; each machine will have its own statements. Normally a summary of the commands will be available at the particular computer centre being used and will vary in scope and power. It is usual for plotter commands to be given as general purpose subroutine subprograms that are called using

330

the CALL command. The plot instructions specific to the
particular program are passed by way of the subroutine
arguments. Typically, a basic set of plotter commands
would include a subroutine to move the pen in the 'up'
position to the point (X,Y); one to move the pen across
the paper from its current position to the point (X,Y),
and one to draw a character of specified type and size at
the current location. Because the (X,Y) co-ordinates are
specified as plotter steps, we must add a fourth routine
that is called at the start of the plotting to specify the
location of the origin (X = ∅, Y = ∅) on the page. On one
system known to the authors these functions are provided by
subroutines called MOVE, DRAW, CENCH and ORIGIN but on
other systems different names might be used.

For most graphical work in geography these routines are
a little too basic; what is needed are routines to perform
common functions such as drawing graph axes, map borders,
labelling, drawing a graticule, joining points by curves,
plotting histograms and threading contours. In the geographical literature, a number of FORTRAN systems that
provide subroutines for these functions have been described
and go by names such as MAPIT (Kern and Rushton, 1969),
KOMPLOT (Kadman, 1971) and GRAFPAC (Rolhf, 1969)(see also
Monmonier, 1969, 1970). In the United Kingdom, many
University systems have implemented a series of routines
called GINO. A very similar system is called GHOST and
can be taken as an example of the *genre*. GHOST is
a FORTRAN graphical output system consisting of 68 subroutines that enable the user to generate and manipulate
graphical information. It is designed to be flexible and
easy to use and allows the user to interface his own programs into the system. GHOST makes full use of the
modular programming methods advocated in Chapter 12.
Apart from two very basic routines to draw a line and to
plot a point that would normally be programmed in the
machine's assembler, all the remaining functions are contained in a hierarchy of standard FORTRAN subroutines.
At some cost in machine efficiency this structure makes
GHOST as machine- and plotter-independent as it is possible
to get.

Figure 13.3 shows a complete program to contour automatically a field of numbers using GHOST routines. The
data input is fully documented in the commentary and consists of the X,Y and Z co-ordinates of a series of up to
100 irregularly spaced control points. Using the simple
inverse distance squared weighting algorithm published and

```
C   A PROGRAM TO ILLUSTRATE THE USE OF PLOTTING FACILITIES AVAILABLE      10
C   ON A PARTICULAR PACKAGE OF GRAPHICAL OUTPUT SUBROUTINES               20
C   FOR THE DIGITAL PLOTTER CALLED G,H,O,S,T.                             30
C   PROGRAM CONTOURS IRREGULARLY SPACED DATA IN TWO STEPS -----           40
C                                                                         50
C   A) IS BASED UPON PROGRAM 6.1, PAGE 317 OF DAVIS,J,C.,1973             60
C      STATISTICS AND DATA ANALYSIS IN GEOLOGY                            70
C      IT PRODUCES A REGULAR GRID OF INTERPOLATED VALUES                  80
C   B) PLOTS THIS USING GHOST ROUTINES FROM THE COMPUTER LIBRARY          90
C                                                                        100
      DIMENSION D(100,3), DIST(100),AM(60,60),H(10),FMT(20),NAME(20)     110
      NIN = 5                                                            120
      NOUT = 6                                                           130
C   READ MAPPING PARAMETERS CARD                                         140
C   NAME = ANY 80 CHARACTER MAP NAME                                     150
C   WIDTH = DESIRED MAP WIDTH IN INCHES                                  160
C   XMIN, YMIN CO-ORDINATES OF BOTTOM LEFT HAND CORNER OF MAP            170
C   YMAX = THE MAXIMUM COORDINATE IN BOTH X AND Y DIRECTIONS             180
C   THE MAP MUST BE SQUARE AND OF SIDE LESS THAN 10 INCHES               190
C   N = NUMBER OF DATA POINTS                                            200
C   NCONT = NUMBER OF CONTOURS THROUGH FIELD                             210
      READ(NIN,200) NAME ,WIDTH,XMIN,YMIN,YMAX,N,NCONT                   220
      IF ( N.GT. 100 .OR. NCONT .GT. 10  .OR. WIDTH .GT. 10.0) STOP      230
C   READ DESIRED CONTOUR VALUES TO MAXIMUM OF 10                         240
      READ(NIN,220)(H(I),I = 1,NCONT)                                    250
C   READ IN DATA FORMAT FOLLOWED BY DATA CARDS                           260
      READ(NIN,230) FMT                                                  270
      DO 90 I =1, N                                                      280
   90 READ(NIN,FMT) D(I,1),D(I,2),D(I,3)                                 290
C   SET UP PLOTTER BY EQUATING THE PLOTTER SPACE WITH THE ACTUAL         300
C   COORDINATE VALUES USED TO CODE UP THE DATA                           310
C   USE LIMITS TO SET LIMITS ON THE OUTER EDGE OF THE PLOTTING AREA      320
      CALL GHOST                                                         330
      CALL PSPACE(1.0,WIDTH,1.0,WIDTH)                                   340
      CALL LIMITS(0.0,10.0,0.0,10.0)                                     350
      CALL MSPACE(1.0,60.0,1.0,60.0)                                     360
C   PUT A BORDER ROUND THIS                                              370
      CALL BORDER                                                        380
C   WRITE OUT MAP PARAMETERS ON LINEPRINTER                              390
      WRITE(NOUT,240) NAME,WIDTH,XMIN,YMIN,YMAX,NCONT,(I,H(I),I=1,NCONT) 400
C   FIND THE REGULAR GRID OF ESTIMATED VALUES                            410
      DY = (YMAX-YMIN) /59.0                                             420
      SMALL = 2* DY*DY / 10000.0                                         430
      Y = YMAX                                                           440
      DO 100 I =1,60                                                     450
      X = XMIN                                                           460
      DO 110 J = 1,60                                                    470
C   FIND DISTANCE FROM THIS NODE TO ALL DATA POINTS                      480
      DO 120 K = 1,N                                                     490
      DIST(K) = ( X - D(K,1))**2 + ( Y - D(K,2))**2                      500
  120 CONTINUE                                                           510
C   FIND SIX NEAREST OF THESE                                            520
      S1 = 0.0                                                           530
      S2 = 0.0                                                           540
      DO 130 K = 1,6                                                     550
      IC = 1                                                             560
      DO 140 L = 2,N                                                     570
      IF ( DIST(L) .LT. DIST(IC )) IC = L                                580
  140 CONTINUE                                                           590
      IF ( DIST(IC) .LT. SMALL ) GO TO 10                                600
      DT = SQRT(DIST(IC))                                                610
      S1 = S1 + D(IC,3)/DT                                               620
      S2 = S2 + 1.0/DT                                                   630
      DIST(IC) = 9.0E+20                                                 640
  130 CONTINUE                                                           650
```

```
          AM(I,J) = S1/S2                                            660
          GO TO 11                                                   670
       10 AM(I,J) = D(IC,3)                                          680
       11 X = X + DY                                                 690
      110 CONTINUE                                                   700
          Y = Y - DY                                                 710
      100 CONTINUE                                                   720
C   WE NOW HAVE A REGULAR GRID OF INTERPOLATED VALUES THAT CAN BE    730
C   CONTOURED .   USE SUBROUTINE CONTRL TO CONTOUR THIS              740
          CALL CONTRL(AM,1,60,60,1,60,60,H,1,NCONT)                  750
C                                                                    760
C   PLOT POINT LOCATIONS OF ORIGINAL INPUT DATA                      770
C   CTRSET GIVES AN APPROPRIATE CHARACTER SET                        780
          CALL CTRSET(4)                                             790
C   CHSIZE SETS THE CHARACTER SIZE                                   800
          CALL CRSIZE(1,5)                                           810
C   PLOT THE POINTS                                                  820
          DO 150 I = 1, N                                            830
          X = D(I,1) / DY                                            840
          Y = D(I,2) / DY                                            850
C   PLOT CHARACTER 63, A CROSS AT THIS POINT                         860
          CALL PLOTNC(X,Y,63)                                        870
      150 CONTINUE                                                   880
C   CHANGE CHARACTER SET AND SIZE                                    890
          CALL CTRSET(1)                                             900
          CALL CRSIZE(1,75)                                          910
C   WIDEN PLOTTING AREA                                              920
          WIDTH = WIDTH + 1.0                                        930
          CALL PSPACE(0.0,WIDTH*2.5,0.0,WIDTH)                       940
C   MOVE PEN TO POSITION FOR TITLE                                   950
          X = WIDTH - 0.5                                            960
          CALL POSITN(X,1.0)                                         970
C   ROTATE ANNOTATION THROUGH 90 DEGREES ON THE PAPER                980
          CALL CTRANG(1,5708)                                        990
          CALL TYPECS(NAME,80)                                       1000
C   MOVE ON PLOTTER AND TERMINATE PLOTTING                           1010
          CALL FRAME                                                 1020
          CALL GREND                                                 1030
          STOP                                                       1040
      200 FORMAT(20A4/4F10.2,2I5)                                    1050
      220 FORMAT(10F10.2)                                            1060
      230 FORMAT(20A4)                                               1070
      240 FORMAT(1H1,20A4//9H WIDTH = ,F10.2,9H XMIN = ,F10.2,       1080
        *  9H YMIN = ,F10.2,9H YMAX = ,F10.2/15H NO CONTOURS = ,I5,  1090
        *     1H ////  11H NO  VALUE/ 10(I3,F10.2))                  1100
          END                                                        1110
```

*Fig 13.3* An example of a GHOST program

described in Davis (1973), the program first produces
a regular 60 by 60 grid of interpolated values.  GHOST
routines are then used to draw this out as a contour map
on the digital plotter using contour values that were
specified in the original data.

The commands CALL GHOST and CALL FRAME load the required
subroutines and advance the plotter paper well clear of any
previous users' work.  Subroutine PSPACE defines the physical area of the plotter on which the plot is to be made,
in this case a square of 7in side located 1in from the
paper margin.  It would be extremely inconvenient always
to have to refer the information to be plotted to this
physical space on the plotter.  To get over this, the
routine called MSPACE is used to equate the space defined
in PSPACE with the mathematical space actually used in the

333

program. The command CALL BORDER draws a border around the plotting area, and all subsequent commands are in the co-ordinates of the mathematical space, not in plotter steps. Contouring, at values specified in the array H, is performed by simple linear interpolation in subroutine CONTIL. For annotation of this plot, a series of statements uses PLOTNC to draw a specified character (CTRSET) shape and size (CRSIZE) at each of the N input control locations. Finally, POSITN and TYPECS are used to put a title on the map. This ends the plotting required to produce our map, but before the program ends it must move on the paper using FRAME and switch off the plotter using GREND.

An example of the output from this simple program is given as Fig 13.4. The input data were the same as used in Davis (1973) and refer to heights on a topographic surface. The output should be compared with the lineprinter maps given by Davis's program.

A second application of plotters that makes use of their flexibility is in drawing perspective representations of complex surfaces for easier visualisation. Again, a number of systems and programs have been described (Kubert, Szabo and Guilien, 1970). One such system is called SYMVU, and like SYMAP it is a FORTRAN program originally devised at the Harvard Computer-graphics Laboratory. Figure 13.5 shows a SYMVU view of the surface for accumulated temperatures below 60°F over London that was used in the SYMAP example. The advantage of this form of representation is mostly that of easy visualisation, and gives a useful means of presenting a complex statistical surface to a lay audience.

Plotter output has the major disadvantage that by computer standards it is ludicrously slow; usually plotters are operated 'off-line' and are driven by a magnetic tape onto which the output from the generating program is stored. There is no waste of CPU time, but this off-line working usually delays the return of the work from the computer room and is wasteful for all except high quality plots intended for direct use in publication.

*Computer-graphics using the cathode ray tube and similar devices.* Much plotting work in geography is of an experimental nature. The researcher is experimenting with different representations of his data in order to obtain the maximum visual impact or is examining his data using different analytic techniques. This type of work is very wasteful if carried out using the graph plotter and

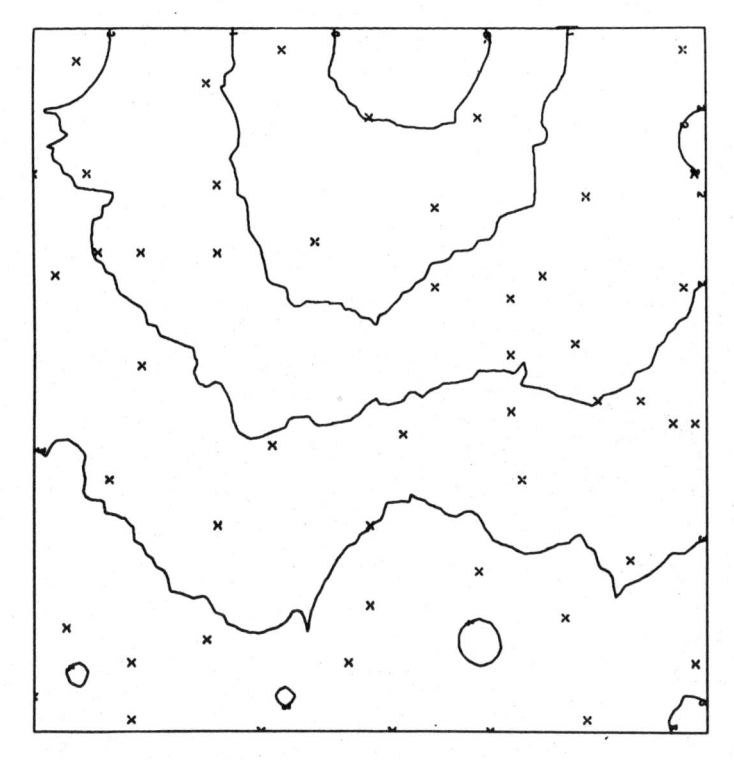

**EXAMPLE OF PLOTTER OUTPUT FROM GHOST**

*Fig 13.4* An example of a GHOST output

is best suited to fast electronic display devices such as the cathode ray tube (CRT) and microfilm plotter. Some examples of CRT output are given by Sprunt (1970) and by Jones (undated).

Although very rapid, electronic output suffers from a number of disadvantages. Unless the screen is photographed a CRT produces no 'hard copy' yet the screens are often curved which introduces distortion into the photographs and the resultant film material may not be stable enough for long-term storage. Secondly, the resolution obtained is poor. Typically, the face of the CRT is divided into addressable locations, or raster points forming a grid with 1024 points in each of the X and Y

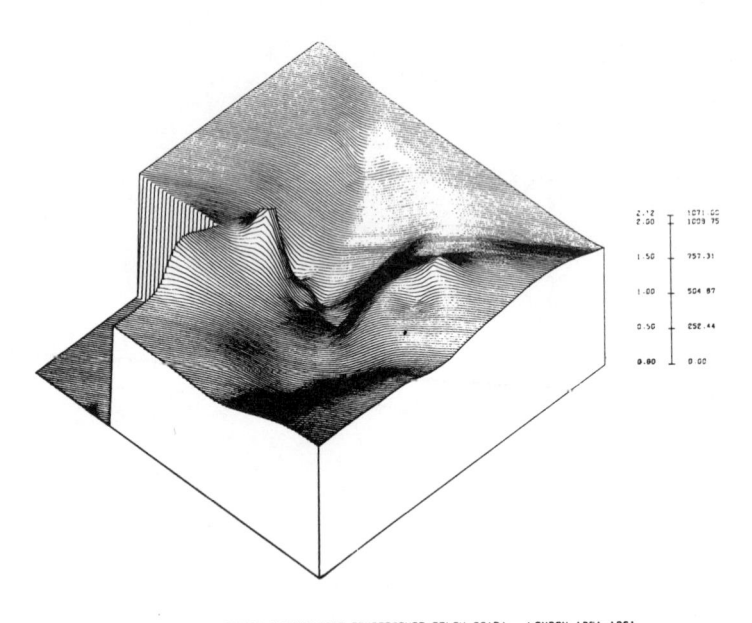

*Fig 13.5* A SYMVU of the London data

directions. This grid limits the resolution of the plot and the physical screen size gives severe limits on the size of plot that can be made. Recently, devices that plot directly onto film, or *microfilm plotters*, have become generally available and are usually faster and more versatile than the CRT.

Despite these physical limitations, the CRT and microfilm plotter do offer fascinating possibilities for geographic research. One possibility is to display a succession of maps and diagrams, filming them as they appear on the screen. On replaying the film the result will be an animated map (Thrower, 1959) or even an animated block diagram (Cutrell, Feeser and Penzien, 1970) that changes as we watch it. These could be used to study diffusion processes by simulation techniques, in the analysis of land-use changes or in any other branch of the subject where an historical perspective is required.

## 13.5 Other Languages

Throughout this book we have restricted ourselves to the high-level programming language FORTRAN. All the examples have used a restricted subset, the 1966 ANSI standard FORTRAN, of the commands that are commonly available. As we have seen, this is the most widely available of the high-level languages in which the vast majority of scientific programs have been written. The reader who has mastered the rudiments of this language may well find it useful to add a second or even a third language to his repertoire. This is not as difficult as it sounds. The real art of programming is in the creation of algorithms, not in the detail of the coding, and this is independent of the particular language used. In practice, too most languages have statements that are fairly similar - computer scientists have even devised programs that will automatically translate programs from language to language. The geographer might wish to program in another language for a number of reasons. He might simply want to translate a program from one language to another, or to make heavy use of a specific facility available in another language, or simply to extend his computing knowledge.

Most universities will have available compilers for many languages other than FORTRAN, but discussion here will be restricted to the other two most commonly used scientific languages, ALGOL and BASIC.

In theory, ALGOL is an extremely important language. It was designed in the early 1960s by an international committee and includes an extremely powerful and attractive series of commands. The designers intended that ALGOL should be used to describe a computational process, independent of any particular machine and without specifying any input/output commands. There is therefore only a single ALGOL that can be used as a 'communications language' between programmers as for example in the algorithms given in each issue of *Communications of the Association of Computing Machinery*. The formal definition of this language, by Backus and others, in terms of a very high-level 'metalanguage' is a classic document in computer science (see the Appendix to Dijkstra, 1962).

In practice, implementations of ALGOL on particular machines fall short of this theoretical ideal. Although

```
CHISQUARE:
"BEGIN"
"COMMENT" A1 ALGOL PROGRAM TO CALCULATE CHISQUARE VALUES COMPARABLE TO
        FORTRAN CODING OF FIGURE 8.3 , USING ICL-ELLIOT 10;
    "INTEGER" NR,NC;
    "READ" NR,NC;
    "BEGIN" "INTEGER" NSAMP,NRS,I,J;
        "REAL" CHISQ,AMULT,EXP;
        "INTEGER""ARRAY" NOS[1:NR,1:NC],NCOL[1:NC],NROW[1:NR];
        "REAL""ARRAY" CHIS[1:NC];
        SAMELINE; ALIGNED(4,3); DIGITS(4);
        "PRINT" ''L?' CONTINGENCY TABLE';
        "COMMENT" READ DATA FORMING ROW AND COLUMN SUMS;
        NSAMP := 0;
        "FOR" J:=1"STEP"1"UNTIL"NC"DO" NCOL[J]:=0;
        "FOR" I:=1"STEP"1"UNTIL"NR"DO"
        "BEGIN" NRS :=0;
            "FOR" J:=1"STEP"1"UNTIL"NC"DO"
            "BEGIN" "READ" NOS[I,J];
                    NRS := NRS + NOS[I,J];
                    NCOL[J] := NCOL[J] + NOS[I,J];
            "END" JS LOOP ;
            NROW[I] := NRS ; NSAMP := NSAMP + NRS ;
        "END" DATA INPUT AND INITIAL SUMMING ;
        "COMMENT" CALCULATE AND PRINT CHISQUARE;
        CHISQ := 0;
        "FOR" I:=1"STEP" 1"UNTIL""R"DO"
        "BEGIN" AMULT := NROW[I]/NSAMP ;
            "FOR"J:=1"STEP"1"UNTIL""C"DO"
            "BEGIN" EXP := NCOL[J] * AMULT;
                "IF" EXP < 5 "THEN""BEGIN"
                            "PRINT"''L'WARNING E ONLY',
                            EXP,' IN CELL',I,J;
                            "IF"EXP=0"THEN""GOTO"L1;
                            "END" INVALID CHISQUARES LOOP ;
                CHIS[J] := (NOS[I,J]-EXP)+2/EXP;
                CHISQ := CHISQ + CHIS[J] ;
            "END";
            "PRINT" ''L''';
            "FOR" J:=1"STEP"1"UNTIL"NC"DO" "PRINT"
                    NOS[I,J],CHIS[J];
        "END" TABLE FORMATION NOW SUMMARISE;
        "PRINT"''L'TOTAL CHISQUARE',CHISQ,' WITH', NSAMP,
                ' CASES';
        "COMMENT" CALCULATE AND PRINT DEGREES OF FREDOM;
        "PRINT"''L'DEGREES FREEDOM',DIGITS(5),(NR-1)*(NC-1);
L1: "END" TIMED BLOCK WITH DYNAMIC STORE ALLOCATION;
"END" OF ALGOL PROGRAM TO DEMONSTRATE CHISQUARES ;
```

*Fig 13.6* A simple ALGOL program

the language is in itself elegant and powerful, only seldom can the same be said of most presently available ALGOL compilers, so that the actual code the machine uses produced by ·the compiler from an ALGOL source is often clumsy and inefficient.  Secondly, every ALGOL compiler has its own unique input/output commands which makes most ALGOL programs very machine dependent.  Figure 13.6 shows

a simple ALGOL program to calculate chi-square values for
a contingency table, similar in most respects to the
FORTRAN coding of Fig 8.4. This particular program is
for one machine, the ICL-Elliot 4130, and could not be
expected to run without substantial modification on other
machines. Apart from the obvious representational
differences (: = for =, ↑ for **, [] for () and so on),
the most interesting features of this ALGOL are

   (i) The use of the commands 'BEGIN' and 'END'; to give
      a nested or 'block' structure to the program.
  (ii) Run-time allocation of store. Notice that the
      sizes of the arrays used are declared using
      variables called NR and NC that are not known until
      they are read as data at execution.
 (iii) The statement terminator is the semi-colon, not the
      end of a card. This allows several statements to
      be packed onto a card and single statements to be
      held on more than one card.
  (iv) Although not illustrated in this example, a par-
      ticularly powerful and elegant ALGOL feature is
      *recursion* which allows a subprogram (or 'PROCEDURE')
      to call itself.

The programming language called BASIC (Beginners All-
purpose Symbolic Instruction Code) has a very different
history from ALGOL. It was developed at Dartmouth College,
New Hampshire, primarily as an extremely easily learnt
language for use in time-sharing systems (Kemeny and Kurtz,
1971). As a result of the simplicity of its commands,
BASIC is rapidly finding favour in schools and colleges as
a first computer language. Figure 13.7 shows a BASIC
program on punched cards to calculate chi-squares comparable to the ALGOL of Fig 13.6 and FORTRAN of Fig 8.4.
The reader should have very little trouble in following
this program, and with the notable exception of its ascending sequence of statement numbers on all statements,
it does not look very different from FORTRAN.
    The real power and flexibility of BASIC lies not in its
application to problems that can be solved using card input and lineprinter output, but in its use in *interactive*
computation. A number of other programming languages,
including an interactive FORTRAN, APL and FOCAL, have been
designed for interactive computing, but BASIC is the most
commonly used. In interactive computation the user sits
at a teletype or visual display unit (VDU) through which

```
10 REM A  BASIC PROGRAM TO CALCULATE CHHISQUARES
20 REM SIMILAR TO THE FORTRAN AND ALGOL EXAMPLES
30 DIM N(10,10),C(10),R(10),S(10)
40 READ K,M
50 PRINT " CONTINGENCY TABLE"
60 REM READ IN DATA FORMING ROW AND COLUMN TOTALS
70 LET S1 = 0
80 FOR J = 1 TO M
90 LET C(J) = 0
100 NEXT J
110 FOR I = 1 TO K
120 LET S2 = 0
130 FOR J = 1 TO M
140 READ N(I,J)
150 LET S2 = S2 + N(I,J)
160 LET C(J) = C(J) + N(I,J)
170 NEXT J
180 LET R(I) = S2
190 LET S1 = S1 + S2

200 NEXT I
210 REM CALCULATE CHISQUARE VALUES ROW BY ROW
220 LET C1 = 0
230 FOR I = 1 TO M
240 LET A = R(I)/S1
250 FOR J = 1 TO K
260 LET E = C(J) * A
280 IF E > 5 THEN 310
290 PRINT " WARNING - EXPECTED ONLY ",E," IN CELL",I,J
300 IF E = 0 THEN 500
310 LET C2 = N(I,J) - E
320 LET C2 = C2 * C2 / E
330 LET S(J) = C2
340 LET C1 = C1 +C2
350 NEXT J
360 FOR J=1 TO K
370 PRINT
380 PRINT N(I,J),S(J)
390 NEXT J
400 NEXT I
405 PRINT
410 PRINT "TOTAL CHISQUARE",C1,"TOTAL NO OF CASES",S1
420 REM CALCULATE AND PRINT DEGREES OF FREEDOM
430 PRINT " DEGREES OF FREEDOM",(M-1)*(K-1)
450 DATA 3,3,1,28,7,9,53,68,28,3,21
500 END
```

*Fig 13.7* A simple BASIC program

he can communicate directly with the computer. The teletype may be physically very remote from the servicing machine. All input and output will pass through the teletype at relatively slow speed, so that the CPU is able to share its time among a number of other users at other teletypes or VDUs. The CPU speed is such that the machine's response to any one user's command is more-or-less instantaneous. Interactive BASIC is available in two types of computing environment. The original Dartmouth College implementation was on a large machine servicing a great many individual users, but increasingly it is available on small mini-computers servicing between one and, say, five users.

Interactive computing has both practical and theoretical advantages. Practically, each statement can be checked for correct syntax as it is typed into the machine so that any program-coding errors are detected immediately, enabling the programmer quickly to develop a correct code. There are no punched cards to bother with and the answers are more-or-less instantaneous. It is thus a very practical and rapid way to learn programming. Theoretically, it allows several useful extensions to the normal range of computer applications. First, it enables the user to experiment freely trying alternative inputs to problems until a reasonable solution is obtained. This experimental approach is often used in design studies in engineering for complex structures such as bridges and motor cars. These systems are so complex that no analytic procedure can be used to produce an optimum design. Instead, the structure is modelled within the computer and its performance under realistic conditions simulated. On the basis of the simulated performance, the design can be altered and the simulation re-run in a cyclical manner until some reasonable performance is achieved. In this work it is extremely useful to be able to combine interactive computing with graphics to develop *interactive computer-graphics* using a CRT to display the results and a light pen (*see Chapter 2*) to input design changes. The hardware to do this is expensive and not often readily available, but there are obvious practical applications to map design in cartography (Newman and Sproull, 1973).

A second theoretical advantage of interactive computing is that programs can be written that guide the user through his analysis by conducting a question (from the program) and answer (from the user) dialogue. At the extreme, programs that take on the role of teacher may be written and used for *Computer Aided Instruction*.

THE CHAGGA FARM GAME
YOU ARE FARMING COFFEE ON MOUNT KILIMANJARO
EACH YEAR YOU MUST DECIDE WHAT PROPORTION OF YOUR
CROP TO SPRAY WITH COPPER. ACCORDING TO WEATHER
CONDITIONS THE PAY OFFS ARE AS GIVEN IN THE TABLE
```
                    HIGH MOISTURE    LOW MOISTURE
       IF SPRAYED       100              -40
     IF NOT SPRAYED      60               40   (MONEY UNITS)
```

AT THE START YOU HAVE 200 MONEY UNITS
   YEAR NUMBER    1
WHAT PERCENT OF YOUR CROPS DO YOU WANT TO SPRAY
?
<u>100</u>
THE MOISTURE CONDITION IS LOW THIS YEAR
THIS STRATEGY YIELDS           -40            MONEY UNITS
CAPITAL IS NOW   160           UNITS
   YEAR NUMBER    2
WHAT PERCENT OF YOUR CROPS DO YOU WANT TO SPRAY
?
<u>100</u>
THE MOISTURE CONDITION IS LOW THIS YEAR
THIS STRATEGY YIELDS           -40            MONEY UNITS
CAPITAL IS NOW   120           UNITS
   YEAR NUMBER    3
WHAT PERCENT OF YOUR CROPS DO YOU WANT TO SPRAY
?
<u>100</u>
THE MOISTURE CONDITION IS HIGH THIS YEAR
THIS STRATEGY YIELDS           100            MONEY UNITS
CAPITAL IS NOW   220           UNITS
   YEAR NUMBER    4
WHAT PERCENT OF YOUR CROPS DO YOU WANT TO SPRAY
?
<u>100</u>
THE MOISTURE CONDITION IS LOW THIS YEAR
THIS STRATEGY YIELDS           -40            MONEY UNITS
CAPITAL IS NOW   180           UNITS
   YEAR NUMBER    5
WHAT PERCENT OF YOUR CROPS DO YOU WANT TO SPRAY
?
<u>100</u>
THE MOISTURE CONDITION IS HIGH THIS YEAR
THIS STRATEGY YIELDS           100            MONEY UNITS
CAPITAL IS NOW   280           UNITS
   YEAR NUMBER    6
WHAT PERCENT OF YOUR CROPS DO YOU WANT TO SPRAY
?

*Fig 13.8* An interactive dialogue – the Chagga Farm Game

Figure 13.8 shows a typical student/program dialogue in the Chagga farm game taken from Abler, Adams and Gould (1971). In it the student takes the role of a Chagga farmer producing coffee on the slopes of Mount Kilimanjaro. Each year he must decide what proportion of his crops to spray with copper solution. The benefits from the spraying depend upon the year's moisture conditions. If the year is wet the sprayed trees give a maximum net yield of 100 money units, but if it is dry the spray inhibits the trees' normal defences against aridity resulting in a loss of 40 units needed to replace the dead trees. If no spraying is attempted, the yield is 60 units in a wet year and 40 in a dry year. At the start of the game the farmer has 200 money units capital and the game lasts for the equivalent of eleven years. The decisions to spray are taken each year without knowledge of the future weather conditions. This decision is taken by the program using a random number generator and a fixed probability of 50 per cent that a year will be wet. The economic outcome in any year is thus related both to the farmer's decision and to the prevailing weather conditions. It is calculated in the program and the farmer's capital adjusted accordingly. The capital obtained at the end of the game is a measure of the farmer's skill in long-range weather forecasting or of his understanding of elementary game theory (Gould, 1963).

In our example, Fig 13.8, the underlined portions represent values typed by the farmer/student. At the command <u>RUN</u>, the machine types out an explanatory message and asks what percentage of the crop the farmer wants to spray. Ours is an inveterate gambler who sprays all his crops. If the weather is wet this will give the maximum return. In the first year he is lucky; the weather is wet and he gains 100 units. Encouraged by this initial success he continues to spray all his crops and ends with a capital of 740 units. In years 4,6,8 and 11 conditions are dry and he loses 40 units in each. Had he been able accurately to forecast the coming weather conditions and always chosen the best strategy he would have ended with 1,140 units. The BASIC program that controlled this dialogue is shown as Fig 13.9, and should be more-or-less self-explanatory. Notice that there are two forms of input statement. The statement 13∅ READ takes its data from the DATA statement labelled 365 whereas 23∅ INPUT A accepts values from the teletype. As given, this exercise is little more than an amusing game, but in following it through a student will have gained practical experience of the problem of decision-making under

```
10  REM THE CHAGGA FARM GAME, ABLER ADAMS AND GOULD,437-8
20  DIM E(2, 2)
30  PRINT "THE CHAGGA FARM GAME"
40  PRINT "YOU ARE FARMING COFFEE ON MOUNT KILIMANJARO"
50  PRINT "EACH YEAR YOU MUST DECIDE WHAT PROPORTION OF YOUR"
60  PRINT "CROP TO SPRAY WITH COPPER. ACCORDING TO WEATHER"
70  PRINT "CONDITIONS THE PAY OFFS ARE AS GIVEN IN THE TABLE"
75  PRINT "              HIGH MOISTURE   LOW MOISTURE"
80  PRINT "     IF SPRAYED     100            -40"
90  PRINT "     IF NOT SPRAYED  60             40 (MONEY UNITS)"
100 PRINT
105 PRINT "AT THE START YOU HAVE 200 MONEY UNITS"
110 RANDOMIZE
120 LET M = 200
130 READ E(0, 0), E(0, 1), E(1, 0), E(1, 1), E1
140 LET E2 = 100 - E1
150 LET I = 1
160 LET C = 0
170 LET C = C + 1
175 PRINT " YEAR NUMBER ", C
180 IF C = 12 THEN 360
190 IF M > 0 THEN 220
200 PRINT " YOU ARE BROKE - TRY FOR A WORLD BANK LOAN - GAME ENDS"
210 GOTO 370
220 PRINT "WHAT PERCENT OF YOUR CROPS DO YOU WANT TO SPRAY"
230 INPUT A
235 LET A = A / 100
240 IF A < 0 THEN 370
245 IF A > 1 THEN 370
250 LET B = 1 - A
260 LET W = RND(0) * 100
270 IF W > E1 THEN 300
280 LET Y = 0
285 PRINT "THE MOISTURE CONDITION IS HIGH THIS YEAR"
290 GOTO 310
300 LET Y = 1
305 PRINT "THE MOISTURE CONDITION IS LOW THIS YEAR"
310 LET G = A * E(0, Y) + B * E(1, Y)
320 PRINT "THIS STRATEGY YIELDS ", G, "MONEY UNITS"
330 LET M = M + G
335 PRINT "CAPITAL IS NOW", M, " UNITS"
340 GOTO 170
360 PRINT " YOU HAVE PLAYED FOR A CHAGGAS LIFETIME - GAME ENDS"
365 DATA 100, -40, 60, 40, 50
370 END
```

*Fig 13.9* The BASIC control program for the Chagga Farm Game

uncertainty. Similar interactive programs can be written to perform simple geographic and statistical analysis at the teletype and it is likely that this form of computation will become increasingly popular (Koch, Link and Schuenemeyer, 1972).

13.6 The Future of Computing in Geography
Much of this book has been concerned with how computers have

helped in the solution of traditional geographical problems and in speeding calculations in general geographical techniques. The help afforded with data manipulation, statistical analysis and in the graphical representation of results is particularly noteworthy. What of the future role of computers in geography? In many ways they will extend existing procedures, allowing more sophisticated statistical analysis, more efficient data sorting of even larger data blocks, and more aesthetically pleasing and accurate mapping. These applications depend upon a continuing growth in the geographer-programmer's art and in the computer scientist's technology. There are however two other uses of computers, which to the geographer are still in their infancy. In both fields geographical developments have been taking place but in a rather tentative fashion yet they offer very considerable potential for the future. The two topics are computer simulation and computer-aided instruction.

Since the mid-1960s many disciplines, not just those in the social sciences, have become aware of the value of computer simulation techniques and an increasing body of general theory related to computer simulation and modelling is developing. Martin (1968) has outlined 27 possible broad areas of learning where simulation models are likely to lead to increased understanding of systems and their processes. Not all are of interest or use to geographers, but several certainly lie well within the scope of mainstream geography. Five such are:

(i) "To analyze the effects of meteorological changes. To model weather and climate. To study weather forecasting. To analyze the effects of random processes in meteorology. To predict weather front movements. To select optimum air, land, and sea routes under existing weather patterns. To analyze the behaviour of hurricanes, typhoons, and tornados.
(ii) "To analyze a wide variety of urban problems, such as land use and transportation. To demonstrate to decision makers the consequences of alternate decisions. To study the effects of metropolitan growth patterns.
(iii) "To assist in demographic studies. To analyze the effects of population growth and population shift. To predict the effects and impacts of population growth. To integrate demographic problems with economics, community planning, zoning, marketing

outlets, resources allocations, and urban renewal. To analyze the effects of population changes on composition of age groups and ethnic groups.
(iv) "To assist in business games and to evaluate business strategies. To analyze business and market trends. To apply cost reduction to business operations. To administer inventory control. To allocate resources and to analyze product and market distribution.
(v) "To study and analyze transportation and traffic problems. To select optimum routing. To select appropriate roadways, seaways, or airways. To analyze and evaluate road and expressway designs. To analyze and evaluate traffic control systems. To assist in transportation planning. To analyze demands for travel. To analyze the effects of demands on air, land, or sea transportation facilities: road networks, airport runways, and sea-lanes. To analyze the effects of traffic delays, re-routing, and perturbation of arterial flow in transportation networks."

Some areas of application have already been explored but as yet not developed. Urban simulation particularly has received attention in a recent summary volume of *Ekistics* (1974), in a theoretical formulation by Wilson (1974), and in detailed case studies by several authors, for example Swerdloff and Stowers (1966) and Batty (1972). Other areas of potential application not mentioned in Martin's summary are in geomorphology, as in Smart, Surkan and Considine (1967), or hydrology, as in Onstad and Brakensiek (1968) and Hufschmidt and Fairing (1966). A computer simulation may be defined as 'a logical-mathematical representation of a concept, system or operation programmed for solution on a high-speed electronic computer' (Martin, 1968). The history and outline of the evolution of the concept as related to the social sciences together with an extensive bibliography is provided by Starbuck and Dutton (1971) as a simulation model of the development of the techniques of simulation. From these various attempts at definition, simulation emerges as a set of techniques which involve setting up a model to determine the influences and relationships among the variables composing the model. In this manner it is hoped to understand more fully the reality which is modelled, and in particular how the real event or system is organised,

functions and perhaps changes. With such a broad
definition of simulation, situations exist for simulation
without a computer. Simulation models can be developed
and studied in a totally non-computerised environment.
Alternatively simulation may be effective by a man-machine
interaction as in Figs 13.8 and 13.9; one further aspect
of this is explored in later paragraphs. Alternatively
the simulation may be totally computer-based and it is this
type which most concerns us.

Simulation of large-scale systems, whether physical
(geomorphological, meteorological, etc), social (economic,
political, etc) or combined (ecological) requires computer
development to keep account of the internal relationships
present in the simulation exercise. In this manner computer based simulation 'permits us to explore the complex,
intuitively holistic systems that will never be understood
by simple additions of partial sub-models' (Hendricks,
1974). In building such a model the geographer is concerned with three phases,

(i) the model concept
(ii) the model implementation
(iii) the results of the model.

The first phase involves the definition of the precise
problem under scrutiny and the definition of a working
hypothesis which is then formalised in terms of its parameters, variables and relationships. The variables may
be random, controlled or uncontrolled while relationships
may be deterministic or stochastic. Precise definition
is required in the conceptualisation phase. The data
required as input for the hypothesis testing procedures
are gathered. In the second phase a flowchart is designed and the mathematical equations in the hypothesis
are explicitly derived. A computer program is written,
possibly in one of a number of programming languages
specially designed for simulation procedures. The final
results phase entails checking of results and their evaluation. Conclusions are drawn as to whether to re-define
the hypothesis or to accept it as valid.

The simulation method is very similar to that outlined
in Chapter 3 as a method of approach to programming.
While the earlier method dealt with rather trivial problems,
we are here concerned with problems that may take 2 or 3
man-years to solve. Yet the basic approach is similar.
The similarity is imposed by the sequential nature of the

analytical process. Activity B cannot be performed before activity A is completed - calculation Y cannot be made before calculation X - the situation at year 9 cannot be determined before the situation at year 8 is known. It is the inbuilt temporal nature of computer methods, together with the computer's ability to manipulate complex relationships, which makes it so useful for developing simulation models in general and dynamic simulation models in particular. With increased geographical interest in computer-based simulation so levels of computer utility will rise.

The second potential growth area for new computer applications in geography is in Computer Aided Instruction. Several of the various techniques involved in CAI are closely related to simulation in which there is man-machine interaction. Computer Aided Instruction studies range from relatively simple question-answer interactive programs (Huke, Fielding and Rumage, 1969) to sophisticated land-use games as reviewed by Taylor (1971) and Patterson (1974) and to full computer simulations designed and written as teaching tools (Smith and Lee, 1970; Marble and Anderson, 1972). The whole field is reviewed by Fielding and Rumage (1972).

Most of the techniques involve interactive communication between a student and a prepared program. The program is designed to ask questions of the student and depending on the student's answer other questions are posed. As yet most of the programs are experimental in that they do not deal with mainstream courses in geography departments, and are concentrated at a very few North American universities, but many geography departments are now experimenting in these techniques. One popular geographical method taught by computer aided instruction is the simulation of a spatial diffusion process. The role of barriers and changing probability fields can be programmed so that a student can decide the types of barriers involved and their location and then generate cycle-by-cycle diffusion pattern. Most of the experiments are ones to teach a specific theory or application of geography, for example Von Thunen's rings or spatial competition and trade areas. Full courses taught by such methods are currently very few but the number is likely to grow considerably over the next decade. As yet most uses seem to be concerned with economic or political geography. The geomorphologist would seem to have a potentially rewarding field for CAI but so far there are only a few experimental programs. Again, growth seems

likely in the near future.
Computer simulation and computer aided instruction do not necessarily need computers with a large store. Amounts of data are often quite small although the programs themselves can be quite complex. Both developments are ideal for implementation on mini-computers and the desire of several university geography departments to obtain their own mini-computers may well be a response to the growing importance of all branches of simulation. These mini-computers are currently getting both smaller and cheaper and perhaps before many years are out student geographers will be expected to be just as competent in the use of both mini- and large computers as they are now expected to be in writing essays.

13.7 Further Reading

Abler, R., Adams, J.S. and Gould, P. *Spatial organisation: the geographer's view of the world* (1971)

Ahnert, F. 'A general and comprehensive theoretical model of slope profile development', *University of Maryland, Occasional Papers in Geography*, 1 (1971)

Baker, L. 'A selection of geographical computer programs', *Geographical papers, London School of Economics*, 6 (1974)

Balfour, A. and Beveridge, W.T. *Basic numerical analysis with FORTRAN* (1972)

Batty, M. 'Recent developments in land-use modelling: a review of British research', *Urban Studies*, 9, 151-71 (1972)

Bertziss, A.T. *Data structures, theory and practice* (1971)

Bickmore, D.P. 'The scope of automatic cartography', *Proceedings, Geological Society of London*, 1642, 205-9 (1967)

___ 'Maps for the computer age', *Geographical Magazine*, 41, 221-7 (1968)

Bisco, R.L. 'Social science data archives: progress and prospects', *Social Science Information*, 6, 39-74 (1967)

___ (ed) *Data bases, computers, and the social sciences* (1970)

Blackman, S. and Goldstein, K.M. *An introduction to data management in the behavioural and social sciences* (1971)

Chandler, T.J. *The climate of London* (1965)

Chorley, R.J. and Haggett, P. 'Trend surface analysis in geographic research', *Transactions, Institute of British Geographers*, 37, 47-67 (1965)

Cuttrell, J.D., Feeser, L.J. and Penzien, J.P. 'Perspective views and computer animation in highway engineering',

*ACSM-ASP Technical Conference Papers*, 74-94 (1970)
Davies, W.K.D. 'Varimax and the destruction of generality: a methodological note', *Area*, 3, 112-8 (1971)
Davis, J.C. *Statistics and data analysis in geology* (1973)
Dijkstra, E.W. *A primer of ALGOL 60 programming* (1962)
Dippel, G. and House, W.C. *Information systems: data processing and evaluation* (1969)
Dixon, W.J. 'Biomedical computer programs', *University of California publications in automatic computation*, 2 (1971)
ECU *Automated cartography and planning* (1971)
Ekistics 'Simulation models', *Ekistics*, 222, 310-73 (1974)
Fielding, G.J. and Rumage, K.W. (eds) 'Computerised instruction in undergraduate geography', *AAG Commission on College Geography, Technical Paper 6* (1972)
Gaits, M. 'Thematic mapping by computer', *Cartographic Journal*, 6, 50-68 (1969)
Garfinkel, D. 'Programmed methods for printer graphical output', *Communications, Association for Computing Mathematics*, 5, 477-9 (1962)
Gould, P.R. 'Man against environment: a game theoretic approach', *Annals, Association of American Geographers*, 53, 290-7 (1963)
Gurr, T. and Panofsky, H. (eds) 'Information retrieval in the social sciences', *American Behavioural Scientist*, 7(10), 3-70 (1964)
Harbaugh, J.W. and Merriam, D.F. *Computer applications in stratigraphic analysis* (1968)
Heap, B.R. and Pink, M.G. 'Three contouring algorithms', *National Physical Laboratory*, DNAM, 81 (1969)
Hendricks, F. 'Problems of large scale simulations', *Ekistics*, 222, 312-5 (1974)
Horwood, E.M. 'Grid co-ordinate geographic identification systems', in Bisco, R.C. (ed) *Data bases, computers, and the social sciences*, -20-37 (1970)
Hufschmidt, M.M. and Fiering, M.B. *Simulation techniques for design of water resource systems* (1966)
Huke, R.E., Fielding, G.J. and Rumage, K.W. (eds) 'Computer assisted instruction in geography', *AAG Commission on College Geography, Technical Paper 2* (1969)
Jaro, M.A. *GRIDS: a computer mapping system* (1972)
Jones, J.G.T. *Diagrams drawn by computer* (undated)
Kadmon, N. 'KOMPLOT: 'Do-it-yourself' computer cartography', *Cartographic Journal*, 8, 139-44 (1971)
Kemeny, J.G. and Kurtz, T.E. *BASIC programming* (1971)
Kern, R. and Rushton, G. 'MAPIT: a computer program for production of flow maps, dot maps and graduated symbol

maps', *Cartographic Journal*, 6, 131-7 (1969)
Kirk, M.V. and Preston, D.A. 'FORTRAN IV programs for computation and printer display of maps and mathematically defined surfaces', *Geocom programs*, 3 (1971)
Koch, G.S., Link, R.F. and Schuenemeyer, J.H. *Computer programs for geology* (1972)
Kubert, B., Szabo, J. and Guillen, S. 'The perspective representation of functions of two variables', *Journal, Association of Computing Machinery*, 15, 193-204 (1968)
LAMSAC *Census analysis* (1974)
McCullagh, M.J. and Sampson, R.J. 'User desires and graphics capabilities in the academic environment', *Cartographic Journal*, 9, 109-22 (1972)
Marble, D.F. and Anderson, B.M. 'LAND USE: a computer program for laboratory use in economic geography courses', *AAG Commission on College Geography, Technical Paper 8* (1972)
Martin, F.F. *Computer modelling and simulation* (1968)
Mather, P.M. 'Varimax and generality', *Area*, 3, 352-9 (1971)
Mendelssohn, R.C. *The manpower information system* (1965)
Monmonier, M.S. 'The production of shaded maps on the digital computer', *Professional Geographer*, 17, 13-14 (1965)
__ 'Computer mapping with the digital incremental plotter', *Professional Geographer*, 20, 408-9 (1969)
__ 'Shaded area symbols for the digital incremental plotter', *Tijdschrift voor Economische en Sociale Geografie*, 61, 374-8 (1970)
Muxworthy, D.T. 'A user's guide to SYMAP and SYMVU', *University of Edinburgh, Program Library service report*, 2 (1972)
National Computer Centre *Computer guide 3: programs for social scientists* (1972)
Newman, W.M. and Sproull, R.F. *Principles of interactive computer-graphics* (1973)
Nie, N.H., Bent, D.H. and Hull, C.H. *Statistical package for the social sciences* (1970)
Olea, R.A. 'Optimal contour mapping using Universal Kriging', *Journal of Geophysics Research*, 79, 695-702 (1974)
Onstad, C.A. and Brakensiek, D.L. 'Watershed simulation by stream path analogy', *Water Resources Research*, 4, 965-71 (1968)
Parslow, R.D. and Green, R.E. *Advanced computer-graphics* (1971)

Patterson, P.D. 'Games as an urban laboratory', *Ekistics*, 222, 349-54 (1974)

Pitts, F.R. 'HAGER III and HAGER IV: two monte-carlo computer programs for the study of spatial diffusion problems', *Office of Naval Research, Geography Branch, Report 4, Task 389-140* (1965)

— 'MIFCEL and NONCEL: two computer programs for the generalisation of Hagerstrand models to an irregular lattice', *Office of Naval Research, Geography Branch, Report 7, Task 389-140* (1969)

Rhind, D.W. 'Automated contouring - an empirical evaluation of some differing techniques', *Cartographic Journal*, 8, 145-58 (1971)

Robertson, J.C. 'The SYMAP program for computer mapping', *Cartographic Journal*, 4, 108-13 (1967)

Rohlf, F.J. 'GRAFPAC, Graphical output subroutines for the GE 635 computer', *Computer Contribution*, 36 (1969)

Rokkan, S. 'Archives for secondary of sample survey data', *International Social Science Journal*, 16, 49-62 (1964)

— (ed) *Data archives for the social sciences* (1966)

Rosing, K.E. 'Computer graphics', *Area*, 3, 2-7 (1969)

— and Wood, P.A. *Character of a conurbation, a computer atlas of Brimingham and the Black Country* (1971)

Sawyer, J.S. 'Graphical output from computers and the production of numerically forecast or analysed synoptic charts', *Meteorological Magazine*, 89, 187-90 (1960)

Scheuch, E.K. and Bruning, I. 'The Zentralarchiv at the University of Cologne', *International Social Science Journal*, 16, 77-85 (1964)

Sentance, W.A. 'The graphic representation of computer output', *Tijdschrift voor Economische en Sociale Geografie*, 60, 180-6 (1969)

Sharp, J.R. *Some fundamentals of information retrieval* (1965)

Shephard, D. 'A two-dimensional interpolation function for irregularly spaced data', *Proceedings, Association of Computing Machinery*, 11, 517-23 (1968)

Smart, J.S., Surkan, A.J. and Considine, J.P. 'Digital simulation of channel networks', *Symposium on river morphology*, Bern, 87-98 (1967)

Smith, D.M. and Lee, T. 'A programmed model for industrial location analysis', *Department of Geography, Southern Illinois University, Discussion Paper*, 1 (1970)

Sprunt, B. 'Geographics: a computer's eye view of terrain', *Area*, 4, 54-9 (1970)

Starbuck, W.H. and Dutton, J.M. 'The history of simulation models', in Dutton and Starbuck (eds) *Computer simulation of human behaviour*, 9-102 (1971)

Swerdloff, C.N. and Stowers, J.R. 'A test of some first generation residual land use models', *Highway Research Record*, 126, 38-59 (1966)

Tarrant, W.R. *Computers in the environmental sciences* (1971)

Taylor, J.L. *Instructional planning systems. A gaming simulation approach to urban problems* (1971)

Thrower, N.J.W. 'Animated cartography', *Professional Geographer*, 11, 9-12 (1959)

Tobler, W.R. 'Automation and cartography', *Geographical Review*, 49, 526-34 (1959)

___ 'Automation in the preparation of thematic maps', *Cartographic Journal*, 2, 32-8 (1965)

___ (ed) *Selected computer programs* (1970)

UNESCO *Data archives for the social sciences: purposes, operations and problems* (1972)

Veldman, D.J. *FORTRAN programming for the behavioural sciences* (1967)

Wilson, A.G. *Urban and regional models in geography and planning* (1971)

# Appendix
# Standard Functions of Ansi Fortran

## (i) Intrinsic Functions

| Function and definition | No of arguments | Symbolic name | Type of argument | Type of function |
|---|---|---|---|---|
| Absolute value, $\|a\|$ | 1 | ABS<br>IABS<br>DABS | real<br>integer<br>double | real<br>integer<br>double |
| Truncation, largest integer $\leq a$ | 1 | AINT<br>INT<br>IDINT | real<br>real<br>double | real<br>integer<br>integer |
| Remaindering, $a_1 \pmod{a_2}$ | 2 | AMOD<br>MOD | real<br>integer | real<br>integer |
| Largest value, $\max(a_1, a_2, \ldots)$ | 2 or more | AMAX0<br>AMAX1<br>MAX0<br>MAX1<br>DMAX1 | integer<br>real<br>integer<br>real<br>double | real<br>real<br>integer<br>integer<br>double |
| Smallest value, $\min(a_1, a_2, \ldots)$ | 2 or more | AMIN0<br>AMIN1<br>MIN0<br>MIN1<br>DMIN1 | integer<br>real<br>integer<br>real<br>double | real<br>real<br>integer<br>integer<br>double |
| Conversion from integer to real | 1 | FLOAT | integer | real |
| Conversion from real to integer | 1 | IFIX | real | integer |
| Transfer of sign, sign of $a_2$ times absolute value of $a_1$ | 2 | SIGN<br>ISIGN<br>DSIGN | real<br>integer<br>double | real<br>integer<br>double |
| Positive difference $a_1 - \min(a_1, a_2)$ | 2 | DIM<br>IDIM | real<br>integer | real<br>integer |
| Obtain most significant part of Double Precision argument | 1 | SNGL | double | real |
| Obtain Real part of complex argument | 1 | REAL | complex | real |

| Function and definition | No of arguments | Symbolic name | Type of argument | Type of function |
|---|---|---|---|---|
| Obtain Imaginary part of complex argument | 1 | AIMAG | complex | real |
| Express Single Precision argument in Double Precision form | 1 | DBLE | real | double |
| Express two Real arguments in complex form $a_1 + a_2\sqrt{(-1)}$ | 2 | CMPLX | real | complex |
| Obtain conjugate of a complex argument | 1 | CONJG | complex | complex |

### (ii) Basic External Functions

| Function and definition | No of arguments | Symbolic name | Type of argument | Type of function |
|---|---|---|---|---|
| Exponential, $e^a$ | 1 | EXP<br>DEXP<br>CEXP | real<br>double<br>complex | real<br>double<br>complex |
| Natural logarithm, $\log_e a$ | 1 | ALOG<br>DLOG<br>CLOG | real<br>double<br>complex | real<br>double<br>complex |
| Common logarithm, $\log_{10} a$ | 1 | ALOG10<br>DLOG10 | real<br>double | real<br>double |
| Trigonometric sine, $\sin(a)$ | 1 | SIN<br>DSIN<br>CSIN | real<br>double<br>complex | real<br>double<br>complex |
| Trigonometric cosine, $\cos(a)$ | 1 | COS<br>DCOS<br>CCOS | real<br>double<br>complex | real<br>double<br>complex |
| Hyperbolic tangent, $\tanh(a)$ | 1 | TANH | real | real |
| Square root, $\sqrt{a}$ | 1 | SQRT<br>DSQRT<br>CSQRT | real<br>double<br>complex | real<br>double<br>complex |
| Arctangent, $\arctan(a)$ | 1 | ATAN<br>DATAN | real<br>double | real<br>double |
| $\arctan(a_1/a_2)$ | 2 | ATAN2<br>DATAN2 | real<br>double | real<br>double |
| Remaindering, $a_1 (\mod a_2)$ | 2 | DMOD | double | double |

| Function and definition | No of arguments | Symbolic name | Type of argument | Type of function |
|---|---|---|---|---|
| Modulus | 1 | CABS | complex | real |

*Note*
Intrinsic functions may not themselves be used as arguments to other functions except where part of an arithmetic expression whereas basic external functions may be so used.

# Index

agricultural geography,
    Chagga farm game, 343
    crop combinations, 207
algorithm, 48, 293, 295, 337
area calculation, 178
arguments,
    of function, 128, 251
    of subroutine, 256, 276
arithmetic, 25
    assignments, 65
    operations, 64
arrays,
    adjustable dimensions, 152
    concept of, 147
    input/output of, 150, 154, 225
    multi-dimensional, 152
    subscript overflow in, 292
    use of, 150

backing store, 40, 315
BACKSPACE statement, 317
binary system, 29
biogeography,
    association in, 112
BMD statistical package, 320
byte, 29

calculators, 15
CALL statement, 257
calling subprograms, 249

cards in FORTRAN,
    comment, 59
    continuation, 60, 219
    identification on, 60
    labels on, 59
    layout of, 58
central processing unit, 27
climatology,
    anomalies in, 180
    aridity indices in, 89
    Casagrande formula, 67
    classification in, 117
    continentality in, 142
    heat islands, 327, 334
    potential evapo-transpiration, 207
    radiation, 252
    temperature scales, 135, 251
    THI index, 143
    wind chill, 131
    wind velocity, 193
coding forms, 51
collinearity, 48
comments in a program, 59, 302
common blocks, 260
    extension of, 264
    where to use, 298
COMMON statement, 259
COMPLEX data type, 295
computers,
    analogue, 31
    appreciation of, 12
    digital, 31

generations of, 16
geographical use, 19, 313, 344
hardware of, 26
micro, 18
mini, 18, 349
revolution, 11
software, 27
university, 313
computer aided instruction, 341, 348
computer graphics,
CRT, 334
general use of, 324
graph plotter, 330
interactive, 341
lineprinter, 326
constants in FORTRAN, 61
alphanumeric, A, 217
complex, 295
double precision, D, 294
exponent-form, E, 62, 218
floating point, F, 61, 294
Hollerith, H, 78
integer, I, 61, 294
logical, L, 296
continuation cards, 60
CONTINUE statement, 100, 123, 128
core storage, 27, 29

data,
banks of, 322
cards, 71
concept of, 25
errors in, 291
files of, 314
format of, 76, 215
type statement, 294
vector, 193
DATA statement, 227, 307
debugging, 53, 289

demography,
life tables, 213
Markov chains, 287
population potential, 204
population pyramids, 228
rate calculations, 66, 143
ratios, 81
transitions, 98
trends in UK, 119
DIMENSION statement, 149
distances between points, 90, 178, 266
documentation of programs, 54, 300
DO statement, 123
DOUBLE PRECISION data type, 294

economic geography,
building activity, 92
connectivity in, 283
diversification index, 207
factory characteristics, 189
factory/warehouse allocation, 198
location quotients, 101, 118
Pareto curves, 247
pie diagrams, 92
shift/share analysis, 162
shop sales, 285
shop sizes, 136
tertiary employment, 143
efficiency in computing, 302
elements of arrays, 147
END statement, 80, 249, 256
ENIAC, 16
EQUIVALENCE statement, 262
errors in computing,
control, 292

diagnostics, 53, 290
naming, 264
overflow, 291
rounding, 62, 293
testing for, 54, 290
truncation, 64, 293
types of, 290
exceptionalism in geography, 13
exponent-mantissa notation, 62
exponentiation in FORTRAN, 64
expressions in FORTRAN,
   arithmetic, 63
   relational, 96, 98, 296
   relative efficiency, 303
   subscripting, 148

F (floating point data), 75, 294
files in computing, 314
floating point overflow, 291
flowcharts, 50
FORMAT statement,
   A(alphanumeric), 215
   alternative reference to, 229
   complex data in, 296
   D(double precision), 295
   E(exponent), 215
   F(floating point), 76
   free form input, 220, 300
   H(Hollerith), 76, 300
   L(logical), 297
   line control using, 222
   repetition of fields, 220
   summary of, 219, 230
   X(space), 215
FORTRAN,
   constants in, 61
   essential commands, 299
   extensions to, 300
   functions in, 128
   graphics capability, 325
   history of, 56
   infrastructure, 56
   portability of programs, 298
   statement types, 56
   variables in, 62, 147
functions in FORTRAN,
   standard, 128
   statements, 250
   subprograms, 253
   subscripting technique, 269

geomorphology,
   bedload, 142
   cirques, 194, 203
   cobble orientation analysis, 193
   cobble shape analysis, 106
   drainage basin description, 90
   drainage basin simulation, 284
   drumlins, 194, 265
   sediment size analysis, 84
   slope deposits, 157
GHOST graphics package, 320, 331
GIGO garbage-in, garbage-out, 319
GINO graphics package, 318, 331
GO TO statement
   unconditional, 94
   computed, 95
graph plotter, 330
GRIDS mapping program, 320

H (Hollerith data type), 78
   direct input of, 216

I (integer data type), 78, 294
  compression to save space, 308
idiography in geography, 13
IF statements, 95
  arithmetic, 96
  logical, 96
information, 14, 21
input,
  basic ideas, 71, 314
  data, 25, 227
  devices for, 35
INTEGER type statement, 294
interactive computation, 339

job control, 53

K (unit of storage capacity), 30
keypunch, 32

libraries,
  geographical programs, 319
  graph plotter programs, 331
  lineprinter mapping programs, 326
  subprograms, 249, 259, 318
linear programming, 199
line control in FORTRAN, 222
LOGICAL data type, 296
loops in FORTRAN, 123
  implied, 150, 154
  nested, 127, 156
  optimising, 305
  value on exit, 127

maps by computer, 20, 324
  animated, 336
  automatically contoured, 325, 331
  choropleth type, 234
  correlation analysis of, 273
  design of, 14
  point location type, 247
  proportional circles, 25
  using graph plotter, 330
  using lineprinter, 326

NAG Numerical Algorithms Group, 318
number crunching, 11
numerical analysis, 293, 318

operating systems, 27, 41, 315
output, 25
  FORTRAN basics, 71, 314
  graphical, 324

parameters of a DO, 123
peripherals, 26, 325
perspective diagrams, 334
portability of programs, 259
problem identification, 46
programming languages, 16, 43
  ALGOL, 337
  BASIC, 339
  FORTRAN, 56
  style in, 289
programs,
  design and portability, 289, 297
  documentation, 289, 300
  efficiency, 289, 302
  example of, 26
  good and bad, 289
  integrity of, 290
  justification for, 21, 41

libraries of, 319
presentation of, 289, 300
segmentation of, 249
writing and running, 46
quantification,
 effect of computers, 19
 requirements, 13
 revolution in geography, 11
questionnaire analysis, 184, 319

radius vector summation, 193
range of a DO, 124
READ statement, 72
 for arrays, 225
 using files, 314
REAL type statement, 294
relational expression, 96, 296
 operators, 97
RETURN statement, 253, 256
REWIND statement, 317
rounding in computer, 31

simulation, 14, 21, 345
software, 41
SPSS statistics package, 321
statements in FORTRAN,
 arithmetic, 63
 arrays, 149
 data type, 227, 294
 executable types, 57
 file handling, 316
 format, 73, 215
 input/output, 71, 214
 labels on, 57, 301
 layout on cards, 58
 loops, 121
 non-executable, 149
 relational, 96, 98

specification, 149, 260, 262
terminal to a DO, 123
terminators, 80
transfer of control, 94
statistical analysis,
 chi-square, 158, 339
 contingency tables, 157, 184
 correlation analysis, 273
 histograms, 231
 Markov chains, 287
 mean centre, 308
 moving means, 115
 multivariate analysis, 12, 22
 nearest neighbour analysis, 265
 normalising data, 130
 packages of programs, 320
 percentages, 150
 programs for factor analysis, 319
 ranking numbers, 178
 reduced major axis, 285
 scatter diagrams, 245
 sorting numbers, 257
 student's t, 142
STOP statement, 80
subprograms, 57, 249, 297, 318
SUBROUTINE statement, 255, 297
SYMAP and SYMVU graphics packages, 320, 326, 334
symbols used in FORTRAN, 58

transfer of control, 94, 121
transport, in Wales, 177
terminator of a DO, 123

variables in FORTRAN, 63
 alphanumeric, 217
 arrays, 147

361

logical, 296
  naming rules, 63
  scalar, 63
  vector, 193

word length, 29
   A format, 217, 300
   consequences, 293
   double length, 294
WRITE statement, 76
   alternative channels, 224
   line control, 222
   lists, 225